50 Springer Series in Solid-State Sciences

Edited by Hans-Joachim Queisser

Springer Series in Solid-State Sciences

Editors: M. Cardona   P. Fulde   H.-J. Queisser

Volume 1 – 39 are listed on the back inside cover

Shih-Lin Chang

Multiple Diffraction of X-Rays in Crystals

With 152 Figures

Springer-Verlag
Berlin Heidelberg New York Tokyo 1984

Professor Shih-Lin Chang, Ph. D.

Instituto de Física, Universidade Estadual de Campinas, Caixa Postal 1170
Campinas, São Paulo 13100, Brazil

Series Editors:

Professor Dr. Manuel Cardona
Professor Dr. Peter Fulde
Professor Dr. Hans-Joachim Queisser

Max-Planck-Institut für Festkörperforschung, Heisenbergstrasse 1
D-7000 Stuttgart 80, Fed. Rep. of Germany

ISBN-13:978-3-642-82168-4   e-ISBN-13:978-3-642-82166-0
DOI: 10.1007/978-3-642-82166-0

Library of Congress Cataloging in Publication Data. Chang, Shih-Lin, 1946-. Multiple diffraction of x-rays in crystals. (Springer series in solid-state sciences ; 50). Bibliography: p. Includes index. 1. X-ray crystallography. 2. X-rays – Diffraction. I. Title. II. Series. QD945.C44   1984   548'.83   84-13900

© Springer-Verlag Berlin Heidelberg 1984
Softcover reprint of the hardcover 1st edition 1984

Typesetting: K & V Fotosatz, 6124 Beerfelden

2153/3140-543210

Dedicated to my wife, Ling-Mei

Preface

The three-dimensional arrangement of atoms and molecules in crystals and the comparable magnitude of x-ray wavelengths and interatomic distances make it possible for crystals to have more than one set of atomic planes that satisfy Bragg's law and simultaneously diffract an incident x-ray beam – this is the so-called multiple diffraction. This type of diffraction should, in principle, reflect three-dimensional information about the structure of the diffracting material. Recent progress in understanding this diffraction phenomenon and in utilizing this diffraction technique in solid-state and materials sciences reveals the diversity as well as the importance of multiple diffraction of x-rays in application.

Unfortunately, there has been no single book written that gives a systematic review of this type of diffraction, encompasses its diverse applications, and foresees future trends of development. It is for this purpose that this book is designed. It is hoped that its appearance may possibly turn more attention of condensed-matter physicists, chemists and material scientists toward this particular phenomenon, and that new methods of non-destructive analysis of matter using this diffraction technique may be developed in the future.

I am very grateful to Professor Ben Post of the Polytechnic Institute of New York (Brooklyn) for having brought me into this field of research and to Professor Hans-Joachim Queisser of the Max-Planck-Institut für Festkörperforschung (Stuttgart) for useful discussions and encouragement in preparing this book. I am also indebted to Dr. H. K. V. Lotsch of Springer-Verlag (Heidelberg), for enthusiasm and much help in the course of writing this book. The assistance and criticism from my colleagues, S. Caticha-Ellis, J. D. Rogers, I. C. L. Torriani, A. M. O. de Almeida, and Y. Kurihara, of the Universidade Estadual de Campinas (Brazil) are also acknowledged. My thanks are also extended to S. L. Gomes and J. A. Fraymann for the preparation of the figures.

Campinas, June 1984   *Shih-Lin Chang*

Acknowledgements

The author is indebted to B. S. Fraenkel, R. Hoier, B. J. Isherwood and P. Mikula for kindly providing their original photographs and figures.

The permission of the following publishers to reproduce their figures (with the original author's consent) in the text is also gratefully acknowledged:

American Institute of Physics

American Physical Society

International Union of Crystallography

Japan Society of Applied Physics

North-Holland Publishing Co.

Physical Society of Japan

Philips Research Laboratories

Royal Society of London

Springer-Verlag, Heidelberg

Verlag, Zeitschrift für Naturforschung

Contents

1. Introduction

Multiple diffraction, in contrast to a simple Bragg – the so-called two-beam (incident and reflected) – reflection, occurs when several sets of atomic planes in a crystal are simultaneously brought into position to diffract an incident x-ray beam. The interaction among the diffracted beams within the crystal gives rise to an increase or a decrease in the intensity of a given two-beam reflection. The decrease in intensity, which brightens the two-beam diffraction images on an x-ray film (a negative), is called 'Aufhellung'. This phenomenon was first discovered by *Wagner* in 1923 [1.1] and extensively investigated by *Berg* [1.2]. In 1928, *Mayer* [1.3] reported the observation of 'Aufhellung' in a four-beam diffraction: in these experiments, three sets of atomic planes, (220), (400) and (2$\bar{2}$0), of a diamond crystal were aligned simultaneously to diffract an incident beam of MoK_α radiation. The diffracted intensity of the two-beam (000) (400) reflection decreased in the vicinity of the four-beam (000) (220) (400) (2$\bar{2}$0) diffraction position, (000) being the incident beam.

Renninger [1.4] in 1937 observed an increase in intensity of the (222) reflection from a diamond crystal for various multi-beam cases. This Renninger effect was named 'Umweganregung' because of the detoured excitation of the (222) reflection by means of multiple diffraction. Since the (222) reflection of diamond is a nearly forbidden reflection, the intensity gained was mainly due to the transfer of diffraction power from a strong reflection into the (222) direction via the coupling between the two reflections. Other investigations on the 'Umweg' diffraction in the 1940s have also been reported [1.5, 6]. An attempt to determine x-ray reflection phases using 'Umweg' reflections was carried out by *Lipscomb* [1.7] in 1949.

Both 'Aufhellung' and 'Umweganregung' involve the Bragg geometry in which both the reflected and the incident beams lie on the same side of the crystal. In contrast to these 'Bragg-(reflection-)type' multiple diffractions, *Borrmann* and *Hartwig* [1.8], in 1965, investigated 'Laue-(transmission-)type' multiple diffraction in which all the diffracted beams involved were transmitted through the crystal. The phenomenon of anomalously high transmission of x-rays through a perfect germanium crystal observed in their experiment is the well-known multi-beam Borrmann effect [1.8, 9].

The possibility of multiple diffraction occurring in crystals was also pointed out by *Bethe* [1.10] for the electron case. The first observation of such diffraction for electrons was reported in 1932 [1.11] and in 1935 [1.12]. *Kambe* and *Miyake* [1.13 – 15] attempted to utilize the intensity variation of three-beam Borrmann diffraction of electrons for reflection-phase determination. A theory with a two-beam approximation for the three-beam case has also been derived by these authors, based on Bethe's theory of electron diffraction [1.10].

Multiple diffraction is a combination of several two-beam diffractions. One can regard multiple diffraction as a generalized two-beam diffraction. Diffraction theories for multi-beam cases have therefore been derived from the two-beam diffraction theories. In discussing the development of the theories for multiple diffraction, the evolution of the diffraction theory of x-rays in crystals for the two-beam cases should be mentioned. Since two-beam diffraction theory has been well established and there are several review articles [1.16 – 22] and books [1.23 – 27] written on this subject, only key contributions are mentioned briefly below.

Soon after the discovery of x-ray diffraction in crystals by *Laue, Friedrich* and *Knipping* [1.28], *Darwin* [1.29] developed a kinematical theory dealing with the diffracted intensities of x-rays from single crystals. Forward and backward multiple scattering among lattice planes in crystals was considered. His approach, which takes into account the interaction between the incident and diffracted beams inside the crystal, is actually more dynamical than kinematical. The concepts of extinction and the crystal mosaic structure, first named by Ewald, were introduced.

The dynamical nature of x-ray diffraction was first considered by Ewald even before the experimental discovery of x-ray diffraction. In 1916 *Ewald* [1.30] developed dispersion theory and the theory of reflection and refraction for crystal optics, in which a crystal was considered to consist of periodic arrays of dipoles. The excitation of these dipoles by an external source, such as x-rays, produces electromagnetic waves. The interaction between the waves of the excited dipoles was automatically taken care of by this self-consistent theory. In 1931, *Laue* [1.31] reformulated the dynamical theory in terms of Maxwell's equations. He treated a crystal as a medium with a periodic, complex dielectric constant for x-rays. Bloch wavefunctions with momentum satisfying Bragg's law were assumed as solutions to Maxwell's equations. This treatment has been more frequently adapted than Ewald's theory by researchers in the field of x-ray diffraction, mainly because of its simple formalism. However, the latter contains more physical insight into the dynamical diffraction of x-rays in crystals. Both Ewald's and Laue's treatments were based on the assumption that the incident waves are plane.

The spherical-wave nature of x-ray diffraction due to the beam divergence and the coherence length of x-rays was experimentally demonstrated by *Kato* and *Lang* [1.32] in 1959. A spherical-wave theory of x-ray diffraction was then derived by *Kato* to account for the experimental results [1.33]. In the early 1960s, the development of a dynamical theory for imperfect crystals was carried out by *Penning* and *Polder* [1.34], *Takagi* [1.35], *Taupin* [1.36], *Kato* [1.37] and many others [1.38 – 44]. Most recently, *Kato* developed a generalized 'extinction theory' [1.45] and 'statistical dynamical theory' [1.46] governing x-ray diffraction from both imperfect and perfect crystals.

The so-called quantum theories of x-ray diffraction have been developed, in parallel to the (classical) dynamical theories, by *Molière* [1.47], *Wagenfeld* [1.48], *Ohtsuki* [1.49], *Ashkin* and *Kuriyama* [1.50] and others [1.51 – 56] for both perfect and imperfect crystals. In these theories, the generalized current density was considered and tensor operators were adapted for the susceptibilities. Excitations such as thermal vibration and Compton scattering were taken into account. A more rigorous treatment has also been given by some of the authors mentioned, based on quantum field theory.

The first attempt to account for the observed diffracted intensities for x-rays in multi-beam cases was given by *Mayer* in 1928 [1.3], employing Ewald's dynamical theory [1.30]. *Renninger* [1.4] presented a semi-quantitative treatment, but useful quantitative relationships could not be found. In the late 1930s, *Ewald* [1.57] and *Lamla* [1.58] discussed the dispersion equation and the excitation of wavefields in crystals for three-beam x-ray dynamical diffractions. For electron diffraction, *Kambe* and *Miyake* [1.13 – 15] adopted Bethe's theory for three-beam cases with at least one of the reflections involved being weak. In 1958, *Kato* [1.59] gave a matrix formulation for solving multi-beam diffraction problems for x-rays. In 1960, *Laue* [1.25] applied his dynamical theory to a three-beam case of x-ray diffraction. The fundamental equations of the wavefield were derived.

The discovery of the multiple Borrmann effect, due to the availability of perfect single crystals in the 1960s, stimulated both experimental and theoretical investigations of anomalous transmission of x-rays through solids in the Laue (transmission) geometry for multi-beam cases. These include work by *Saccocio* and *Zajac* [1.60], *Hildebrandt* [1.61], *Joko* and *Fukuhara* [1.62], *Ewald* and *Heno* [1.63], *Heno* and *Ewald* [1.64] and many others [1.65 – 67]. The work of Ewald and Heno gave a detailed analytical expression for the dispersion relation of wavevectors in reciprocal space and discussed the relationship between the dispersion surface and reflection phases. A relevant theory was also formulated by *Penning* and *Polder* [1.68, 69]. Experimental verification of the intensity enhancement in various three-, four-, six- and eight-beam Laue-type diffractions was also carried out by many investigators [1.70 – 85].

Interpretation of Bragg-type multiple diffraction, Aufhellung and Umweganregung, was made in the early work of Mayer and by *Moon* and *Shull* [1.86] in 1964 for neutron diffraction; modification for x-rays was made by *Zachariasen* [1.87] in 1965, within the framework of the kinematical theory. The latter bore a strong relation to *Zachariasen*'s extinction theory [1.88]. More exact expressions for the diffracted intensities in three- and four-beam cases were reported by *Caticha-Ellis* [1.89]. Recently *Chang* [1.90] reformulated the kinematical calculation procedure in a dynamical-theory formalism for multiple diffracted powers, basing this on the *Zachariasen* treatment [1.87].

A dynamical approach, accounting for the diffracted intensity in Bragg-type multi-beam cases, has been pursued by *Colella* [1.91], *Kohn* [1.92] and *Chang* [1.93]. The difference between Bragg-type and transmission-type multiple diffraction is given in Pinsker's book [1.27].

Multiple diffraction, resulting from the scattering of x-rays through a three-dimensional array of atoms and molecules, should, in principle, provide three-dimensional information about the structure of matter, i.e., the lattice constants of a crystal unit cell, lattice mismatches (the relative variation in lattice constants) in layered materials, reflection phases, and the relative positions of atoms and molecules in crystal unit cells. Efforts have been made in these directions, including: the determination of lattice constants, for example, of diamond by *Lonsdale* [1.94], of silicon by *Isherwood* and *Wallace* [1.95], of gallium arsenide by *Spooner* and *Wilson* [1.96], and of germanium, silicon and diamond by *Post* [1.97] and *Hom* et al. [1.98]; characterization of three-dimensional lattice mismatches between liquid phase epitaxial layers and substrates by *Isherwood* et al. [1.99, 100] and *Chang* [1.101]; experimental determination of x-ray reflection phases reported by *Hart* and *Lang* [1.102], *Post* [1.103], *Chapman* et al. [1.104], *Chang* [1.105], *Juretschke* [1.106] and their first application, in conjunction with direct methods [1.107], to crystal-structure determination carried out by *Chang* and *Han* [1.105, 108 – 110]. Applications to x-ray optics have also been reported for multi-beam monochromators [1.111, 112], a multi-beam x-ray interferometer [1.113] and multi-beam x-ray topography [1.114]. Also, due to the fact that the x-ray standing waves formed by dynamical diffraction can generate fluorescence from the foreign atoms in a host crystal lattice, it can be foreseen that the technique of x-ray standing-wave excitation [1.115, 116] may be advanced through the use of multiple diffraction to detect more precisely the positions of impurity atoms on crystal surfaces.

Multiple diffraction depends not only on the structure of matter but also on the radiation employed. The sensitivity of this type of diffraction to the variation in wavelength of the radiation provides a possibility for using multiple diffraction for spectroscopic characterization of radiation sources, such as plasmas. Plasma diagnosis, based on this idea, has been carried out very recently [1.117].

This book is written following closely the above historical survey of the development of the multiple-diffraction technique. It deals first with the experimental and theoretical aspects of multiple diffraction and then with applications to perfect and imperfect crystals and related problems. In Chap. 2, the geometrical conditions and the existing techniques for generating and for indexing multiple diffraction lines, the Kossel patterns [1.118 – 122], and reflected intensities, the Renninger diagrams [1.4], are described. The relation between the geometry of multiple diffraction and crystal symmetry is briefly discussed.

To interpret experimental images and intensities, diffraction theories, both kinematical and dynamical, are necessarily employed. The kinematical theory, valid for diffraction in small crystals, involves a set of differential equations governing the transfer of diffraction power among the diffracted beams [1.86, 87, 89]. Various solutions to these equations form the central part of Chap. 3. Some geometrical factors, such as Lorentz-polarization factors, which are closely related to diffraction powers, or intensities, are included. A generalized calculation procedure for diffraction in multi-layered crystals [1.90] is also given in this chapter.

In Chap. 4, the fundamentals of the dynamical theory [1.30, 31] are reviewed for simple two-beam diffraction and are extended for multi-beam cases. The cofactors of matrix algebra are adopted to simplify the presentation of the fundamentals of the theory for multiple diffraction. These include the fundamental equation of wavefields, the dispersion relation between the wavevectors and the angular position of the crystal, the direction of energy flow and modes of wave-propagation, the boundary conditions for wavefields at the crystal surfaces, and the spherical nature of x-rays.

Discussion on the dynamical theory is extended, in Chap. 5, to include the two-beam approximation for three-beam cases slightly off the exact three-beam point. With this approximation, an analytical expression for the integrated intensity near a 3-beam position is derived. This two-beam approximation was first used by *Kambe* [1.15] for electron diffraction. For x-ray cases, the two-beam approximation was adopted by *Kohler* et al. [1.123] to interpret the interaction between x-rays and acoustoelectrically generated phonons, and by *Juretschke* [1.106] to account analytically for the experimental method of *Chang* [1.105] for phase determination. In Chap. 5, this two-beam approximation is generalized to both transmission- and reflection-type multiple diffraction. Following the two-beam treatment of *Zachariasen* [1.23], approximate analytical expressions for the integrated reflection for various experimental conditions are obtained. Numerical calculation, which is indispensable for calculating intensities at the exact multi-beam diffraction position and for providing a general feature of the diffraction line profile is presented to complement the two-beam approximation. The calculation procedure is based on the theory described in Chap. 4. The quantum mechanical approach to intensity interpretation, which is important for cases involving radiation near critical absorption edges, is also mentioned in Chap. 5. Other phenomena similar to multiple diffraction, such as x-ray diffraction with specular reflections [1.124 – 128] and x-ray-phonon interaction [1.123, 129 – 138] are discussed.

Chapter 6 is designed to illustrate the validity of the theoretical approaches given in Chaps. 3 – 5 in interpreting experimental observations – mainly the diffracted intensities. In addition, attention is also given to the calculated dispersion surface, linear absorption coefficients and excitation of modes of propagation. Some of the calculations, which are not yet available in the literature, are also presented for cases involving total reflection. These provide readers with a preparatory background to understanding the experimental solution to the x-ray phase problem using Bragg-type multiple diffraction, described in Chap. 7. Calculated results for two-beam cases are included to give a check for the asymptotic behavior of multi-beam cases, when the crystal is set off the exact multi-beam positions.

Chapter 7 covers the applications just mentioned. Emphasis is given to one of the important subjects in x-ray physics and crystallography, the x-ray phase problem. A brief description of this problem and an analysis [1.93, 139] on the shortcomings of the existing proposed experimental ways of solving this problem are presented in the beginning of this chapter. The presentation of the operational method given by *Chang* [1.105] follows. It shows that this method bears strong relation to the discussions given in the previous chapters on the geometry and dynamical interaction of wavefields of Bragg-type multiple diffractions. Examples on how to apply this method to crystal-structure determination are demonstrated and considerations on solving the phase problem for acentric crystals are included. Other applications of either *Kossel* [1.118 – 122] or *Renninger* [1.4] diffraction techniques to the determination of lattice constants and wavelengths of radiation sources, and applications in x-ray topography and interferometry fill in the rest of this chapter. At the end of this book, the possible development of multiple diffraction techniques in the near future is tentatively explored.

Multiple diffraction also takes place in polycrystalline materials. This fact was pointed out by *Renninger* [1.4]. There are two types of multiple diffraction in polycrystals, the external and internal grain diffractions [1.140 – 142]. Multiple diffraction is called internal if the first (primary) and the second reflections take place in the same grain, and external if the first is in one grain and the second in another. Since the internal grain multiple diffraction is the same as that in single crystals and the external reflection resembles the diffractions from mosaic crystals, multiple diffractions from polycrystals are therefore excluded from this book. Multiple diffraction accompanied by small-angle x-ray scattering will also not be discussed because its origin is the same diffraction mechanism as in the normal cases. Furthermore, successive reflection, which is often referred to as multiple reflection [1.143], is left out due to the fact that successive reflections are composed of only two-beam reflections. This type of reflection therefore lies outside the scope of multiple diffraction.

2. Geometry, Peak Indexing, and Experimental Techniques

The occurrence of multiple diffraction depends on the geometrical aspects of the crystal itself and the relative arrangement of the crystal with respect to the incident radiation. These geometrical factors are the interatomic distances, the space group to which the crystal belongs, the wavelength of the radiation, and the experimental arrangement. This chapter describes the general aspect of the geometry of multiple diffraction and provides a detailed account of each experimental technique by which multiple diffractions are generated. A description of the methods of identifying multiple diffraction peaks and lines follows.

Although the Lorentz-polarization factor for multiple diffraction involves the geometry a great deal, it is more closely connected with the kinematical theory for the consideration of the diffracted intensity. The Lorentz-polarization factor is therefore discussed in Chap. 3.

2.1 Geometry of Multiple Diffraction

2.1.1 Real-Space and Reciprocal-Space Representations

No Bragg reflection occurs when a crystal is set so that Bragg's law is not satisfied for any set of atomic planes. This situation can be represented as shown in Fig. 2.1a in reciprocal space, where only the origin O, with Miller indices (000), of the reciprocal lattice is situated on the surface of the Ewald sphere. The vector k_0, from the center C of the sphere to point O, is the wavevector of the incident beam. The radius of the Ewald sphere is $1/\lambda$, λ being the wavelength inside the crystal. The real-space representation is also shown in the same figure. This case is often referred to as one-beam diffraction, for only the incident beam is involved. If the crystal is tilted so that an additional reciprocal lattice point A is brought onto the surface of the Ewald sphere at point G_1, Bragg's law is then satisfied (Fig. 2.1b):

$$k_0 + g_1 = k_1, \quad k_0 \cdot g_1 = -g_1^2/2, \tag{2.1}$$

where the vectors k_0 $(= CO)$ and k_1 $(= CG_1)$ are the wavevectors of the incident and the reflected beams and g_1 $(= OG_1)$ is the reciprocal lattice vector of the $(hkl)_1$ planes. The angle θ_1 is the Bragg angle of the G_1 reflection.

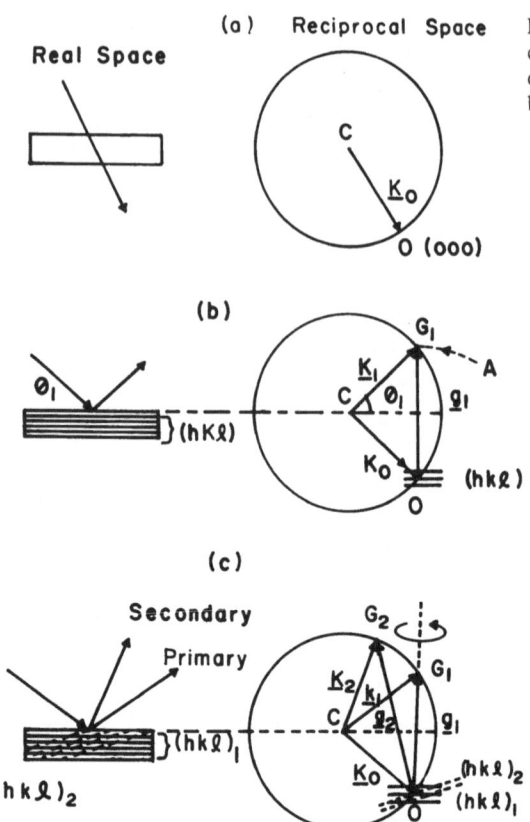

Fig. 2.1a–c. Real-space and recipro-cal-space representations of x-ray diffraction for (a) one-beam, (b) two-beam and (c) three-beam cases

The crystal surface can also be represented in this geometry as a section plane at the center C. For example, the plane perpendicularly bisecting the vector OG_1 stands for the crystal surface which is parallel to the $(hkl)_1$ planes. This is a two-beam symmetric Bragg reflection since the two wavevectors k_0 and k_1 are symmetric about the crystal surface. If the crystal surface is parallel to OG_1, this is a symmetric Laue reflection, because the two diffracted beams associated with the wavevectors k_0 and k_1 are symmetric about the crystal surface normal and transmit through the crystal.

If now the crystal is rotated around OG_1 without disturbing G_1 to bring $N-2$ additional reciprocal lattice points onto the surface of the Ewald sphere, $N-1$ sets of planes with reciprocal lattice vectors $g_1, g_2, \ldots, g_{N-1}$ satisfying simultaneously Bragg's conditions, and N-beam diffraction (including the incident beam 0) occurs. A three-beam case, O, G_1, G_2, is illustrated in Fig. 2.1c. For convenience, the reflection G_1 is called the 'primary' reflection and G_2, the 'secondary' reflection. The vector G_2G_1 ($= g_2 - g_1$) indicates the reciprocal lattice vector for the coupling reflection $G_2 - G_1$, namely, $(h_2 - h_1, k_2 - k_1, l_2 - l_1)$. The geometrical conditions for this N-beam diffraction are the following:

$$k_0 \cdot g_i = -g_i^2/2 \, ,$$
$$k_i \cdot (g_j - g_i) = -|g_j - g_i|^2/2 \, ,$$

(2.2)

for all i and j $(i \neq j)$.

If all the wavevectors $k_1, k_2, \ldots, k_{N-1}$ are transmitted through the crystal, one has an N-beam Laue (transmission) diffraction, or an N-beam Borrmann diffraction [2.1]. The case when the N-beam diffraction involves at least one Bragg reflection is called Bragg-type multiple diffraction. The terms Bragg-Bragg and Bragg-Laue multiple diffraction are sometimes used to distinguish between Bragg-type diffraction involving all Bragg reflections, and that involving at least one transmission. When one of the diffracted beams is directed along the crystal surface, the case is called Bragg surface diffraction. When an N-beam Bragg-type multiple diffraction involves a symmetric Bragg reflection G_1, these three Bragg cases can be distinguished by the conditions

$$g_i \cdot g_1 \gtreqless g_1^2/2 \, ,$$

(2.3)

where the equality is for the Bragg surface case, and the greater than and less than signs are for the Bragg-Bragg and Bragg-Laue cases, respectively. For a more general case, the direction cosine $\gamma \, (= \hat{n} \cdot k/k)$ of the diffracted vector k with respect to the inward-pointing crystal surface normal $\hat{n}$ is used. For $\gamma = 0$, the diffraction is a surface case. For $\gamma < 0$ and $\gamma > 0$, the cases are Bragg and Laue, respectively.

2.1.2 Persistent and Coincidental Multiple Diffractions

Most multiple diffractions are usually either lattice-vector coplanar, or wave-vector coplanar. In the lattice-vector coplanar case, all the reciprocal lattice vectors (or reciprocal lattice points) lie in a plane which intersects the Ewald sphere. All the reciprocal lattice points are circumscribed by a circle, the so-called reflection circle, whose radius r_0 is smaller than that of the Ewald sphere. The conditions for this kind of multiple diffraction are

$$r_0 = g_i g_j |g_i - g_j|/(2|g_i \times g_j|) \, ,$$
$$(g_i \times g_j) \cdot g_l = 0 \, ,$$

(2.4)

for $i \neq j \neq l \neq i$. Since the radius is independent of the wavelength, this type of diffraction occurs for all wavelengths of radiation. It is called persistent or systematic multiple diffraction and depends only on the crystal symmetry. All the reflection planes involved in this case have a common zone axis z, i.e., all the g vectors are perpendicular to z. When r_0 is equal to $1/\lambda$, not only the end points of the reciprocal lattice vectors, but also those of the wavevectors lie on the great circle of the Ewald sphere. This type of diffraction is called coincidental diffraction [2.2]. It occurs for only one specific wavelength. There is, however, another type of coincidental diffraction, the non-coplanar multiple diffraction, for which the reciprocal lattice points involved do not lie in a plane but lie on the surface of the Ewald sphere. On particular occasions, both persistent and coincidental diffractions can occur simultaneously for a given wavelength. The eight-beam (000) $(0\bar{2}2)$ (022) (004) $(3\bar{1}1)$ (311) $(3\bar{1}3)$ (313) diffraction of diamond for CuK_{α_1} is a case for which the reflections $(0\bar{2}2)$, (022) and (004) have a common zone axis $\langle 100 \rangle$ and persistent multiple diffraction occurs, while the rest of the reflections are involved in a non-coplanar coincidental diffraction [2.2].

2.1.3 Lattice-Symmetry Dependence – Intrinsic Multiple Diffraction

The number of reciprocal lattice points, including the origin of the lattice, involved in multiple diffraction determines the number of diffracted beams. In the coplanar case, this number depends on the symmetry of the reciprocal lattice plane (two-dimensional plane lattice) [2.5, 61] containing these reciprocal lattice points and on the relative position of the reflection circle with respect to the lattice points in this two-dimensional plane. In principle, it is possible to derive general conditions under which possible N-beam diffractions take place for a given plane lattice. It is, however, difficult in practice to deduce such conditions, since the variable position of the reflection circle in a plane lattice provides a great variety of conditions under which multiple diffraction occurs.

Of particular interest is the case where one of the reflections involved has its reciprocal lattice vector, with magnitude equal to the diameter of the reflection circle, lying in a horizontal plane which contains the incident beam. Under this condition, whenever this particular reflection is aligned for diffraction in the horizontal plane, the other reflection planes are automatically in position to diffract an incident beam. In other words, this particular reflection is always accompanied by multiple diffraction. This case is called intrinsic multiple diffraction [2.3]. Figure 2.2 shows examples for three-, four-, five-, six- and eight-beam intrinsic diffractions. Since the diameter of the reflection circle must coincide with one of the reciprocal lattice vectors, the position of the reflection circle is fixed. The conditions for the occurrence of N-beam cases, in relation to the symmetry of the plane lattice, can therefore be obtained.

It was proposed by *Burbank* [2.3] that, for a given plane lattice, the possibilities of having various N-beam diffractions can be seen graphically by constructing reflection circles of various sizes in the plane lattice. General conditions for multiple diffractions can thus be systematically obtained.

There are, according to *Burbank* [2.3], six plane lattices: oblique, primitive rectangular, centered rectangular, rhombohedral rectangular, hexagonal,

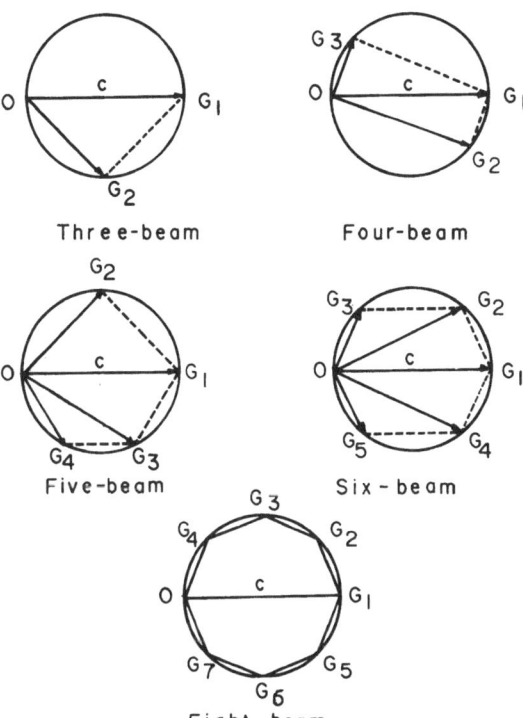

Fig. 2.2. Intrinsic multiple diffractions

and square. These six lattices can be described in terms of two basis vectors, v_1 and v_2. All the reciprocal lattice points in a plane lattice are defined by linear sums of the two basis vectors, $p_1 v_1 + p_2 v_2$. The indices (p_1, p_2) identify a reflection. These two basis vectors can be the three-dimensional reciprocal lattice vectors. The definitions of these six plane lattices and the generalized conditions for multiple diffraction are:

1) *Oblique:* $v_1 \neq v_2$ and the angle α between the two vectors is neither 60° nor 90°. Only two-beam diffraction occurs for any lattice point with both p_1 and p_2 not simultaneously equal to zero.

2) *Primitive rectangular:* $v_1 \neq v_2$, $\alpha = 90°$. Four-beam diffraction takes place for non-zero p_1 and p_2.

3) *Centered rectangular:* primitive rectangular lattice with an additional lattice point at $(v_1 + v_2)/2$. Half of the reflections have integral p_1 and p_2, half have non-integral p's. The four-beam case occurs only for non-zero integral p_1 and p_2.

4) *Rhombohedral rectangular:* primitive rectangular lattice plus two additional lattice points at $(v_1 + v_2)/3$, $2(v_1 + v_2)/3$ or at $-(v_1 + v_2)/3$, $-2(v_1 + v_2)/3$. One-third of the reflections have integral p_1 and p_2, two-thirds have non-integral p's. Four-beam diffraction occurs only for non-zero integral p_1 and p_2.

5) *Hexagonal:* $v_1 = v_2$, $\alpha = 60°$. Four-beam diffraction occurs for odd non-zero p_1 and p_2 and $p_1 \neq p_2$. Six-beam diffraction occurs for even p_1, or $p_2 = 0$, or $p_1 = p_2$. Twelve-beam diffraction takes place for even non-zero p_1 and p_2 and $p_1 \neq p_2$.

6) *Square:* $v_1 = v_2$, $\alpha = 90°$. The four-, eight-, twelve- and sixteen-beam cases have been noted [2.3] for even p's, and the four-, six-, eight- and twelve-beam cases for odd p's. Generalized conditions for p's have not been deduced because of the great variety of multiple diffractions involved in the square lattice.

The two-dimensional plane lattices, which are sections of the three-dimensional lattices, should be related to the seven crystal systems: triclinic, monoclinic, orthorhombic, tetragonal, cubic, rhombohedral, hexagonal. The relation between the two-dimensional plane lattices and the sections of three-dimensional lattices projected normal to the crystallographic axes are listed as follows:

1) *Oblique:* all reflection plane normals for triclinic; $\langle 010 \rangle$, $\langle 0kl \rangle$, $\langle hk0 \rangle$ and $\langle hkl \rangle$ for monoclinic; $\langle hkl \rangle$ for orthorhombic, tetragonal and cubic; and $\langle hkil \rangle$ for hexagonal.

2) *Primitive rectangular:* monoclinic $\langle 100 \rangle$, $\langle 001 \rangle$, $\langle h0l \rangle$; orthorhombic $\langle 100 \rangle$, $\langle 010 \rangle$, $\langle 001 \rangle$, $\langle 0kl \rangle$, $\langle h0l \rangle$, $\langle hk0 \rangle$; tetragonal $\langle 100 \rangle$, $\langle 010 \rangle$, $\langle hhl \rangle$, $\langle h0l \rangle$, $\langle 110 \rangle$, $\langle hk0 \rangle$; cubic $\langle 100 \rangle$, $\langle 110 \rangle$, $\langle hk0 \rangle$, $\langle hhl \rangle$; and hexagonal $\langle 10\bar{1}0 \rangle$, $\langle 11\bar{2}0 \rangle$, $\langle hki0 \rangle$, $\langle h0\bar{h}l \rangle$, $\langle hh\overline{2h}l \rangle$.

3) *Centered rectangular:* the conditions are the same as in 2).

4) *Rhombohedral rectangular:* cubic $\langle hhh \rangle$ and rhombohedral $\langle hhh \rangle$.

5) *Hexagonal:* cubic $\langle 111 \rangle$, $\langle hhh \rangle$ and hexagonal $\langle 000l \rangle$.

6) *Square:* cubic $\langle h00 \rangle$, $\langle 0k0 \rangle$, $\langle 00l \rangle$ and tetragonal $\langle 00l \rangle$.

General rules governing the occurrence of three-, four-, ..., N-beam diffraction in non coplanar multiple diffraction cases for a given three-dimensional lattice are difficult to acquire because of the difficulty in constructing graphically the Ewald sphere for a three-dimensional lattice. Nevertheless, from the discussion of two-dimensional plane lattices, the cubic lattice should have the highest possibility of generating non-coplanar multiple diffractions.

2.1.4 Multiple-Diffraction Possibilities

The number N_{G_1} of possible multiple diffractions for a given crystal and radiation depends on the crystallographic space group, and on the sizes of the Ewald sphere and the crystal unit cell. According to *Prager* [2.4], this number can be estimated from the number of reciprocal lattice points of secondary reflections in the volume v^* swept out by the surface of the Ewald sphere during a full rotation around the primary reflection vector OG_1 (Fig. 2.3). Since each

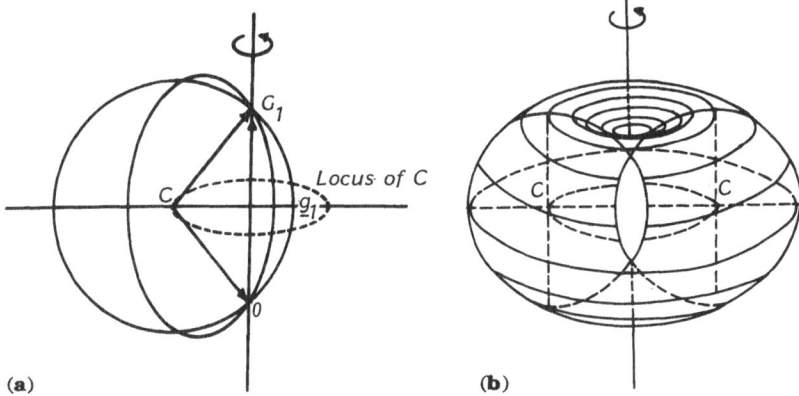

Fig. 2.3. (a) Rotation of Ewald sphere about the reciprocal lattice vector of the primary reflection. (b) Volume v^* swept out by the surface of the Ewald sphere during a full rotation (after *Prager* [2.4])

reciprocal lattice point occupies a reciprocal unit cell of volume V^*, the number N_{G_1} is therefore proportional to the ratio v^*/V^*, i.e.,

$$N_{G_1} = 2a_{G_1} v^*/V^* = 4\pi^2 a_{G_1} V\lambda^{-3}\cos\theta_{G_1}, \qquad (2.5)$$

where the volume v^* is approximately equal to the area $\pi(1/\lambda)^2$ times the locus of the rotation of the center C, $2\pi(\cos\theta_{G_1}/\lambda)$. The angle θ_{G_1} is the Bragg angle of the G_1 reflection and V is the volume of the unit cell. The factor 2 is due to the secondary reciprocal points within the volume v^* entering and leaving the Ewald sphere. The term a_{G_1} takes care of the systematic extinction of reflections imposed by the space group. For example, in the diamond structure, which belongs to the space group $Fd3m$, (hkl) reflections with $h+k+l = 4n\pm 2$ are forbidden reflections [2.5]. For a given three-beam, O, G_1, G_2, diffraction, if two of the three reflections G_1 (primary), G_2 (secondary) and $G_1 - G_2$ (coupling), are forbidden by the space group, no relative intensity variation can be detected, see (3.20). If the primary reflection is a reflection with $h+k+l = 4n\pm 2$, the secondary reflections must involve all odd Miller indices so that the corresponding coupling reflections do not belong to the class $h+k+l = 4n\pm 2$. In this case, a_{G_1} must be 1/8, since there is only a one-half chance of having odd numbers for each index. However, care must be taken to include the space group forbidden reflections, like (222) and (442), whose structure factors are not zero due to the asymmetrically distributed valence electron densities about the nuclei. When N_{G_1} is large, the expression for N_{G_1} can be derived from consideration of the probability density $\varrho(r)$ of finding operative secondary reciprocal lattice points as a function of the distance r between the origin and the secondary reciprocal lattice points G_2:

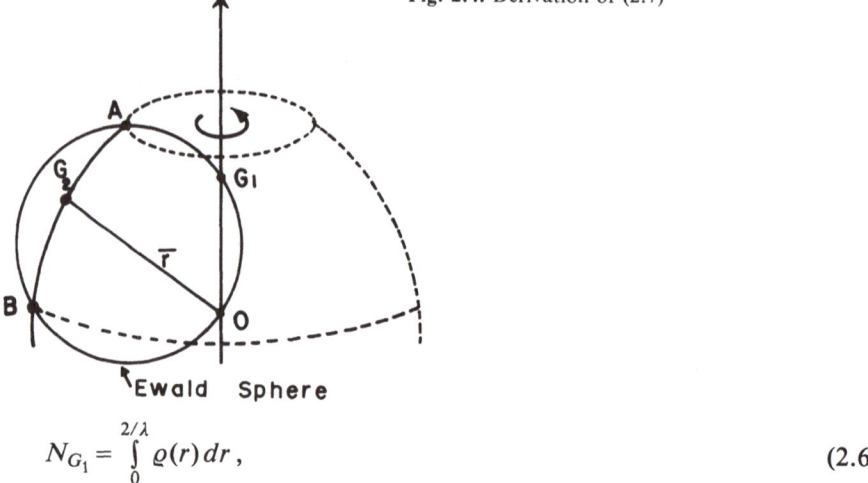

Fig. 2.4. Derivation of (2.7)

$$N_{G_1} = \int_0^{2/\lambda} \varrho(r)\,dr,\qquad(2.6)$$

where the probability density is equal to the area swept out by the arc AB (Fig. 2.4) during a full rotation about OG_1, namely,

$$\varrho(r) = 8\pi r^2 a_{G_1} V\cos\theta_{G_1}\cos\theta_{G_2},\qquad\text{and}\qquad(2.7)$$

$$r = 2\sin\theta_{G_2}/\lambda,\qquad(2.8)$$

which is the magnitude of the reciprocal lattice vector g_2.

2.1.5 Decomposition of Multiple Diffraction

High-order multiple diffraction with $N > 3$ frequently takes place in high-symmetry crystal systems, for example, in cubic systems. When the crystal is subjected to external disturbances such as temperature, pressure or composition gradients, a crystal system can be transformed from a cubic system to other systems. The dimensions of the reciprocal lattice are then changed accordingly. Under this circumstance, the N reciprocal lattice points in a given N-beam case may no longer lie simultaneously on the surface of the Ewald sphere. The N-beam case is then decomposed into several lower-order multiple diffractions. To illustrate this decomposition, let us consider those cases shown in Fig. 2.2. The four-beam case may be decomposed into two three-beam cases, O, G_1, G_2, and O, G_1, G_3, provided that the angle $<G_2OG_3$ is not $90°$. If $<G_2OG_3 = 90°$, this case remains as a four-beam diffraction, because the four reciprocal lattice points forming a rectangle always lie on a circle. The five-beam case can be decomposed into a three-beam, O, G_1, G_2, and a four-beam, O, G_1, G_3, G_4, case, or into three three-beam cases, i.e., O, G_1, G_2; O, G_1, G_3; and O, G_1, G_4. Similarly, higher-order multiple diffraction can be decomposed into three-, four- or five-beam cases.

2.2 Experimental Techniques for Obtaining Multiple Diffraction

2.2.1 Collimated-Beam Technique

The first systematic investigation on multiple diffraction was carried out by *Renninger* [2.6] who used a collimated x-ray beam and an ionization chamber for intensity detection. The experimental set-up is schematically shown in Fig. 2.5. Similar arrangements were proposed by *Williamson* and *Fankuchen* [2.7] and *Cole* et al. [2.8]. The former used a double-crystal spectrometer to improve the linewidth of multiple diffraction peaks, while the latter utilized a 2 m long evacuated pipe with collimators to confine the beam divergence to 1 to 2 minutes of arc. The advantage of the latter arrangement is that it provides K_{α_1}, K_{α_2} and K_β radiation sources without the experimental setting having to be changed. In these methods, a crystal is first aligned for a simple Bragg (two-beam) reflection, say reflection G_1. The crystal is then rotated around the reciprocal lattice vector of the reflection G_1 without disturbing the setting for this reflection, to bring additional sets of atomic planes into positions where they satisfy the Bragg condition. A detector is always placed in position to monitor the G_1-reflected beam. The interaction between the diffracted beams within the crystal modifies the intensity of the G_1 reflection. The intensity variation is transmitted through electronic devices to a chart recorder. A multiple diffraction pattern is therefore obtained.

Figure 2.6 shows the multiple diffraction pattern of germanium for CuK_{α_1} radiation with the (222) reflection as the primary reflection. The vertical and the horizontal axes indicate the intensity variation on the (222) reflected background and the azimuthal angle of rotation about the direction $\langle 222 \rangle$. Although (222) is a forbidden reflection, its diffracted intensity is not null. This is due to scattering from the aspherical distribution of valence electrons in germanium and thermal diffuse scattering. The Miller indices of the secondary reflections are labeled. Most of the multiple diffractions are three-beam cases.

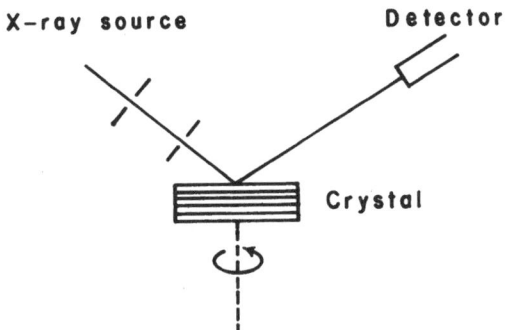

Fig. 2.5. Experimental set-up for the collimated-beam technique

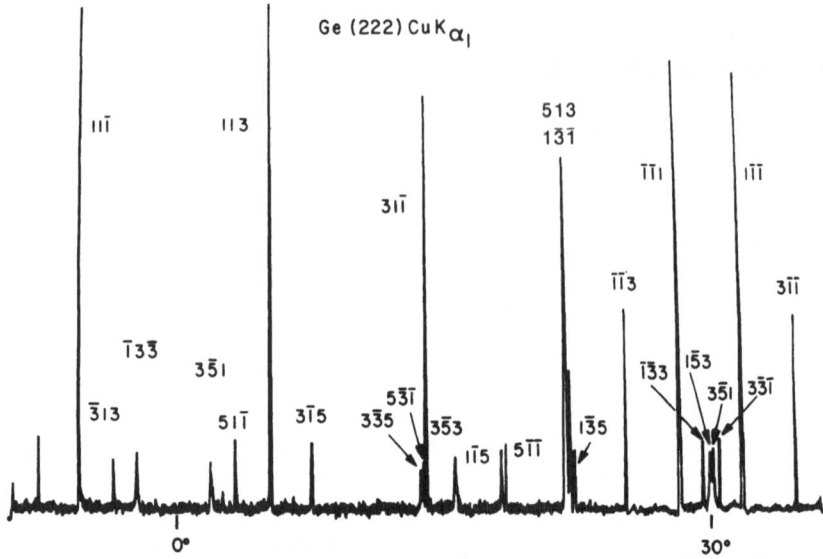

Fig. 2.6. Multiple diffraction pattern of germanium (222) reflection for CuK_{α_1} radiation

Only one four-beam case, (000) (222) ($1\bar{3}\bar{1}$) ($5\,1\,3$), exists. Bragg-Bragg and Bragg-Laue can be easily distinguished, according to (2.3), by $h+k+l>3$ and $h+k+l<3$, respectively. There is no Bragg surface diffraction.

There exist other rotating-crystal methods utilizing collimated beams [2.9], such as rotation, oscillation (Weissenberg), and precession. In these methods, the intensity measurement of a simple Bragg (two-beam) reflection is the objective for crystal-structure determination. Multiple diffractions are generated as undesired by-products of the crystal rotation [2.7, 10, 11]. The presence of multiple diffraction definitely causes inaccuracy in the intensity measurement of the two-beam diffraction.

In the Weissenberg methods, the crystal is rotated around a zone axis, which is inclined with respect to the plane normal to the incident x-ray beam. Figure 2.7 shows the geometry, where X, Y and Z are three mutually orthogonal axes. The Y axis is coincident with the incident beam and the rotation axis lies in the horizontal YZ plane. The rotation angle is u. If $u=0$, the method is called normal-beam Weissenberg; if u is non-zero, it is called equi-inclination Weissenberg. When u is equal to the Bragg angle θ for the rotation vector, the geometry is exactly the same as that for the collimated-beam method described above. The plane containing reciprocal lattice points, including the origin of the lattice, perpendicular to the rotation axis is called the zero-level. Other planes parallel to the zero-level are called the higher-levels.

Multiple diffraction occurs, like persistent diffraction, when the reciprocal lattice points that reflect simultaneously lie in a vertical plane perpendicular to

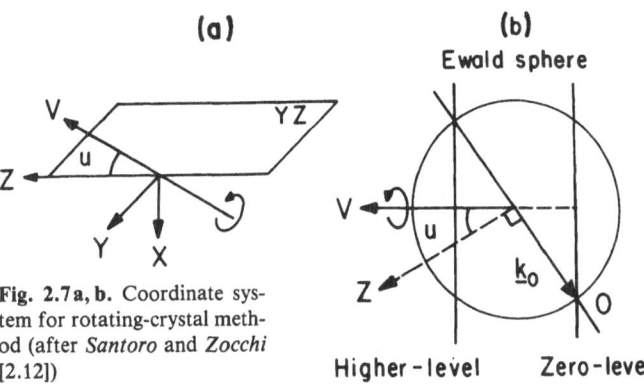

Fig. 2.7a, b. Coordinate system for rotating-crystal method (after *Santoro* and *Zocchi* [2.12])

the *YZ* plane. Multiple diffractions generated with $u \neq 0$ and $u = 0$ are shown in Fig. 2.8a and b, respectively. The difference between these two rotating-crystal techniques and the collimated-beam method is that the intersection of the rotation axis with the reflection circle in Fig. 2.8a does not coincide with a reciprocal lattice point or there is no such intersection (Fig. 2.8b). In these two cases, the occurrence of multiple diffraction depends very much on the choice of the rotation axes, which is closely related to the crystal symmetry.

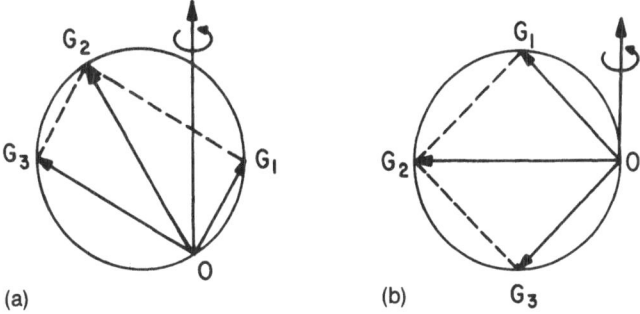

Fig. 2.8 a, b. Multiple diffraction with only the origin of the reciprocal lattice lying on the rotation axis

Considering the most commonly used axes, rotations about $\langle 100 \rangle$ in monoclinic crystals, and about $\langle 100 \rangle$, $\langle 010 \rangle$ and $\langle 001 \rangle$ in orthorhombic, tetragonal and hexagonal crystals generate three- and four-beam diffraction in the normal-beam and equi-inclination Weissenberg methods, respectively [2.10]. For cubic crystals, rotation about any zone axis always gives rise to multiple diffraction. Some special conditions for simultaneous diffraction in cubic and hexagonal crystals have been given in [2.12].

Many reflections in the normal-beam and equi-inclination methods for non-primitive Bravais lattices are free from multiple diffractions. These include, in the equi-inclination method, the odd-layer reflections of the rotations about $\langle 010 \rangle$ for C-centered monoclinic, about $\langle 100 \rangle$ and $\langle 010 \rangle$ for C-centered orthorhombic and about $\langle 100 \rangle$, $\langle 010 \rangle$ and $\langle 001 \rangle$ for body- and face-centered orthorhombic, and about $\langle 001 \rangle$ for body-centered tetragonal crystals. The same situation happens for rotations about $\langle 100 \rangle$ and $\langle 010 \rangle$ for tetragonal body-centered crystals, except that reflections of the $(hh0)$ type are excluded. For the rotation about $\langle 111 \rangle$ in rhombohedral crystals, reflections from the $3n^{th}$ layers are exempted from multiple diffraction in both normal-beam and equi-inclination arrangements [2.12].

There are other types of rotating-crystal methods such as flat-cone Weissenberg, equal-cone Weissenberg [2.9], and the rotation method, which are similar or equivalent to the normal-beam and equi-inclination Weissenberg. The possibilities for generating multiple diffractions with these methods are similar to those discussed in this section.

In the precession methods [2.3, 9], the crystal precesses about the incident beam at an angle of inclination u. The geometry in reciprocal space is shown in Fig. 2.9. The zero- and higher-levels are indicated. The projection of the section lattice plane, for example, the $0A$ plane, that precesses about the incident beam is shown in Fig. 2.9b, the side view of Fig. 2.9a. There are two possibilities for multiple diffraction to take place: (1) more than two reciprocal lattice points of the same layer form a reflection circle; (2) more than two reciprocal lattice points from different layers lie on the surface of the Ewald sphere. For the first possibility, the chance of forming multiple diffractions depends, as in the case of intrinsic multiple diffraction (Sect. 2.1.3), on the type of two-dimensional lattice plane, the plane OA. Since the OA plane precesses around the incident beam, the chance of multiple diffraction occurring is greatly increased compared to the intrinsic case. For example, if the OA plane is an oblique lattice, three-beam diffraction occurs for any pair of reciprocal lattice points, together with the origin O. For higher-symmetry lattices like the square lattice, the chance of multiple diffraction is drastically in-

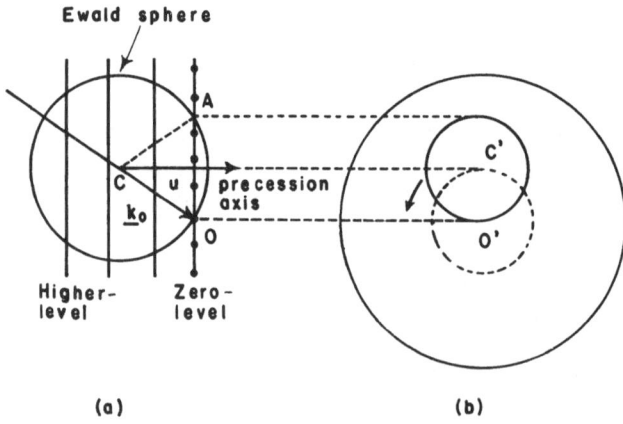

Ewald sphere

A

C'

C
k_0
u   precession
    axis
O'
O

Higher-   Zero-
level   level

(a)   (b)

Fig. 2.9. (a) Schematic representation of precession method in reciprocal space. (b) Side view of (a)

creased. A spectacular case, reported in [2.3], showed that 128 lattice points undergo sixteen-beam diffraction during one passage of the OA plane across the Ewald sphere.

2.2.2 Divergent-Beam Techniques

The collimated-beam method mentioned above involves crystal rotation for generating multiple diffraction. Another method utilizing a rotation of the x-ray source can be used as an alternative for the same purpose. The most convenient way of simulating the rotation of the source is to use a very divergent incident beam. Diffractions of this divergent source by sets of crystal planes form conic sections on a film. These conic diffraction images imply the rotation of a tiny x-ray source about the normal to the crystal planes, since the same diffraction cone can be obtained with an aligned rotating x-ray source. The intersections of these conics locate multiple diffractions.

The application of this divergent-beam technique dates far back to 1914, when *Rutherford* and *Andrade* [2.13] investigated the intensity for both reflection and transmission (absorption) of γ rays in the two-beam diffraction from a rock-salt crystal. This kind of investigation was actually first carried out for a collimated x-ray beam by *W. H. Bragg* [2.14] in the same year, using an ionization spectrometer. In the following years, *Seemann* [2.15 – 17] and later *Gerlach* [2.18, 19] developed Cu x-ray tubes to provide wide-angle beams. Divergent-beam photographs were obtained for a rock-salt crystal with a back-reflection geometry [2.20, 21], and for gypsum and mica, with a transmission arrangement [2.22, 23]. *Linnik* [2.24, 25], in 1929, also suggested the use of the divergent beam for obtaining diffraction conics photographically and discussed the similarity between the divergent-beam diffraction pattern of x-rays and the Kikuchi patterns of electron diffraction obtained by *Kikuchi* [2.26]. All these cases involved an x-ray source which is placed outside the crystal under investigation.

In the so-called Kossel method [2.27 – 32], a single crystal is utilized as anode in the x-ray tube. Divergent characteristic x-rays were generated within the crystal itself, and were diffracted from various sets of atomic planes before emerging from the crystal. Intense (black) and deficient (white) conics, due to reflection and absorption respectively, were obtained. Alternatively, *Borrmann* [2.33, 34] placed the crystal outside the x-ray tube and obtained a divergent beam via fluorescence excitation of the high-energy x-ray beam from the tube. Both discrete Laue diffraction spots (Laue method) of high-frequency x-rays and continuous conics of lower-frequency x-rays were observed.

The names 'Kossel' and 'pseudo-Kossel' were used to distinguish the methods with x-ray sources inside (Kossel) from those with x-ray sources outside the crystal (pseudo-Kossel). *Lonsdale* attained pseudo-Kossel patterns using a

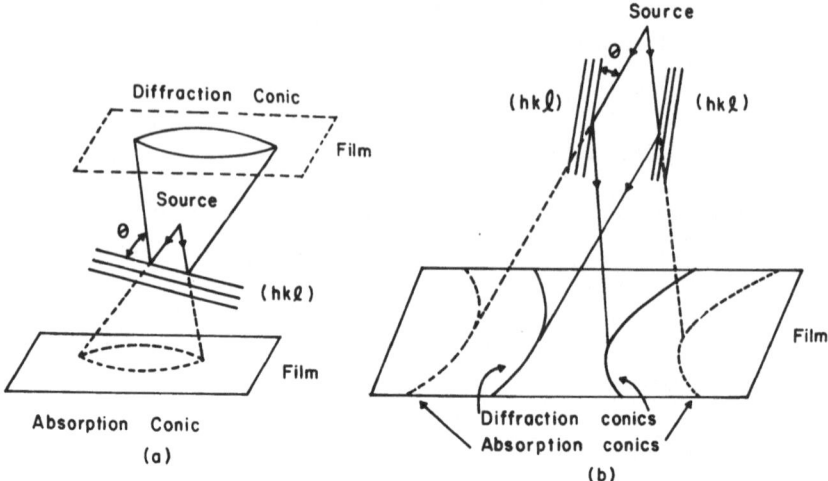

Fig. 2.10a, b. Formation of Kossel conics: (a) back reflection, (b) transmission

special x-ray tube designed by *Müller* [2.2]. The use of an electron probe microanalyzer for obtaining Kossel patterns was suggested by *Castaing* and *Guinier* [2.35]. The focused electrons impact on the crystal surface in an area of few micrometers in diameter and produce x-rays. Because of the small size of the x-ray source, the Kossel patterns obtained showed high resolution and clear contrast for the diffraction lines. As an alternative to direct electron bombardment on the crystal, it is possible to generate x-ray sources by electron bombardment on a thin metal foil which covers the crystal. Producing pseudo-Kossel patterns with a desired wavelength is thereby achieved [2.36 – 38]. Pseudo-Kossel patterns can also be obtained with a commercially available microfocus x-ray generator. An incident-beam size of about 10 μm in diameter is obtainable.

The geometry for obtaining Kossel patterns is shown in Fig. 2.10a and b. In Fig. 2.10a, a divergent characteristic x-ray beam from a point source is reflected by the atomic planes with Miller indices (hkl) and forms a diffraction cone (back-reflection Kossel). Meanwhile, the transmitted beam suffers crystal absorption and reflection (which transfers energy towards the diffracted beam), and produces a deficient absorption conic on an x-ray film (transmission Kossel). The semi-epical angle α of the conic is $90° - \theta$, where θ is the Bragg angle of the (hkl) planes. When both the (hkl) and $(\bar{h}\bar{k}\bar{l})$ planes satisfy the Bragg condition, both diffraction and absorption conics can be observed on one film (Fig. 2.10b). The diffraction conics always lie inside the two absorption conics. The deficient absorption lines are not always observable. They depend on the linear absorption coefficient μ which is a function of the radiation used and the crystal thickness t. The best condition for obtaining a clear image of the absorption conics is $\mu t \simeq 1$ [2.2, 2.39]. The number of

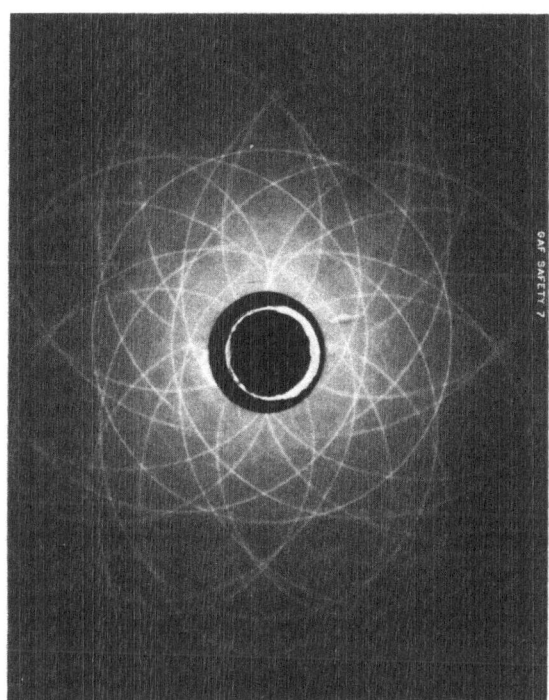

Fig. 2.11. Back-reflection Kossel pattern of (1 00) KCl for Cu radiation

families of planes taking part in the divergent-beam diffraction is limited by the Bragg reflection condition, i.e., $2d < \lambda$, by the position of the film, and by the crystal structure. The quantity d is the interatomic distance for a given plane. Figure 2.11 shows a back-reflection pseudo-Kossel pattern of a KCl crystal for Cu radiation. The film is parallel to the (1 00) plane of KCl.

The geometric condition for forming Kossel lines and their intersections can be derived from the Bragg law (2.2). We define a double wavevector c_i [2.40] as

$$c_i = -2k_i, \tag{2.9}$$

such that

$$c_i \cdot g_i = g_i^2 \tag{2.10}$$

for all i. The magnitude of c_i is $2/\lambda$. For a two-beam, O and G_1, case, a sphere of reflection with radius $2/\lambda$, centered at the reciprocal lattice point O, can be constructed (Fig. 2.12a). The plane perpendicular to the reciprocal lattice vector g_1 at G_1 is known as the Kossel plane. For a divergent incident beam, a diffraction cone is formed. The intersection of this cone with a film gives rise to an image of a conic. This conic is the projection of the intersection of the Kos-

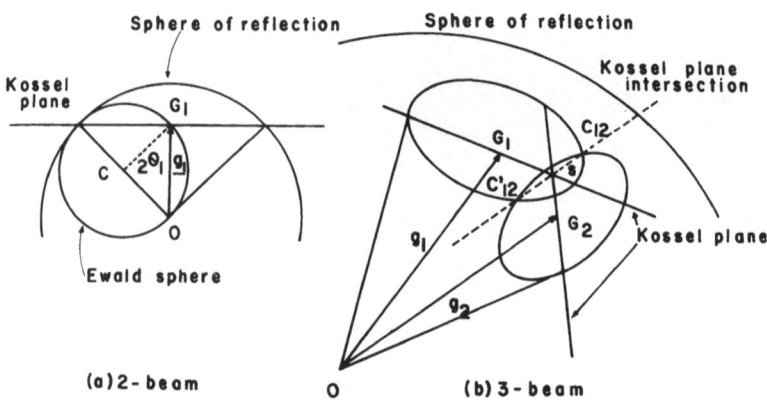

Fig. 2.12a, b. Double-wavevector representation of x-ray diffraction in (a) two-beam, (b) three-beam cases (after *Isherwood* and *Wallace* [2.40])

sel plane with the sphere of reflection. When such conics intersect each other, multiple diffractions occur at the intersection points.

Figure 2.12b shows the geometry of a three-beam, O, G_1 and G_2, multiple diffraction in the double-wavevector construction. The two points C_{12} and C'_{12} are the intersections of the two Kossel conics of the reflections G_1 and G_2. The line $\overline{C_{12}C'_{12}}$ also represents the intersection of the Kossel plane with the reflections G_1 and G_2. The double wavevectors at these intersection points for the three-beam case are $c_{12} (= OC_{12})$ and $c'_{12} (= OC'_{12})$. The angle between the two is denoted as ψ. The magnitudes of these two vectors c_{12} and c'_{12} can be written as

$$c_{12} = \sqrt{(OS)^2 + (SC_{12})^2} = 2/\lambda , \tag{2.11}$$

and

$$c'_{12} = \sqrt{(OS)^2 + (SC'_{12})^2} . \tag{2.12}$$

Since the vectors OS, g_1 and g_2 lie in the same plane, which is perpendicular to SC_{12}, then SC_{12} (with magnitude equal to SC'_{12}) is therefore in the direction of $g_1 \times g_2$. Because the angles $<SG_1O$ and $<SG_2O$ are equal to 90°, the four points O, S, G_1 and G_2 lie along the same circle, and $\overline{OS}$ is the diameter of that circle. By simple manipulation, it can be shown that

$$\overline{OS} = g_1 g_2 |g_1 - g_2| / |g_1 \times g_2|, \tag{2.13}$$

and $\overline{SC_{12}}$ can be expressed in terms of $\overline{OC_{12}}$ and the angle ψ as

$$\overline{SC_{12}} = \overline{SC'_{12}} = \overline{OC_{12}} \sin \frac{\psi}{2} , \tag{2.14}$$

where ψ is the angle between $\overline{OC_{12}}$ and $\overline{OC'_{12}}$ (Fig. 2.12).

From (2.9 – 14), the condition for multiple diffraction is obtained:

$$\frac{1}{\lambda^2} = \frac{g_1^2 g_2^2 |g_1 - g_2|^2}{2|g_1 \times g_2|^2 (1 + \cos\psi)}, \quad \text{or} \tag{2.15}$$

$$\frac{1}{\lambda} = \frac{|g_1 - g_2|}{4\sin\phi\cos(\psi/2)}, \tag{2.16}$$

where ϕ is an angle such that $|g_1 \times g_2| = |g_1||g_2|\sin\phi$. When the two Kossel conics are tangential to each other, there is only one intersection point S and the angle ψ is zero. Therefore, (2.15) becomes

$$\frac{1}{\lambda} = \frac{g_1 g_2 |g_1 - g_2|}{2|g_1 \times g_2|}. \tag{2.17}$$

This implies that this three-beam case is a coplanar coincidental diffraction (Sect. 2.1.2), since the double wavevector at the point S and the two reciprocal lattice vectors g_1 and g_2 are coplanar and the radius r_0 of the reflection circle in (2.4) is $1/\lambda$.

For four-beam diffraction, O, G_1, G_2 and G_3, the three Kossel conics of the reflections G_1, G_2 and G_3 intersect at a point T such that (Fig. 2.13)

$$OT \cdot g_i = g_i^2. \tag{2.18}$$

Since the intersection lines of the three pairs of Kossel planes also intersect at T, the double wavevector OT can be written in terms of the three directions $g_1 \times g_2$, $g_2 \times g_3$ and $g_3 \times g_1$, which are along the three intersection lines:

$$OT = \frac{g_1^2(g_2 \times g_3) + g_2^2(g_3 \times g_1) + g_3^2(g_1 \times g_2)}{g_1 \cdot (g_2 \times g_3)}, \tag{2.19}$$

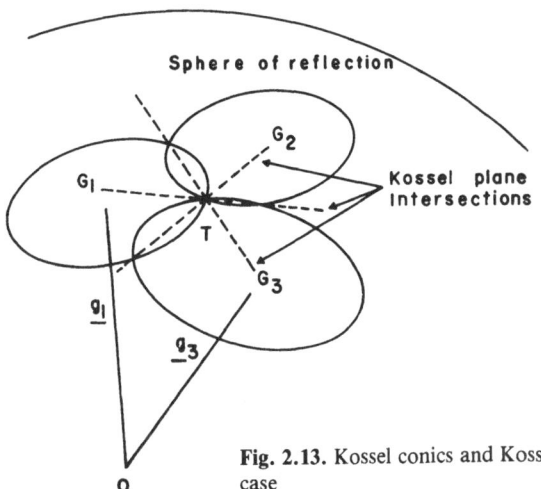

Fig. 2.13. Kossel conics and Kossel plane intersection for a four-beam case

which also satisfies (2.15). The condition for four-beam diffraction is then

$$\overline{OT} = \frac{2}{\lambda}. \tag{2.20}$$

For a general N-beam case, a similar condition with a more complicated vectorial expression for the double wavevector at the N-beam intersection point can be derived.

It should be noted that pseudo-Kossel lines are not exactly conics. Only ellipse-like lines are observed [2.41], while the back-reflection Kossel pattern gives hyperbolae [2.42 – 44].

Section pseudo-Kossel is most useful for revealing the details of the intersections of Kossel lines, i.e., the images of multiple diffractions in transmission [2.45] and reflection geometry [2.46, 47]. In this method, the beam divergence is limited to 5 – 10 degrees. The experimental arrangements for transmission and reflection geometries are shown in Figs. 2.14a, b, respective-

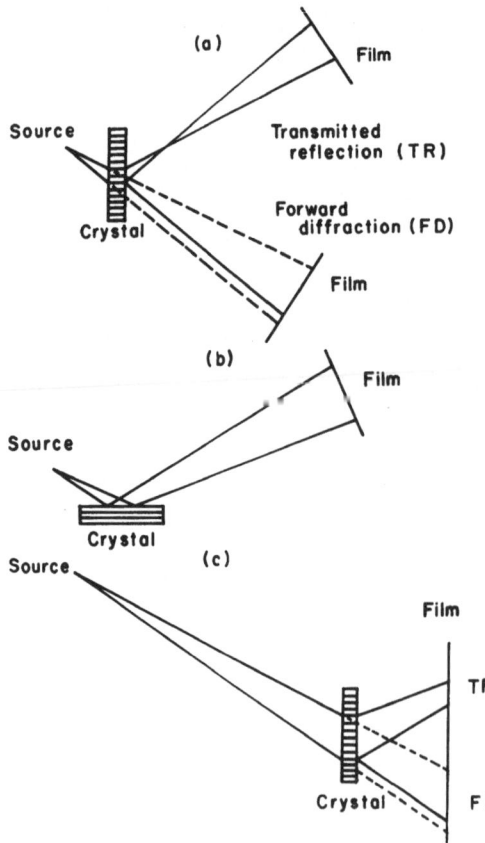

Fig. 2.14a – c. Experimental set-ups for (a) transmission, (b) reflection and (c) section-transmission divergent-beam technique

ly. The size of the point focus from a microfocus generator is 50×50 μm^2 or 100×100 μm^2. The crystal is placed about $5 - 10$ cm away from the source. The distance between the crystal and the film is $1 - 2$ m. The angular resolution at 1 m from the crystal is the angle subtended at the focus, which is about 10^{-4} rad. With this resolution, a detailed intensity distribution in the vicinity of the multiple diffraction can be obtained. In the transmission case (Fig. 2.14a), the forward-diffracted (FD) and the transmitted-reflected (TR) images [2.45] are recorded on separate films. To record both the FD and TR images on one film, the experimental arrangement shown in Fig. 2.14c is employed [2.48]. In this set-up the distances from the source to the crystal and from the crystal to the film are $1 - 2$ m and a few centimeters, respectively. The long source-to-crystal distance, shown in Fig. 2.14c, limits the beam divergence. This makes the crystal alignment difficult.

The procedure of aligning the crystal for a desired multiple diffraction is the same as that described in the collimated-beam method. If the crystal is plate-like and if all the reciprocal lattice points involved lie in the plane parallel to the crystal plate, the crystal can be aligned in a particularly simple way. The crystal plate is first placed normal to the incident beam, supposing that the incident beam lies as usual in a horizontal plane; the crystal is then rotated around its surface normal so that the center of the reflection circle circumscribing the reciprocal lattice points stays in the horizontal plane; then it is tilted horizontally towards the source by an angular amount α such that

$$(2/D^*) \sin \alpha = \lambda \, , \tag{2.21}$$

where D^* is the diameter of the reflection circle.

2.3 Indexing Multiple Diffraction Patterns

2.3.1 Reference Vector Method

There are several ways of indexing multiple diffraction peaks. We shall describe the procedure given by *Cole* et al. [2.8] since it is easily adapted for computer programming.

During the rotation of a crystal in the generation of multiple diffraction, each secondary reciprocal lattice point enters and then leaves, or vice versa, the Ewald sphere. This is demonstrated in Fig. 2.15a. Figure 2.15b is the projection of Fig. 2.15a on the plane perpendicular to the reflection vector of the primary reflection OG. The points of entry and exit are denoted as "IN" and "OUT" in the figure. For simplicity, vectors g and p are used as the reciprocal lattice vectors for the primary and the secondary reflections. The vectors p_n and p_p are the components of the vector p, perpendicular and parallel to g.

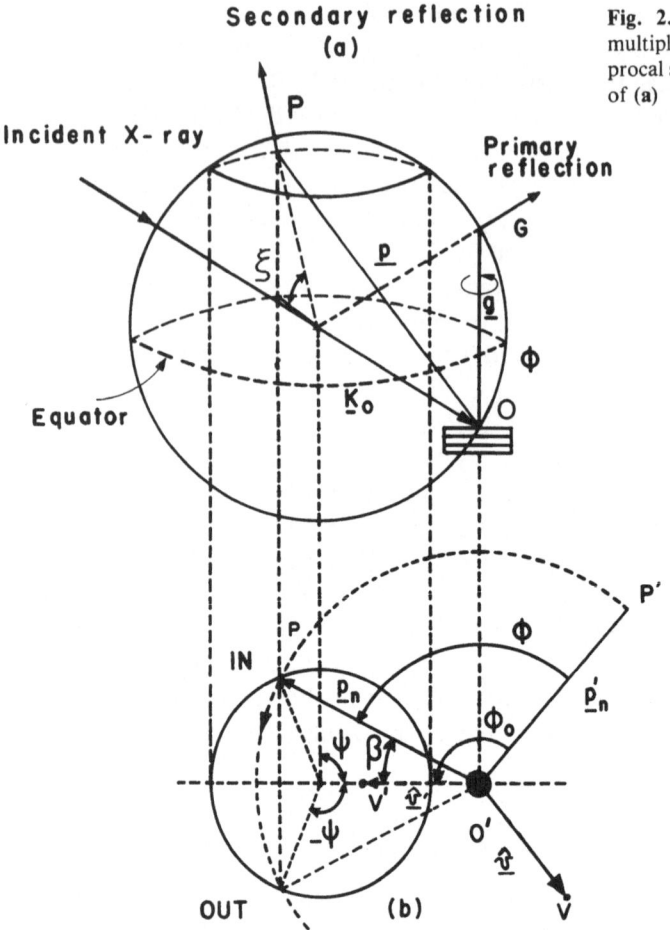

Secondary reflection
(a)

Fig. 2.15. (a) Geometry of multiple diffraction in reciprocal space. (b) Bottom view of (a)

Vector $\hat{v}$ is a unit vector in a reference direction which is usually chosen to be perpendicular to g. Hereafter, vector $\hat{u}$ denotes the unit vector of any vector u. Suppose that initially the reference unit vector $\hat{v}'$ lies in the plane of incidence of the G reflection along the line $\overline{O'V'}$ and that p_n' lies along $\overline{O'P'}$ as shown in Fig. 2.15b. The angle ϕ_0 between $O'V'$ and $O'P'$ is equal to the angle between $O'V$ and $O'P$, because the crystal rotation does not change the angle between any pair of reciprocal lattice vectors. Points V and P are the positions of V' and P' after the crystal has been rotated by an angle ϕ so as to give multiple diffractions. Hence $\phi_0 = \cos^{-1}(\hat{p}_n' \cdot \hat{v}')$. The azimuthal angles ϕ, the amounts by which the crystal needs to be rotated, are therefore

$$\phi = \phi_0 - \beta, \qquad \phi = \phi_0 + \beta, \tag{2.22}$$

for the IN and OUT positions, respectively, where

$$\beta = \cos^{-1}(\hat{p}_n \cdot \hat{v}'), \quad \text{and} \tag{2.23}$$

$$\hat{v}' = \frac{k_0 + (g/2)}{|k_0 + (g/2)|}, \quad \text{or} \tag{2.24}$$

$$\cos \beta = \frac{\sqrt{p^2 - p_p g}}{2p_n \sqrt{(1/\lambda^2) - (g^2/4)}}. \tag{2.25}$$

Equation (2.25) can be written as

$$\sqrt{\frac{1}{\lambda^2} - \frac{g^2}{4}} \cos \beta = \frac{\sqrt{p^2 - p_p g}}{2p_n} = C_{GP}, \tag{2.26}$$

where C_{GP} depends only on the reflections G and P.

From (2.26) there are two constraints, $0 \leqslant \cos \beta \leqslant 1$ and $1/\lambda^2 \geqslant g^2/4$, for a positive C_{GP}. When $\beta = 0$, the diffraction is coplanar and coincidental, and there exists a smallest sphere such that

$$\left(\frac{1}{\lambda}\right)^2 = \frac{g^2}{4} + C_{GP}^2. \tag{2.27}$$

The radius of the Ewald sphere is infinite when $\beta = 90°$ for a positive C_{GP}. If $C_{GP} = 0$,

$$p^2 - p_p g = 0, \tag{2.28}$$

which implies that the point P lies on the circle with diameter equal to g, passing through O and G. Under this condition, if $\cos \beta \neq 0$, then $1/\lambda = g/2$. This means that the point P lies on the great circle which intersects the Ewald sphere at the center C with $\beta \neq 90°$. If (2.28) holds and $\beta = 90°$, multiple diffraction occurs for all wavelengths.

For negative C_{GP}, $180° \geqslant \beta \geqslant 90°$. When $\beta = 180°$, the multiple diffraction is a coplanar and a coincidental diffraction.

The appearance of mirror symmetries in a multiple diffraction pattern depends on the symmetry of the rotation vector. An expression can be used as a general rule for predicting the angular repetition of the mirror symmetries, the n_M-fold mirror symmetry, in the form [2.5]:

$$n_M = n_r(p) \cdot R, \tag{2.29}$$

where $n_r(p)$ is the n_r-fold symmetry of the rotation vector. R is the rotation operator, which is 2 in this case to take into account the entering and leaving situations. For two-, three- and four-fold rotation axes, the corresponding multiple diffraction patterns show four-, six- and eight-fold mirror symmetries. For example, when the rotations are about the axes $\langle 220 \rangle$ and $\langle 400 \rangle$

for the diamond structure, mirror symmetries appear in the multiple diffraction patterns every 90° and 45°. The multiple diffraction pattern for the rotation about $\langle 111 \rangle$ shows twelve-fold symmetry because the symmetry of the projected plane normal to $\langle 111 \rangle$ is six-fold. However, it should be emphasized that the mirror symmetries divide a multiple diffraction pattern into n_M asymmetric portions. A complete symmetric portion of the pattern actually repeats as it has an $n_r(p)$-fold symmetry. For the rotations about $\langle 220 \rangle$, $\langle 111 \rangle$ and $\langle 400 \rangle$, a symmetric portion of the multiple diffraction pattern appears every 180°, 60°, and 90° in the azimuths, respectively.

A computer program for multiple diffraction peak indexing can be easily written by considering (2.25). However, it may be somewhat difficult to write such a program for crystal systems, such as hexagonal, rhombohedral, monoclinic, and triclinic, whose basis vectors are not orthogonal to each other. Nevertheless, by choosing convenient orthogonal coordinates as the basis vectors and by transforming the Miller indices according to the new system, the indexing problem can be treated as easily as for cases involving cubic, tetragonal, and orthorhombic crystal systems. For example, the new reciprocal orthogonal axes, A^*, B^* and C^*, may be chosen as

$$A^* = a^*,$$
$$B^* = c^* \times a^*,$$
$$C^* = A^* \times B^*,$$

(2.30)

where a^*, b^* and c^* are the non-orthogonal axes. The relationships between the indices (HKL) of the orthogonal and (hkl) of the non-orthogonal axes are

$$H = h + \frac{A^* \cdot b^*}{(A^*)^2} k + \frac{A^* \cdot c^*}{(A^*)^2} l,$$

$$K = \frac{B^* \cdot a^*}{(B^*)^2} h + k + \frac{B^* \cdot c^*}{(B^*)^2} l,$$

(2.31)

$$L = \frac{C^* \cdot a^*}{(C^*)^2} h + \frac{C^* \cdot b^*}{(C^*)^2} k + l.$$

As an illustration of the indexing procedure, let us consider the (222) multiple diffraction pattern of germanium for CuK_{α_1} (Fig. 2.6). Germanium has a diamond structure and belongs to the space group $Fd3m$. The $\langle 222 \rangle$ direction has three-fold symmetry. Rotation about this axis generates six-fold symmetry. Therefore, multiple diffraction peaks involving equivalent secondary and coupling reflections should appear every sixty degrees in the azimuthal angle ϕ. Because a pair of tetrahedrons are different by sixty degrees

Table 2.1. Indexing of the multiple diffraction pattern of Ge (222) taken with CuK_{α_1} radiation referred to origins $\langle 1\bar{1}0\rangle$ and $\langle 0\bar{1}1\rangle$, 60° apart

Azimuthal angle [°]	Origin on $\langle 1\bar{1}0\rangle$ Secondary/Coupling	Origin on $\langle 0\bar{1}1\rangle$ Secondary/Coupling
2.087	$3\bar{5}1/\bar{1}71$	$1\bar{1}7/13\bar{5}$
3.484	$51\bar{1}/\bar{3}13$	$3\bar{3}1/\bar{1}51$
5.424	$113/11\bar{1}$	$\bar{1}11/311$
7.735	$3\bar{1}5/\bar{1}3\bar{3}$	$\bar{3}13/53\bar{1}$
13.820	$3\bar{3}5/\bar{1}5\bar{3}$	$\bar{3}15/53\bar{3}$
13.976	$5\bar{3}\bar{1}/\bar{3}53$	$3\bar{3}5/\bar{1}5\bar{3}$
14.161	$31\bar{1}/\bar{1}13$	$3\bar{1}\bar{1}/\bar{1}31$
15.720	$3\bar{5}3/\bar{1}7\bar{1}$	$\bar{1}\bar{1}7/33\bar{5}$
18.302	$1\bar{1}5/13\bar{3}$	$\bar{3}13/51\bar{1}$
18.561	$5\bar{1}\bar{1}/\bar{3}33$	$3\bar{3}3/\bar{1}5\bar{1}$
21.869	$\begin{cases} 1\bar{3}\bar{1}/153 \\ 513/\bar{3}1\bar{1} \end{cases}$	$\begin{array}{l} 315/\bar{1}13 \\ \bar{1}31/351 \end{array}$
22.081	$\bar{5}11/\bar{3}11$	$1\bar{3}1/151$
22.326	$1\bar{3}5/15\bar{3}$	$\bar{3}15/51\bar{3}$
25.180	$\bar{1}\bar{1}3/33\bar{1}$	$133/\bar{3}\bar{1}\bar{1}$
28.212	$\bar{1}\bar{1}1/331$	$133/1\bar{1}\bar{1}$
29.522	$\bar{1}\bar{1}3/35\bar{1}$	$\bar{1}35/3\bar{1}\bar{3}$
29.883	$1\bar{5}3/17\bar{1}$	$\bar{1}17/315$

upon rotation about $\langle 222\rangle$, symmetry mirrors in the diffraction pattern repeat every 30° (Fig. 2.6). Figure 2.6 is such an asymmetric portion of the multiple diffraction pattern. The angles $\phi = 0$ and $\phi = 30°$ correspond to $\langle 1\bar{1}0\rangle$ and $\langle 121\rangle$, respectively. The former, $\langle 1\bar{1}0\rangle$, is used as the reference vector for indexing the multiple diffraction peaks.

Ambiguity is often encountered in indexing multiple diffraction patterns, especially for crystals involving high symmetry. The diffraction pattern just discussed is a particular case. If one chooses two equivalent directions as the reference vectors, two possible sets of indices can be assigned to the pattern. Table 2.1 shows the two possible sets of indices, with either $\langle 1\bar{1}0\rangle$ or $\langle 0\bar{1}1\rangle$ as the origin for ϕ. These two directions are sixty degrees apart. Because of the indeterminate origin of the azimuths, a correct set of indices cannot be assigned to the observed diffraction peaks. In Table 2.1, each peak of the three-beam case can be indexed as either (A) (000) (222) $(hkl)/(2-h, 2-k, 2-l)$ or (B) (000) (222) $(2-l, 2-h, 2-k)/(lhk)$, where the reflections before and after the slashes are the secondary and the coupling reflections, respectively. The first two reflections are the incident and the primary reflections.

The difficulty is to identify the two cases (A) and (B). To solve this problem, *Chang* and *Caticha-Ellis* [2.49] proposed an experimental way of distinguishing between the two cases. One of these cases involves a Bragg-type secondary reflection, say case A (Bragg-Bragg), and the other a transmission (Laue) reflection, case B (Bragg-Laue). It is easy to distinguish the Bragg-

Bragg case from the Bragg-Laue case by detecting the secondary reflection: if the situation conforms to case A, the secondary reflection would be measurable on the same side of the crystal as the incident beam, while in case B, the secondary reflection would be eventually detected on the other side of the crystal. In each case a scintillation counter can be placed in the precalculated positions ψ and ξ, defined in Fig. 2.15 for the secondary reflection as

$$\xi = \sin^{-1}(2\sin\theta_2\cos\gamma - \sin\theta_1),$$

$$\psi = \cos^{-1}\left(\frac{\cos^2\xi + \cos^2\theta_1 - 4\sin^2\theta_2\sin^2\gamma}{2\cos\xi\cos\theta_1}\right),$$

(2.32)

where θ_1 and θ_2 are the Bragg angles for the primary (222) and the secondary reflections, and γ is the angle between the reciprocal lattice vectors of the two reflections. The maximum values of ξ and ψ are 90° and $\pm 180°$, respectively. The counter and the incident beam in the Bragg-Bragg case always lie on the same side of the crystal. In that case signals should be detected either at the position (ξ, ψ) or at $(\xi, -\psi)$ depending on whether the secondary reciprocal lattice point is going into or out of the Ewald sphere (Fig. 2.15b). If no signals are detected at these two positions, the multiple diffraction therefore involves a transmitted secondary reflection. Since the selected secondary reflection is a strong one, there is little chance of error. In order to avoid all possibility of error, a further check can be made by rotating the crystal 60° around $\langle 222\rangle$ so that the Bragg-Bragg case is produced; then, signals have to be found at one of the positions $(\xi, \pm\psi)$ of the detector. In this way, the two strong three-beam cases, (000) (222) ($\bar{1}\bar{1}1$)/(331) and (000) (222) (133)/(11$\bar{1}$), are identified by the presence of the diffracted intensity of (133) at $\xi = 38.96°$ and $\psi = 30.26°$. The correct indexing for the diffraction pattern is therefore obtained. The method described above can also be used to solve the same ambiguity in the case of Aufhellung "peaks", and it is equally applicable to centrosymmetric and non-centrosymmetric crystals.

2.3.2 Orientation Matrix Method

The method proposed above to distinguish the Bragg-Bragg from the Bragg-Laue case is limited to the cases in which three-beam diffraction can be found. Cases involving even-order multiple diffraction are difficult to handle since secondary Laue and secondary Bragg reflections usually appear simultaneously at the same ξ angle. In addition, a 180° rotation is needed to reveal the symmetry mirrors in the multiple diffraction pattern for monoclinic crystals. For an absolute indexing of a multiple diffraction pattern of any primary reflection, *Han* and *Chang* [2.50] replaced the reference vector by an orientation matrix [2.51] which specifies the initial position of a crystal on a four-

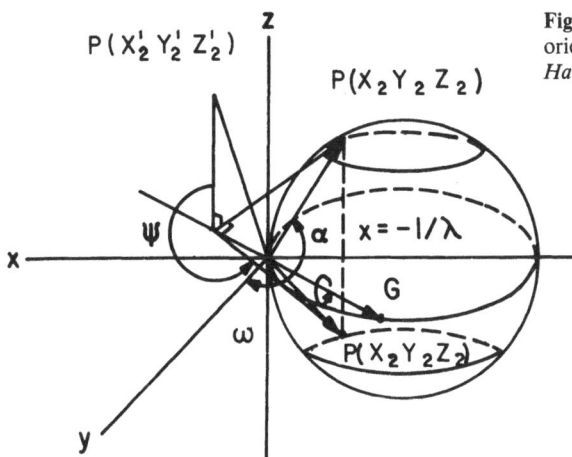

Fig. 2.16. Coordinate system for the orientation matrix method (after *Han* and *Chang* [2.50])

circle x-ray single-crystal diffractometer. In a four-circle diffractometer, the rotation around a reciprocal lattice vector is called a ψ-scan [2.51] which is a combined movement of three circles of the diffractometer, i.e., the rotation angles ϕ, χ and ω (Fig. 2.16). This can be described by the product of three rotation matrices $M_{r\phi}$, $M_{r\chi}$, $M_{r\omega}$, about the ϕ, χ, and ω rotation axes [2.51−53]:

$$M_r = M_{r\omega} M_{r\chi} M_{r\phi}$$

$$= \begin{bmatrix} \sin\phi\cos\chi\sin\omega & -\cos\phi\cos\chi\sin\omega & -\sin\chi\sin\omega \\ +\cos\omega\cos\phi & +\cos\omega\sin\phi & \\ -\sin\phi\cos\chi\cos\omega & \cos\phi\cos\chi\cos\omega & \sin\chi\cos\omega \\ +\cos\phi\sin\omega & +\sin\phi\sin\omega & \\ \sin\phi\sin\chi & -\cos\phi\sin\chi & \cos\chi \end{bmatrix}. \quad (2.33)$$

Supposing the initial orientation matrix is M_{or}, with ϕ, χ and ω equal to zero, the product of M_r and M_{or} yields a new orientation matrix M'_{or}:

$$M'_{or} = M_r M_{or}. \quad (2.34)$$

The initial position (X'_2, Y'_2, Z'_2) of the secondary reciprocal lattice point can then be determined by using the matrix M'_{or} as an operator. Figure 2.16 shows the coordinates and the angular relations between ψ, χ and ω when P is brought onto the surface of the Ewald sphere. The position (X_2, Y_2, Z_2) of P can be determined from

$$X_2^2 + Y_2^2 + Z_2^2 = \overline{OP}^2,$$
$$(X_2 + 1/\lambda)^2 + Y_2^2 + Z_2^2 = 1, \quad (2.35)$$
$$\frac{Y_2 - Y'_2}{X_2 - X'_2} = \tan\omega, \quad \text{where}$$

$$X_2' = -p \cos \alpha \sin \omega,$$

$$Y_2' = p \cos \alpha \cos \omega, \tag{2.36}$$

$$Z_2' = p.$$

The double solutions for (2.35) are

$$X_2 = -\lambda p^2/2,$$

$$Y_2 = Y_2' + (X_2 - X_2') \tan \omega, \tag{2.37}$$

$$Z_2 = \pm \sqrt{p^2 - X_2^2 - Y_2^2},$$

for the entering (positive Z_2) and leaving (negative Z_2) situations. From (X_2', Y_2', Z_2') and (X_2, Y_2, Z_2), the angular distance between the starting position and the multiple diffraction peak is obtained.

2.4 Indexing Kossel Patterns

The intersection of two or more Kossel lines indicates the occurrence of multiple diffraction. As long as the Kossel lines involved are indexed, the corresponding multiple diffractions are identified.

The first step in indexing a Kossel pattern is to identify the symmetry elements of the diffracting crystal from the symmetry of the diffraction pattern, as in the Laue method [2.9]. The rotation and mirror symmetries appearing in the pattern reflect the symmetry possessed by the crystal. For example, if the diffraction pattern shows four-, three- and two-fold symmetries, then the pole axes are $\langle 001 \rangle$, $\langle 111 \rangle$, $\langle 011 \rangle$, respectively.

When the symmetry elements are identified, the Kossel lines may be indexed by several available methods. The techniques most frequently used are described below.

2.4.1 Stereographic Projection Method

Identification of Kossel lines is made by comparing a Kossel photograph with a diagram obtained from projections, such as gnomonic, Kossel, cylindrical and stereographic projections. There are several disadvantages with the first three methods; the stereographic projection is relatively easy to construct in comparison with the others [2.2, 2.54] (Fig. 2.17). The usual procedure of stereographic projection is the following: (1) The Bragg angles of all the possible reflections for a given crystal and radiation are calculated; (2) a circle of the radius corresponding to $90° - \theta$ is inscribed on the stereographic projec-

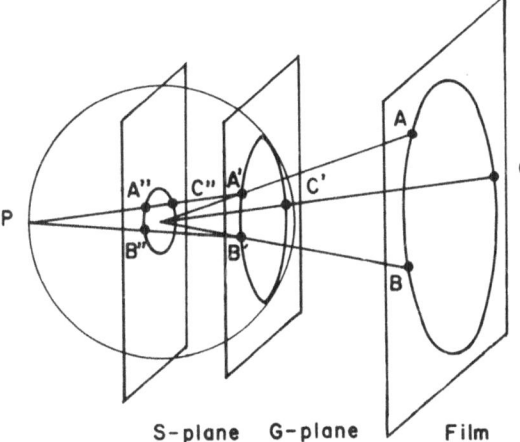

Fig. 2.17. Gnomonic and stereographic projections (after *Tixier* and *Waché* [2.55])

S-plane   G-plane   Film

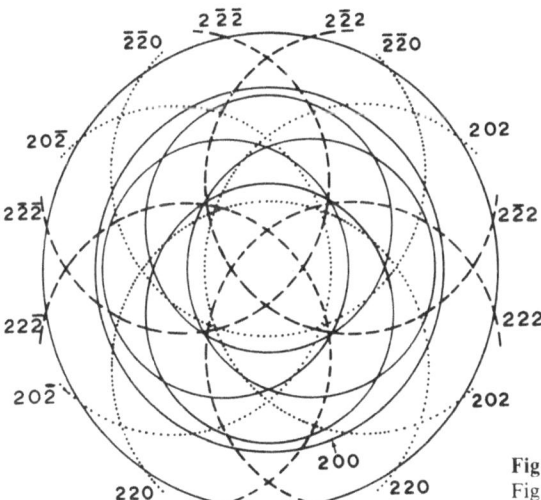

Fig. 2.18. Stereographic projection of Fig. 2.11 (only a few strong reflections are shown)

tion about the pole of each diffraction plane. It should be noted that although the stereographic projection of a circle is still a circle, the center of the projected circle is not necessarily coincident with that of the original circle. With the aid of a computer, the coordinates of the center of the pattern and the radius of each Kossel circle on a stereographic projection can be calculated, and the complete stereographic projection can be plotted [2.55 – 57]. Figure 2.18 gives the stereographic projection of Kossel circles for Fig. 2.11.

2.4.2 Combined Gnomonic-Stereographic Projection Method

This method, proposed by *Peters* and *Ogilvie* [2.58], uses two graphic projections. Three points, A, B, C, on a Kossel circle are first projected (gnomonic) onto the *G* plane (Fig. 2.17) as the points A', B', C' with a suitable scalar factor to take into account the magnification of the diffraction pattern on the film. In a gnomonic projection, meridians and parallels on a sphere appear as radial straight lines and circles concentric about the center of projection. The angular coordinates of points A', B' and C', are measured from the gnomonic polar chart with its center coincident with the center of the Kossel pattern. These points, A', B' and C', are transferred into a stereographic Wulff net as points A'', B'' and C'' shown in Fig. 2.17. A circle circumscribing these three points is therefore constructed graphically on the Wulff net chart. The angle subtended by this circle, the meridian, which can be read off from the length of the segment intersecting the circle through its center, is equal to $90° - \theta$. The indices hkl of this angle θ are therefore identified. The center of the circle also determines the crystal orientation. There are other approaches which also utilize, in part, either gnomonic or stereographic projections [2.59, 60].

3. Kinematical Theory of Diffraction

Crystal lattice points serve in x-ray diffraction as the scattering centers for incident waves. The spherically scattered wavefronts are detected by their intensities – the amplitude squared at a point of observation far away from the crystal. To account for the diffracted intensity and to understand the diffraction mechanism, diffraction theories were developed under a few simplifying assumptions which often rely on the nature of the crystal. For example, in *Laue*'s geometrical theory of x-ray diffraction [3.1], the following assumptions were made: (1) The velocity of the x-ray beam traveling through the crystal is the velocity of light. The interaction between the incident and diffracted beams is neglected. (2) The multiple scattering effect is excluded. This means that the scattered waves are not subjected to re-scattering at other lattice points. (3) Absorption plays no role in the diffraction. This theory is therefore valid for diffraction in small crystals.

Kinematical theory was developed to account for the observed intensity from a real crystal. In developing the kinematical theory of x-ray diffraction, the mosaic crystal model was adopted by *Darwin* [3.2] for real crystals. The atoms in a mosaic crystal structure are assumed to be arranged in blocks. Each block is a perfect crystal. Adjacent blocks are slightly misaligned with each other. Quantitatively, the distribution $W(\phi)$ is employed to describe this misorientation where ϕ is the angle made by two adjacent blocks. This function is called the mosaic spread distribution. Because the crystal blocks are usually of microscopic size, the irradiated region of the crystal, which diffracts an incident beam, is composed of a great number of such blocks. Continuous distribution functions may therefore be used to represent the distribution of blocks, with the angular misalignment deviating from the mean value of the mosaic spread. Lorentzian and Gaussian distribution functions are often employed to describe the mosaic structure in real crystals. For the sake of convenience, a Gaussian distribution function is adopted here for $W(\phi)$:

$$W(\phi) = \frac{1}{\sqrt{2\pi}\eta} \exp\left(-\frac{\phi^2}{2\eta^2}\right), \qquad (3.1)$$

where η, the standard deviation, is the characteristic mosaic spread of a given crystal.

The other assumptions of the kinematical theory are: (1) the scattering process is coherent. Changes in wavelength of the radiation while traveling from one medium to another are not considered and Compton-type scattering is excluded. In addition, the interaction between the scattered x-ray photons and the nuclei is ignored. (2) Both incident and scattered waves have plane wavefronts, and (3) the scattered wave amplitude is a small fraction of the incident wave. This implies that the interaction of the electromagnetic waves comprising the incident and diffracted beams is ignored as in the geometrical theory. Only the reflection powers, or intensities, are considered. The incident reflection power is attenuated due to diffraction and ordinary absorption, i.e., $\exp(-\mu t)$, where μ is the linear absorption coefficient and t is the crystal thickness. Although some of the assumptions violate physical laws, for example, the conservation of energy, and also the validity of the mosaic structure assumed for real crystals, the theory is still justifiable for small crystals, weak reflections, and short-wavelength radiation. In the following, we shall discuss the theory in the multi-beam regime.

3.1 Equation of Power Transfer for Multi-Beam Cases

In this section we shall derive, based on the assumptions for the kinematical theory, the equation of power transfer to describe the transfer of diffraction power among the diffracted beams in multi-beam cases.

For simplicity, let us consider a general four-beam diffraction for a plate-like mosaic crystal of thickness T. Suppose the ideal crystal blocks are also plane-parallel plates nearly parallel to the large crystal. The thickness of each individual block is assumed to be so small that x-ray absorption in the block may be neglected. Because of the misalignment of the blocks, no definite phase relationship exists between the scattered beams from different blocks. This implies that each individual block therefore diffracts independently. The mosaic distribution in the crystal is $W(\phi)$ as defined in (3.1).

A well-collimated and monochromatic x-ray beam O is incident on the crystal. The crystal is so adjusted that four-beam diffraction occurs (Fig. 3.1). In Fig. 3.1, O and L_1 are the Laue transmissions, while M_1 and M_2 are Bragg

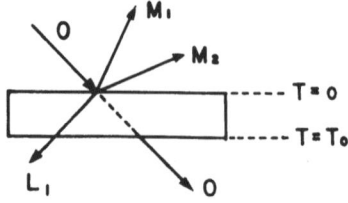

Fig. 3.1. Four-beam diffraction from a plane-parallel crystal plate

reflections. The direction cosines of the diffracted beams with respect to the inward crystal normal are γ_i, where $i = O, L_1, M_1$ and M_2. Consider the diffraction from a crystal layer of thickness dx with an incident beam i and a reflected beam j. This layer is assumed to be composed of many ideal crystal blocks with a mosaic spread distribution $W(\phi)$ as defined by (3.1). The diffraction power of a single block for a reflection j is [3.3]

$$\Delta P_s = \int W(\phi) P_j(\theta - \theta_B + \phi) d\phi \simeq W(\theta - \theta_B) R_{ij}(\theta) , \qquad (3.2)$$

where $R_{ij}(\theta)$ is the integrated reflecting power, defined as the product of the reflection power and the angular deviation, $\theta - \theta_B + \phi$, of a single block, and P_j is the power ratio between the reflected beam j and the incident beam. Consequently, the reflection power of the layer dx can be written in the form

$$\Delta P_{ij} = Q_{ij} dx = W(\theta - \theta_B) R_{ij}(\theta) n_0 , \qquad (3.3)$$

where n_0 is the number of single crystal blocks. The quantity ΔP_{ij} can also be considered as the amount of reflection power transferred from the incident beam i to the reflected beam j. Q_{ij} therefore serves as a linear reflection coefficient.

The variation, either increase (gain) or decrease (loss), in the power of the various diffracted beams is due to normal absorption and diffraction. The power loss of beam i results from the absorption when the ith beam traverses the crystal, and from the diffraction from beam i to other reflected beams, say beam j. The power increase is obtained by means of power transfer from the other j reflected beams to beam i. The differential equations describing these power variations can be obtained [3.4] by generalizing the power-transfer equation of two-beam cases [3.5], see (3.8). A compact form for N-beam cases can be written [3.6]:

$$\pm \frac{dP_i}{dx} = - \frac{\mu_0 P_i}{\gamma_i} + \sum_j \left(\frac{Q_{ji} P_j}{\gamma_j} - \frac{Q_{ij} P_i}{\gamma_i} \right) , \qquad (3.4)$$

where the positive sign is for the transmissions O and L_1, and the negative sign is for the reflections M_1 and M_2. The term P_i is the power in beam i and μ_0 is the ordinary linear absorption coefficient. The first and the last terms on the right-hand side of (3.4) are the decrease in power due to absorption and diffraction. The second term is the power increase of the ith beam gained from the jth beam via the diffraction $j - i$. Equation (3.4) involves n linear differential equations of the form [3.5]

$$\frac{dP_i}{dx} = - \sum_j a_{ij} P_j . \qquad (3.5)$$

The solution will be of the type

$$P_i = \sum_j A_{ij} \exp(Z_j T),$$

(3.6)

where the Z's are the roots of the secular equation of (3.5), and T is the thickness of the crystal plate. The coefficients A_{ij} correlating the i and j diffractions can be determined from the following boundary conditions:

$$P_0(T) = P_0(0), \quad P_{L_1}(T) = 0, \quad \text{at } T = 0;$$

$$P_{M_1}(T) = P_{M_2}(T) = 0, \quad \text{at } T = T_0.$$

(3.7)

Considering two-beam Bragg and two-beam Laue diffractions ($N = 2$), the differential equations are

$$\frac{dP_0}{dx} = -\mu_0 \frac{P_0}{\gamma_0} - Q_{01} P_0 + Q_{10} P_1,$$

$$\pm \frac{dP_1}{dx} = -\mu_0 \frac{P_1}{\gamma_1} - Q_{10} P_1 + Q_{01} P_0,$$

(3.8)

where the subscripts 0 and 1 indicate the incident and the diffracted beams. For the Bragg case, $\gamma_1 < 0$ and the negative sign is used for P_1. These two cases have been extensively studied [3.2, 3]. We shall only present the results. The exact solutions are [3.3]:

for symmetric Bragg, $\gamma_0 = -\gamma_1$,

$$\frac{P_M(0)}{P_0(0)} = \frac{Q_{01} + (\mu_0/\gamma_0) - U}{Q_{01}}$$

$$- \frac{U[Q_{01} + (\mu_0/\gamma_0) - U] \exp(-UT_0)}{Q_{01}\{[Q_{01} + (\mu_0/\gamma_0)] \sinh(UT_0) + U \cosh(UT_0)\}},$$

(3.9)

for symmetric Laue, $\gamma_0 = \gamma_1$,

$$\frac{P_L(T_0)}{P_0(0)} = \sinh(Q_{01} T_0) \exp\left[-\left(\frac{\mu_0}{\gamma_0} + Q_{01}\right) T_0\right],$$

(3.10)

where

$$U = \sqrt{[Q_{01} + (\mu_0/\gamma_0)]^2 - Q_{01}^2}.$$

(3.11)

Expressions for thin and thick crystals can be obtained from the above equations, with the condition that $Q_{ij} T_0$ be less than and greater than unity, respectively [3.3].

The diffracted power, as mentioned above, decreases due to ordinary absorption and the attenuation caused by diffraction. Considering the incident beam, (3.8) shows that the effective absorption μ_e can be defined as

$$\mu_e = \mu_0 + \frac{Q_{01}}{\gamma_0}, \tag{3.12}$$

where μ_0 is the ordinary linear absorption coefficient. The second term on the right side represents the fractional power loss due to diffraction per unit path length. It is called the secondary extinction coefficient when the reflection power loss takes place among the mosaic blocks. The attenuation in power due to the reflection within a single ideal crystal block is named primary extinction [3.2]. Primary extinction depends on the thickness of the block and the strength of the reflection, i.e., the structure factor of the reflection. Secondary extinction thus depends on the primary extinction and the mosaic distribution.

3.2 Approximate Solutions to the Equation of Power Transfer

To find the exact solution for a general N-beam ($N > 2$) case is a formidable task. However, a proper approximate solution can easily be obtained in accordance with the experimental conditions. *Moon* and *Shull* [3.4] and *Zachariasen* [3.5] used a Taylor series expansion for the exponential term in (3.6), when mosaic crystals with moderate or small extinction were considered.

Small extinction signifies that the width of the mosaic spread is much larger than that of the diffraction from a perfect crystal block, and that the incident beam divergence is much smaller than the mosaic spread but much larger than the intrinsic diffraction width from a perfect crystal. In other words, the power loss due to diffraction is very small, such that

$$\mu_e l \ll 1,$$

or, from (3.12),

$$\mu_0 l_i \ll 1, \quad Q_{ij} l_i \ll 1, \tag{3.13}$$

where l_i is the path length of the diffracted beam i in the crystal, defined as

$$l_i = \frac{T}{\gamma_i}. \tag{3.14}$$

The quantity T is the thickness of the crystal. Under these conditions, the reflectivity Q_{ij} can be expressed as [3.4]

$$Q_{ij} = W(\Delta\theta_{ij}) \left(\frac{\lambda^3 N_0^2 |F_s|^2}{\sin 2\theta} \right)_{ij}, \tag{3.15}$$

where $\Delta\theta_{ij}$ is the deviation from the Bragg angle of the $(i-j)$ reflection, N_0 is the number of unit cells per volume and F_s is the structure factor.

The assumed solution of (3.4) is in the form of a Taylor series expansion about the point $x = 0$, the entrance surface of the crystal:

$$P_i(T) = P_i(0) + T\frac{dP_i(0)}{dx} + \frac{1}{2}T^2\frac{d^2P_i(0)}{dx^2} + \dots , \tag{3.16}$$

for all the diffracted beams i. For simplicity, only terms up to second order are included. By substituting (3.16) and the boundary conditions (3.7) into (3.4), the equation of power transfer can be written in the form

$$\sum_L a_{OL}P_L(T) + \sum_M a_{OM}P_M(0) = P_O(0) , \tag{3.17a}$$

$$\sum_L a_{KL}P_L(T) + \sum_M a_{KM}P_M(0) = 0 , \tag{3.17b}$$

for all the transmitted beams K;

$$\sum_L a_{K'L}P_L(T) + \sum_M a_{K'M}P_M(0) = 0 , \tag{3.17c}$$

for all the reflected beams K'. The summations are taken over all the transmitted beams L and reflected beams M, respectively. $P_O(0)$ is the incident beam power, which is assumed to be unity. There are N unknown variables which consist of L transmitted powers at the exit surface $P_L(T)$, and M reflected powers at the entrance surface $P_M(0)$. The coefficients are functions of the coefficients of the P's in (3.4) and the crystal thickness.

For illustration, the following three three-beam cases are discussed – case A: three transmission reflections, case B: two transmission reflections plus one Bragg reflection, and case C: one transmission plus two Bragg reflections.

The boundary conditions in case A are $P_O(0) = 1$ for the incident beam O, and $P_1(0) = P_2(0) = 0$ for the other two transmitted beams, 1 and 2. After a simple manipulation, the power of the ith diffracted beam is obtained as

$$P_1(T) = Q_{01}l_0 - \tfrac{1}{2}Q_{01}l_0[\mu_0l_0 + \mu_0l_1 + Q_{01}l_0 + Q_{10}l_1 + \sum_i (Q_{0i}l_0 + Q_{1i}l_1)]$$
$$+ \tfrac{1}{2}\sum_i Q_{0i}l_0Q_{i1}l_i . \tag{3.18}$$

If there is no secondary reflection i, one has a simple two-beam Laue (transmission) case. The diffracted power of the primary reflection 1 can be obtained by setting the terms involving the i reflection equal to zero:

$$P_1'(T) = Q_{01}l_0[1 - \tfrac{1}{2}(\mu_0l_0 + \mu_0l_1 + Q_{01}l_0 + Q_{10}l_1)] . \tag{3.19}$$

The same result can be obtained from the exact solution for the two-beam Laue case (3.10) by assuming that $\mu_0l \ll 1$ and $Ql \ll 1$. The difference ΔP between P_1 and P_1' gives the variation in the primary reflected power caused by the presence of the secondary reflection i:

$$\Delta P_1(T) = \tfrac{1}{2} \sum_i \left(-Q_{01} l_0 Q_{0i} l_0 - Q_{01} l_0 Q_{1i} l_1 + Q_{0i} l_0 Q_{i1} l_i \right). \tag{3.20}$$

It is this difference which causes the multiple diffraction pattern to show peaks for $\Delta P_1 > 0$ and dips for $\Delta P_1 < 0$ with respect to the two-beam background $P_1'(T)$.

Fig. 3.2a, b. Vector representations for (a) $j - i$ reflection and (b) $i - j$ reflection

The subscripts ij in Q_{ij} indicate the reflection $(j - i)$, where i and j are the incident and the diffracted reflections (Fig. 3.2a) and $(i - j)$ is a back reflection of $(j - i)$ (Fig. 3.2b). For example, Q_{01}, Q_{0i} and Q_{1i} in (3.20) represent the reflectivities of the primary reflection $(1 - 0) = 1$, the secondary reflection $(i - 0) = i$, and the coupling reflection $(i - 1)$ between the secondary and the primary reflections. If any pair of these three reflections is zero, $\Delta P_1 = 0$. This means that no multiple diffraction peak or dip can be observed when any two of the primary, the secondary and the coupling reflections are forbidden reflections. Diffraction peaks are always observable when the primary reflection is a forbidden reflection, since ΔP_1 depends only on the last term of (3.20).

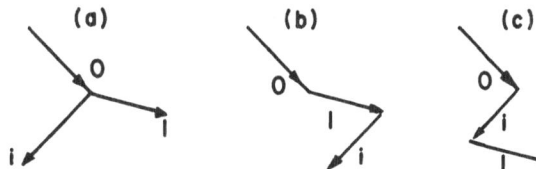

Fig. 3.3a – c. Second-order successive reflections

The three terms in (3.20) involve three successive reflections, which are represented graphically in Fig. 3.3. The first term, involving $(-Q_{0i})(Q_{01})$, implies that the incident beam is then reflected toward the direction of the primary reflection, 1. The minus sign is due to the power loss in the i reflection (Fig. 3.3a). Figure 3.3b, representing the second term, shows that the power loss for the primary reflected beam 1 results from the $(i - 1)$ reflection, the backward reflection of $1 - i$ (Fig. 3.3c). Because of the involvement of the $(i - 1)$ and $(1 - i)$ reflections, the multiple scattering (forward and backward reflections) effect is implicitly included in (3.20).

Cases B and C, which involve at least one Bragg reflection, can be considered jointly. Assume that the primary reflection, 1, is of the Bragg type. The boundary conditions are:

at $x = 0$:

$$P_0 = P_0(0) \, ,$$

$$P_1 = P_1(0) \neq 0 \, ,$$

$$P_i = P_i(0) \begin{cases} = 0 & \text{for secondary transmitted beam} \, , \\ \neq 0 & \text{for secondary reflected beam} \, , \end{cases} \tag{3.21}$$

at $x = T$:

$$P_1(T) = 0 \, ,$$

$$P_i(T) = 0 \quad \text{for all secondary reflected beams} \, .$$

Using these conditions together with the Taylor series expansion up to second-order terms, the following relation between $P_1(0)$ and the other P's is obtained [3.6]:

$$P_1(0)[1 + A_1 l_1 + \tfrac{1}{2}(A_1^2 l_1^2 - Q_{10}^2 l_0 l_1) + \tfrac{1}{2} \sum_r Q_{1r}^2 l_1 l_r - \tfrac{1}{2} \sum_t Q_{1t}^2 l_1 l_t]$$

$$= Q_{01} l_0 [1 + \tfrac{1}{2}(A_1 l_1 - A_0 l_0)] + \sum_r P_r(0) Q_{r1} l_r$$

$$+ \tfrac{1}{2} \sum_r P_r(0) l_r (Q_{r0} Q_{01} l_0 + A_1 Q_{r1} l_1)$$

$$- \tfrac{1}{2} \sum_r Q_{r1} l_r [Q_{0r} l_0 - P_r(0) A_r l_r + \sum_{j \neq r} P_j(T) Q_{jr} l_j]$$

$$+ \tfrac{1}{2} \sum_t Q_{t1} l_t [Q_{0t} l_0 + \sum_r P_r(0) Q_{rt} l_r] \, , \tag{3.22}$$

where

$$A_0 = \mu_0 + Q_{01} + \sum_i Q_{0i} \, ,$$

$$A_1 = \mu_0 + Q_{10} + \sum_i Q_{1i} \, , \tag{3.23}$$

$$A_r = \mu_0 + Q_{r0} + Q_{r1} + \sum_{j \neq r} Q_{rj} \, .$$

Equation (3.22) is valid for all multiple-beam cases with a Bragg reflection as the primary reflection. With the aid of this relation, the reflection powers in cases B and C can be obtained separately.

Case B: Suppose the secondary reflection 2 is of the transmission type with $Q_{1r} = 0$. The reflection power $P_1(0)$ is obtained as

$$P_1(0) = \frac{N_2}{D_2} \, , \quad \text{where} \tag{3.24}$$

$$N_2 = Q_{01}l_0 + \tfrac{1}{2}Q_{01}l_0(A_1l_1 - A_0l_0) + \tfrac{1}{2}Q_{02}Q_{21}l_0l_2 ,$$
$$D_2 = 1 + A_1l_1 + \tfrac{1}{2}(A_1^2l_1^2 + Q_{10}^2l_0l_1 - Q_{12}^2l_1l_2) .$$

(3.25)

Following the same procedure as in case A, the diffraction power of two-beam Bragg reflection is

$$P_1'(0) = \frac{N_2'}{D_2'} , \qquad \text{where}$$

(3.26)

$$N_2' = Q_{01}l_0 + \tfrac{1}{2}Q_{01}l_0(\mu_0l_1 + Q_{10}l_1 - \mu_0l_0 + Q_{01}l_0) ,$$
$$D_2' = 1 + (\mu_0l_1 + Q_{10}l_1) + \tfrac{1}{2}[(\mu_0 + Q_{10})^2l_1^2 - Q_{10}^2l_0l_1] \simeq 1 + \mu_0l_1 + Q_{10}l_1 .$$

(3.27)

The difference $\Delta P_1(0) = P_1(0) - P'(0)$ can be obtained accordingly.

Case C: The secondary reflection is of the Bragg type. The reflection powers $P_1(0)$ and $P_2(0)$ have the forms

$$P_1(0) = \frac{a_1b_1 + b_1c_1}{\Delta} , \qquad P_2(0) = \frac{a_2b_1 + b_2c_1}{\Delta} ,$$

(3.28)

where

$$\Delta = c_1c_2 - a_1a_2 ,$$
$$a_1 = Q_{12}l_2 + (Q_{01}Q_{02}l_0l_2 + A_1Q_{21}l_1l_2 + A_2Q_{21}l_2^2) ,$$
$$b_1 = Q_{01}l_0[1 + \tfrac{1}{2}(A_1l_1 - A_0l_0)] - \tfrac{1}{2}Q_{20}Q_{21}l_0l_2 ,$$
$$c_1 = 1 + A_1l_1 + (A_1^2l_1^2 - Q_{10}^2l_0l_1) + \tfrac{1}{2}Q_{12}^2l_1l_2 ,$$

(3.29)

and a_2, b_2 and c_2 can be obtained from the expressions for a_1, b_1 and c_1 by interchanging subscripts 1 and 2.

The reflection powers $P_1'(0)$ of the two cases and the difference $\Delta P_1(0)$ can be derived similarly, as in case B.

Since $\mu_0l \ll 1$ and $Ql \ll 1$, the second-order terms involved in the denominators of (3.24, 26, 28) can be neglected. Equations (3.24, 28) for $P_1(0)$ are reduced to (3.18) and the expression (3.20) for $P_1(0)$ is therefore valid under the second-order approximation, for all multiple diffractions in a plate-like crystal, regardless of the boundary conditions. In other words, the reflected beams can be of either the transmission type or the reflection type and/or any mixed Bragg-Laue type. This conclusion can also be directly drawn from consideration of the successive diffractions shown in Fig. 3.3, since the three types of successive diffractions are the only possible diffraction processes involved in the second-order approximation. All types of multiple diffraction, either Bragg-Bragg or Bragg-Laue or a mixture, involve the same three second-order successive diffractions. Hence, it is not surprising that (3.20) is valid for all cases.

The validity of the approximate solutions relies on the assumptions $\mu_0 l \ll 1$ and $Ql \ll 1$. The assumption $Ql \ll 1$ is usually fulfilled in x-ray and neutron cases. The first assumption, $\mu_0 l \ll 1$, however, is not quite satisfied in x-ray cases when highly absorbing crystals are being considered. In such cases, with $\mu_0 l \simeq 1$, higher-order approximations then need to be included in the solution to the power-transfer equation.

The reflection power of the primary reflection in case B is given as an example of the third-order approximation [3.6]:

$$P_1(0) = \frac{N_3}{D_3}, \quad \text{where} \tag{3.30}$$

$$
\begin{aligned}
N_3 &= N_2 + \tfrac{1}{6}\bigl(-A_0^2 Q_{01} l_0^3 - A_0 A_1 Q_{01} l_0^2 l_1 - A_0 Q_{02} Q_{12} l_0^2 l_2 \\
&\quad + A_1^2 Q_{01} l_0 l_1^2 - Q_{01}^3 l_0^2 l_1 + Q_{01} Q_{02}^2 l_0^2 l_2 - Q_{12}^2 Q_{01} l_0 l_1 l_2 \\
&\quad + A_1 Q_{21} Q_{02} l_0 l_1 l_2 - A_2 Q_{21} Q_{02} l_0 l_2^2 \bigr), \\
D_3 &= D_2 + \tfrac{1}{6}\bigl[(A_0 l_0 - 2 A_1 l_1) Q_{01}^2 l_0 l_1 - 2 Q_{01} Q_{02} Q_{12} l_0 l_1 l_2 \\
&\quad + A_1^3 l_1^3 - 2 A_1 Q_{12}^2 l_1^2 l_2 + A_2 Q_{12}^2 l_1 l_2^2 \bigr].
\end{aligned}
\tag{3.31}
$$

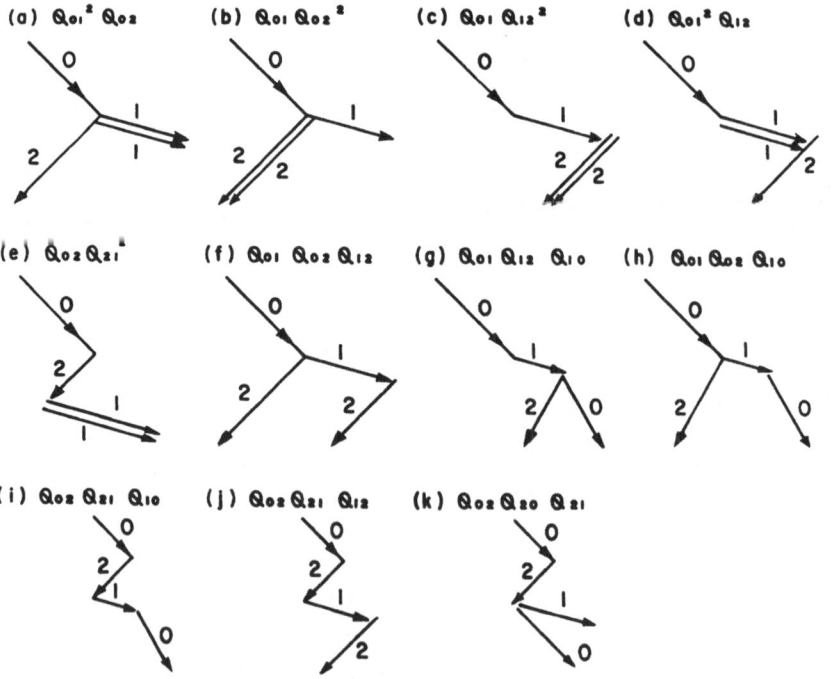

Fig. 3.4a – k. Third-order successive reflections

If the condition $\mu_0 l \ll 1$ still holds, $\Delta P_1(0)$ involves, in addition to the second-order successive reflections (Fig. 3.3), the following reflections: (1) third-order successive reflections which can be represented by the following eleven triple products of the reflectivities: $Q_{01}^2 Q_{02}$, $Q_{02}^2 Q_{01}$, $Q_{01} Q_{02} Q_{12}$, $Q_{12}^2 Q_{01}$, $Q_{10} Q_{12} Q_{01}$, $Q_{01} Q_{02} Q_{10}$, $Q_{01}^2 Q_{12}$, $Q_{10} Q_{21} Q_{02}$, $Q_{12} Q_{21} Q_{02}$, $Q_{20} Q_{21} Q_{02}$, $Q_{21}^2 Q_{02}$; (2) second-order successive reflections plus the ordinary absorption, i.e., $\mu_0 Q_{01} Q_{02}$, $\mu_0 Q_{01} Q_{12}$ and $\mu_0 Q_{02} Q_{21}$. Figure 3.4 shows a graphic representation of third-order successive reflections. However, these eleven reflections do not cover all the possible third-order successive reflections. For three-beam transmission diffraction (case A) the following eleven third-order reflections are involved: $Q_{01}^2 Q_{02}$, $Q_{01}^2 Q_{12}$, $Q_{01} Q_{12}^2$, $Q_{02}^2 Q_{21}$, $Q_{02} Q_{21}^2$, $Q_{01} Q_{02} Q_{21}$, $Q_{01} Q_{02} Q_{20}$, $Q_{01} Q_{12} Q_{21}$, $Q_{02} Q_{21} Q_{10}$, $Q_{02} Q_{21} Q_{12}$ and $Q_{02} Q_{21} Q_{20}$. They are not all the same as the eleven successive reflections mentioned in case B. Therefore, (3.31) is valid only for cases involving three-beam Bragg-Laue diffraction.

3.3 Integrated Intensity and the Lorentz-Polarization Factors

The reflection power given in (3.4) depends on the reflectivity defined in (3.15). In order to compare this with experimental results, the peak intensity or integrated intensity is often calculated. Since multiple diffraction is generated by crystal rotation, the Lorentz factor, which is defined as the duration of the stay of the reciprocal lattice point on the surface of the Ewald sphere, should be included in the reflectivity. In (3.15), the angle $\Delta \theta_{ij}$ is closely related to the azimuth angle $\Delta \varepsilon$ of rotation about an arbitrary axis. The geometry of this rotation is shown in Fig. 3.5. k_i and k_j are wavevectors of the reflections i and j. These two vectors can be expressed in terms of trigonometric functions of the angles ψ, ξ, and χ. The derivative of the scalar product of k_i and k_j with respect to θ_{ij} gives the relation

$$\Delta \theta_{ij} = \left(\frac{\sin \psi \cos \chi \cos \xi}{\sin 2\theta_{ij}} \right) \Delta \varepsilon = K_{ij}^\varepsilon \Delta \varepsilon, \qquad (3.32)$$

where $\Delta \varepsilon$, the deviation from the mean value of ψ, is equal to $\Delta \psi / 2$ [3.3].

The mosaic distribution may now be written in terms of the rotation angle $\Delta \varepsilon$ and renormalized to unity:

$$W(\Delta \varepsilon) = \frac{K_{ij}^\varepsilon}{\sqrt{2\pi\eta}} \exp \left(-\frac{(K_{ij}^\varepsilon \Delta \varepsilon)^2}{2\eta^2} \right). \qquad (3.33)$$

The term involving ψ, χ and ξ in (3.32, 33) is the reciprocal of the Lorentz factor L_f:

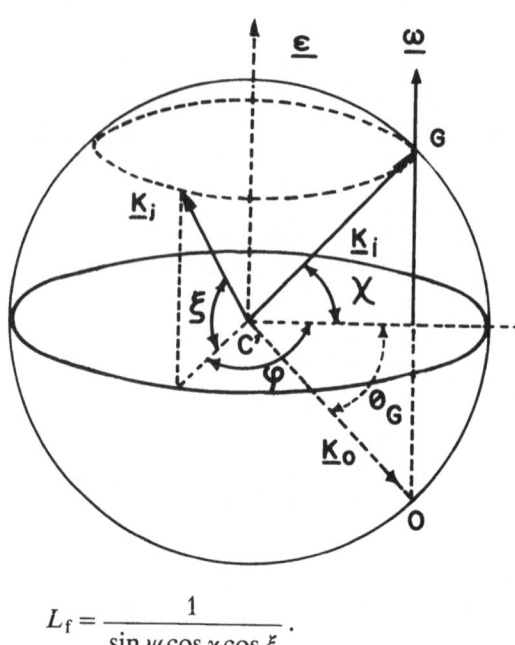

Fig. 3.5. Definition of angles involved in a crystal rotation

$$L_f = \frac{1}{\sin \psi \cos \chi \cos \xi} . \tag{3.34}$$

This expression can be easily obtained by considering the time which a secondary reciprocal lattice point takes to pass through the surface of the Ewald sphere. This time duration is inversely proportional to the speed of the rotation around the primary reflection vector. Referring to Fig. 2.15, the speed v_r is equal to the component of the velocity of rotation in the direction towards the center of the Ewald sphere, i.e.,

$$|v_r| = \frac{(\omega \times p) \cdot r}{|r|} = \omega p_n \cos \alpha , \tag{3.35}$$

where α is the angle between r and the direction of motion of the reciprocal lattice point (Fig. 3.6) and ω is the angular velocity of the rotation. In Fig. 3.6, referring to Fig. 3.5 for the definition of χ, ψ and ξ,

$$\cos \alpha = \frac{h_0}{r} , \quad \sin \beta = \frac{h_0}{r \cos \chi} , \quad \text{and} \tag{3.36}$$

$$p_n \sin \beta = \frac{\sin \psi \cos \xi}{\lambda} , \tag{3.37}$$

where β has been defined in (2.25). From (3.36, 37), we obtain

$$\cos \alpha = (\sin \psi \cos \chi \cos \xi)/p_n \lambda . \tag{3.38}$$

This leads to

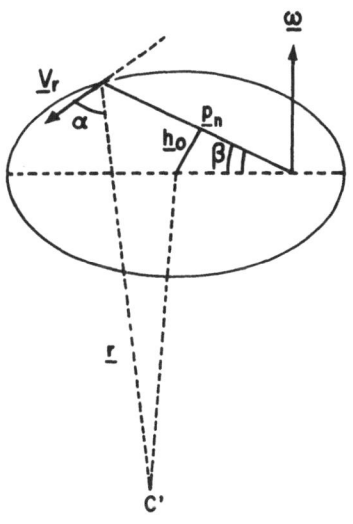

Fig. 3.6. Geometry for defining Lorentz factors

$$L_f \propto \frac{1}{v_r} \simeq \left(\frac{\lambda}{\omega}\right) \frac{1}{\sin\psi\cos\chi\cos\xi}, \tag{3.39}$$

where λ/ω is a constant for the given speed ω of rotation and the radiation (λ) used.

It should be noted that the expression for K_{ij}^ε in (3.32) is not valid when the rotation axis is parallel to the reciprocal lattice vector of a coupling reflection $(j-i)$. Such instances are often encountered in cases of even-order multiple diffraction, for example, four-, six- and eight-beam diffractions. In this situation, $K_{ij}^\varepsilon = 0$ because $\psi = 0$. The angle $\Delta\theta_{ij}$, according to (3.32), is therefore zero. This is incorrect because $\Delta\theta_{ij}$ is not zero when the two lattice points i and j are not on the surface of the Ewald sphere. A suitable geometric factor K_{ij}^ε for this special case can be derived from Bragg's law:

$$2\sin\theta_{ij} = \lambda|\boldsymbol{g}_i - \boldsymbol{g}_j|, \tag{3.40}$$

where $\boldsymbol{g}_i$ and $\boldsymbol{g}_j$ are the reciprocal lattice vectors of the reflections i and j. By squaring both sides of (3.40) and utilizing Bragg's law for each individual reflection, the following relation is obtained [3.6]:

$$\sin^2\theta_{ij} = \sin^2\theta_i + \sin^2\theta_j - 2\sin\theta_i\sin\theta_j\cos\alpha_{ij}, \tag{3.41}$$

where α_{ij} is the angle between $\boldsymbol{g}_i$ and $\boldsymbol{g}_j$. Differentiating (3.41) with respect to the angles θ and using the relationships

$$\Delta\theta_i = K_{i0}^\varepsilon\Delta\varepsilon, \quad \Delta\theta_j = K_{j0}^\varepsilon\Delta\varepsilon, \tag{3.42}$$

with K_{i0}^ε and K_{j0}^ε defined in (3.32), we obtain

$$\Delta\theta_{ij} = K_{ij}^\varepsilon\Delta\varepsilon, \quad \text{where} \tag{3.43}$$

$$K_{ij}^{\varepsilon} = \frac{1}{\sin \theta_{ij} \cos \theta_{ij}} [K_{i0}^{\varepsilon}(\sin \theta_j \cos \theta_i - \cos \theta_i \sin \theta_j \cos \alpha_{ij})$$

$$+ K_{j0}^{\varepsilon}(\sin \theta_j \cos \theta_j - \sin \theta_i \cos \theta_j \cos \alpha_{ij})] \, . \tag{3.44}$$

By combining the Lorentz factor and the mosaic spread, the reflectivity can be written

$$Q_{ij} = \frac{\lambda^3 N_0^2 |F_s|^2}{\sin 2\theta_{ij}} \frac{C_{ij}(i-j)}{\sqrt{2\pi\eta}} \exp\left(-\frac{(K_{ij}^{\varepsilon} \Delta \varepsilon)^2}{2\eta^2}\right), \tag{3.45}$$

where C_{ij} is the polarization factor.

The Lorentz factor discussed here is associated with each individual reflection. The resultant Lorentz factor for multiple diffraction should take the successive reflections into account. Since the integrated intensity is proportional to the product of the diffraction peak height and the peak width, and since the peak width depends on the Lorentz factor, it is therefore straightforward to determine the resultant Lorentz factor from the integrated intensity. In a multiple diffraction experiment, there are two angles involved, the Bragg angle θ of the primary reflection and the azimuth angle ϕ. Integrated intensities with respect to ϕ and θ for the difference between the peak and the background should be considered (Fig. 3.7) [3.4]:

$$R_{\phi} = \frac{\int \Delta P_1(\theta_1, \phi) d\phi}{\int P_1(\theta, \phi_A) d\theta}, \quad \text{and} \tag{3.46}$$

$$R_{\theta} = \frac{\int \Delta P_1(\theta, \phi_1) d\theta}{\int P_1(\theta, \phi_A) d\theta}, \tag{3.47}$$

where the quantity R_{ϕ} represents the integrated intensity above background of the primary reflection, divided by the background (see Fig. 3.7). The quantity

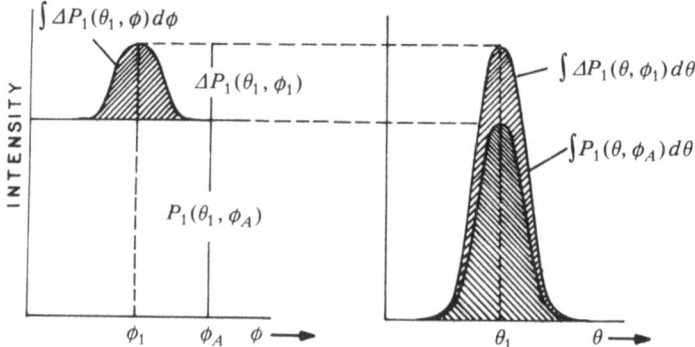

Fig. 3.7. Definition of peak intensities and integrated intensities for multiple diffraction (after *Moon* and *Shull* [3.4])

R_θ is proportional to the difference between the integrated intensity at the multiple diffraction position ϕ_1 and the integrated intensity of the primary reflection at the position ϕ_A where only two-beam diffraction takes place. θ_1 is the Bragg angle of the primary reflection. For simplicity, let us use (3.20) for ΔP_1. The integrated intensities are then given by

$$R_\phi = \frac{1}{2} \frac{Q_{01} I_0}{\sqrt{2\pi\eta}} \sum_i \left(-\frac{1}{K_{0i}^\phi} \frac{Q_{0i}}{Q_{01}} - \frac{1}{K_{1i}^\phi} \frac{Q_{1i}}{Q_{01}} \frac{l_1}{l_0} + \frac{(Q_{0i}Q_{i1}/Q_{01}^2)(l_i/l_0)}{\sqrt{(K_{0i}^\phi)^2 + (K_{i1}^\phi)^2}} \right),$$

(3.48)

and

$$R_\theta = \frac{1}{2} \frac{Q_{01} I_0}{\sqrt{2\pi\eta}} \sum_i \left(-\frac{Q_{0i}/Q_{01}}{\sqrt{1 + (K_{0i}^\theta)^2}} - \frac{(Q_{1i}/Q_{01})(l_i/l_0)}{\sqrt{1 + (K_{1i}^\theta)^2}} \right.$$
$$\left. + \frac{(Q_{0i}Q_{i1}/Q_{01}^2)(l_i/l_0)}{\sqrt{(K_{0i}^\theta)^2 + (K_{i1}^\theta)^2}} \right).$$

(3.49)

The resultant Lorentz factors for an nth-order successive reflection can be written in the form

$$L_n^{\phi,\theta} = \left[\sum_{i=1}^n (K_{i-1,i}^{\phi,\theta})^2 \right]^{-1/2},$$

(3.50)

for both ϕ and θ rotations. Note that $K_{01}^\theta = 1$ and $K_{01}^\phi = 0$. When the Lorentz factor is excluded in (3.48) and $\Delta\varepsilon$ is set equal to zero, the peak intensity ratio

$$R_p = \frac{\Delta P(\theta_1, \phi_1)}{P_1(\theta_1, \phi_A)}$$

(3.51)

can be obtained. A schematic representation for R_p is also shown in Fig. 3.7.

Because multiple diffraction involves many successive reflections, the polarization factors should be adjusted accordingly to include these reflections. In addition, if a monochromator is used in the experiment, the polarization effect caused by the monochromator should be considered.

Let us consider the case of nth-order successive reflections with a monochromator. As is shown in Fig. 3.8, the electric fields of an unpolarized incident beam can be resolved into two components, E_σ and E_π, which are perpendicular to each other and to the direction of the incident beam. The incident beam is first reflected by the monochromator at the angle α. The plane containing the incident beam and the reflected beam is the plane of incidence. The components of the reflected x-rays, E'_σ and E'_π, are related to E_σ and E_π as in a simple reflection:

$$E' = \begin{pmatrix} E'_\sigma \\ E'_\pi \end{pmatrix} = \begin{pmatrix} 1 & 0 \\ 0 & \cos(2\alpha) \end{pmatrix} \begin{pmatrix} E_\sigma \\ E_\pi \end{pmatrix} = \underset{\sim}{C}_M E,$$

(3.52)

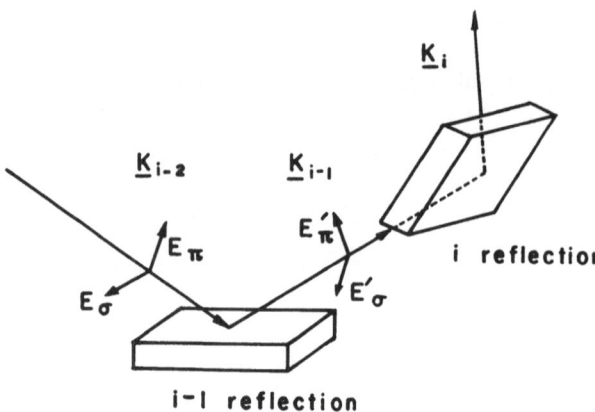

where $\underset{\sim}{C}_M$ is the polarization matrix. The polarization factor for a mosaic crystal is then equal to the sum of the squares of the elements in $\underset{\sim}{C}_M$, i.e.,

$$C_M = \frac{1 + \cos^2 2\theta}{2} . \tag{3.53}$$

The factor $1/2$ is the normalization constant and $\theta = \alpha$. E_σ and E'_σ are perpendicular and E_π and E'_π are parallel to the plane of incidence.

Suppose that the incident beam is $i-1$ and the diffracted beam i; the polarization matrix can then be written as

$$\underset{\sim}{C}_i = \begin{pmatrix} 1 & 0 \\ 0 & \cos 2\theta_i \end{pmatrix} , \tag{3.54}$$

where $2\theta_i$ is the angle between the $(i-1)^{\text{th}}$ beam and the ith beam. Since the plane of incidence of the ith reflection is, in general, not the same plane of incidence as for the jth reflection, the angle $\psi_{i-1,i}$ between these two planes of incidence should be included in the polarization matrix. This angle can be regarded as the rotation angle by which the coordinate system for the σ and π components of the electric field defined by the $(i-1)^{\text{th}}$ reflection should be rotated to give the coordinate system defined by the ith reflection. This rotation of the coordinate system can be described by the transformation matrix [3.7]

$$\underset{\sim}{T}_i = \begin{pmatrix} \cos \psi_{i-1,i} & \sin \psi_{i-1,i} \\ -\sin \psi_{i-1,i} & \cos \psi_{i-1,i} \end{pmatrix} , \tag{3.55}$$

where

$$\psi_{i-1,i} = \cos^{-1}[(\hat{k}_{i-2} \times \hat{k}_{i-1}) \cdot (\hat{k}_{i-1} \times \hat{k}_i)] . \tag{3.56}$$

Using the relation

$$(A \times B) \cdot (C \times D) = (A \cdot C)(B \cdot D) - (A \cdot D)(B \cdot C),$$

we obtain

$$\cos \psi_{i-1,i} = \frac{\cos 2\theta_m - \cos 2\theta_{i-1} \cos 2\theta_i}{\sin 2\theta_{i-1} \sin 2\theta_i}, \tag{3.57}$$

where m is the coupling between $(i-1)$ and i. The vectors $\hat{k}$ in (3.56) are the unit vectors of the diffracted beams in the successive reflections $(i-1)$ and i. We denote the transformation matrix for the reflection from the monochromator as $T_M(\phi)$, ϕ being the angle between the normals to the monochromator and the first plane of incidence. Thus, the polarization matrix for the nth-order successive reflection is [3.7]

$$\underset{\sim}{C} = \underset{\sim}{C_M} \underset{\sim}{T_M} \underset{\sim}{C_1} \underset{\sim}{T_1} \cdots \underset{\sim}{C_{n-1}}. \tag{3.58}$$

The corresponding polarization factor is therefore equal to one-half of the sum of the squared matrix elements of $\underset{\sim}{C}$.

As an example, let us consider second-order successive reflections $(n = 2)$ with a monochromator. The polarization factor for the successive reflections i and j is [3.6]:

$$C_{ij}(m) = \frac{1}{2} \Bigg[\left(1 - \frac{(\cos 2\theta_m - \cos 2\theta_i \cos 2\theta_j)^2}{\sin^2 2\theta_i} \right) (1 - \sin^2 2\alpha \cos^2 \phi)$$

$$\times \cos^2 2\theta_i + \left(\frac{(\cos 2\theta_m - \cos 2\theta_i \cos 2\theta_j)^2}{\sin^2 2\theta_i} + \cos^2 2\theta_j \right)$$

$$\times (\sin^2 2\alpha \cos^2 \phi + \cos^2 2\alpha) \Bigg], \tag{3.59}$$

where $m = j - i$. The angle ϕ between the planes of incidence of the monochromator and of the first reflection i depends on the experimental arrangement. When there is no monochromator, $\alpha = 0$. The polarization factor becomes [3.5]

$$C_{ij}(m) = \tfrac{1}{2} [\cos^2 2\theta_i + \cos^2 2\theta_j + (\cos 2\theta_m - \cos 2\theta_i \cos 2\theta_j)^2]. \tag{3.60}$$

Equation (3.59) can also give the familiar expression for the polarization for a single reflection with a monochromator:

$$C_{0i} = \tfrac{1}{2}(1 + \cos^2 2\alpha \cos^2 2\theta_i). \tag{3.61}$$

The condition $\phi = \pi/2$ has been employed.

3.4 Path Lengths of X-Ray Beams in Crystals

The path lengths of diffracted beams in a very thin plate-like crystal or for a low absorbing crystal are determined by the crystal thickness T, as defined in (3.14). If the crystal is of arbitrary shape, the average path length may be defined as

$$\langle l \rangle = \left\langle \frac{x}{\gamma} \right\rangle = A \frac{dA^*}{d\mu},$$ (3.62)

where $A^* = A^{-1}$ and A is the appropriate absorption function. For a plate-like crystal,

$$A = \exp(-\mu_0 T)$$ (3.63)

for a transmitted beam [3.5], or

$$A = \frac{\gamma}{\mu_0 T} \left[1 - \exp\left(-\frac{\mu_0 T}{\gamma} \right) \right]$$ (3.64)

for a reflected beam [3.6], where $1/\gamma = 1/\gamma_0 + 1/\gamma_i$. For a highly absorbing crystal plate, the total path length is equal to

$$\langle l \rangle = \frac{1}{\mu_0}, \qquad l_i = \frac{1}{\mu_0} \frac{\gamma}{\gamma_i}.$$ (3.65)

If the primary reflection is symmetric, such that $\gamma_0 = \gamma_i$ and $\gamma/\gamma_i = 1/2$, then

$$l_i = \frac{1}{2\mu_0}$$ (3.66)

for all the secondary reflections i. The same result (3.66) can be obtained for transmitted beams.

3.5 Exact Solution to the Power-Transfer Equation

Equation (3.4) involves n linear differential equations of the form given in (3.5, 6). By substituting (3.6) into (3.5), the linear equation

$$(a_{ii} - Z_m) A_{im} + \sum_{j \neq i} a_{ij} A_{jm} = 0$$ (3.67)

can be obtained. In comparison with (3.4),

$$a_{ii} = -\frac{\mu_0 + \sum\limits_j Q_{ij}}{\gamma_i}, \qquad a_{ij} = \frac{Q_{ji}}{\gamma_j}. \tag{3.68}$$

The equation can be solved as an eigenvalue equation. The eigenvector gives the ratio between A_{im} and A_{jm} for each eigenvalue Z_m. A_{im} and A_{jm} can be determined completely from the boundary conditions mentioned in the previous section.

It is, however, not always possible to find analytical expressions for Z_m. When the multiple diffraction is a symmetric transmission diffraction, for which the reciprocal lattices involved form a symmetric polygon, it is possible to solve (3.67) analytically. *Zachariasen* [3.5] gave the reflection powers for three-, four- and six-beam symmetric multiple diffractions of the transmission type. As an illustration, the symmetric three-beam case, O, G_1, G_2, with $|g_1| = |g_2|$ and $\gamma_0 = \gamma_1 = \gamma_2$ is considered. The reflectivities of the reflections G_1, G_2 and $G_2 - G_1$ are Q_1, Q_2 and Q_3. The condition for a non-trivial solution of (3.67) to exist is

$$\begin{vmatrix} -\mu_0 - 2Q_1 - Z & Q_1 & Q_1 \\ Q_1 & -\mu_0 - Q_1 - Q_3 - Z & Q_3 \\ Q_1 & Q_3 & -\mu_0 - Q_1 - Q_3 - Z \end{vmatrix} = 0. \tag{3.69}$$

The reflection powers are obtained at the exit surface of the crystal:

$$P_0(T) = \tfrac{1}{3}[1 + \exp(-3Q_1 T)]\exp(-\mu_0 T), \tag{3.70}$$
$$P_1(T) = P_2(T) = \tfrac{1}{3}[1 - \exp(-3Q_1 T)]\exp(-\mu_0 T).$$

The integrated reflections therefore take the form

$$R_i = \int P_i(\Delta\varepsilon)\,d(\Delta\varepsilon). \tag{3.71}$$

To facilitate the integrations, series expansion of the exponent $\exp(Z_j T)$ is necessary. Suitable expressions for the integrated reflections can be written as

$$R_0(T) = \exp(-\mu_0 T)(1 - 2Q_1' T + 3qQ_1'^2 T^2 + \ldots), \tag{3.72}$$
$$R_1(T) = T\exp(-\mu_0 T)(Q_1' - \tfrac{3}{2}qQ_1'^2 T + \ldots),$$

where

$$Q_1' = \int Q_1\,d(\Delta\varepsilon),$$
$$q = \int W^2\,d(\Delta\varepsilon), \tag{3.73}$$
$$\int Q_i Q_j\,d(\Delta\varepsilon) = qQ_i'Q_j'.$$

The condition similar to (3.69) and the reflection powers for four-beam and six-beam symmetric reflections can be summarized [3.5]:

i) *Four-beam case, O, G_1, G_2 and G_3 ($g_1 = g_2, |g_1-g_2| = g_3, |g_1-g_3| = g_2$):*

$$\begin{vmatrix} -\mu-Z & Q_1 & Q_2 & Q_3 \\ Q_1 & -\mu-Z & Q_3 & Q_2 \\ Q_2 & Q_3 & -\mu-Z & Q_1 \\ Q_3 & Q_2 & Q_1 & -\mu-Z \end{vmatrix} = 0; \tag{3.74}$$

$$\mu = \mu_0 + Q_1 + Q_2 + Q_3,$$

$$\begin{aligned} P_0(T) &= \tfrac{1}{4}B_0(1 + B_1 + B_2 + B_3), \\ P_1(T) &= \tfrac{1}{4}B_0(1 + B_1 - B_2 - B_3), \\ P_2(T) &= \tfrac{1}{4}B_0(1 - B_1 + B_2 - B_3), \\ P_3(T) &= \tfrac{1}{4}B_0(1 - B_1 - B_2 + B_3); \end{aligned} \tag{3.75}$$

$$\begin{aligned} R_0(T) &= B_0[1 - (Q_1 + Q_2 + Q_3)T + q(Q_1^2 + Q_2^2 + Q_3^2 + Q_1 Q_2 \\ &\quad + Q_1 Q_3 + Q_2 Q_3)T^2 + \ldots], \\ R_1(T) &= B_0 T[Q_1 + q(Q_2 Q_3 - Q_1^2 - Q_1 Q_2 - Q_1 Q_3)T + \ldots], \\ R_2(T) &= B_0 T[Q_2 + q(Q_1 Q_3 - Q_1 Q_2 - Q_2^2 - Q_2 Q_3)T + \ldots], \\ R_3(T) &= B_0 T[Q_3 + q(Q_1 Q_2 - Q_1 Q_3 - Q_2 Q_3 - Q_3^2)T + \ldots]; \end{aligned} \tag{3.76}$$

where

$$\begin{aligned} B_0 &= \exp(-\mu_0 T), \\ B_1 &= \exp[\;2(Q_2 + Q_3)T], \\ B_2 &= \exp[-2(Q_1 + Q_3)T], \\ B_3 &= \exp[-2(Q_1 + Q_2)T]. \end{aligned}$$

ii) *Six-beam case, O, G_1, G_2, G_3, G_4 and G_5 ($g_1 = g_2, g_3 = g_4$),* $|g_1-g_3| = |g_2-g_4| = |g_3-g_5| = |g_4-g_5| = g_5,$ $|g_1-g_2| = |g_1-g_5| = |g_2-g_5| = g_3, |g_1-g_4| = |g_2-g_3| = g_5$):

$$\begin{vmatrix} -\mu'-Z & Q_1 & Q_1 & Q_3 & Q_3 & Q_5 \\ Q_1 & -\mu'-Z & Q_3 & Q_1 & Q_5 & Q_3 \\ Q_1 & Q_3 & -\mu'-Z & Q_5 & Q_1 & Q_3 \\ Q_3 & Q_1 & Q_5 & -\mu'-Z & Q_3 & Q_1 \\ Q_3 & Q_5 & Q_1 & Q_3 & -\mu'-Z & Q_1 \\ Q_5 & Q_3 & Q_3 & Q_1 & Q_1 & -\mu'-Z \end{vmatrix} = 0; \tag{3.77}$$

$$\mu' = \mu_0 + 2Q_1 + 2Q_3 + Q_5 ,$$

$$P_0(T) = \tfrac{1}{6}B_0(1 + 2B_4 + 2B_5 + B_6) ,$$

$$P_1(T) = P_2(T) = \tfrac{1}{6}B_0(1 - B_4 + B_5 - B_6) , \tag{3.78}$$

$$P_3(T) = P_4(T) = \tfrac{1}{6}B_0(1 - B_4 - B_5 + B_6) ,$$

$$P_5(T) = \tfrac{1}{6}B_0(1 + 2B_4 - 2B_5 - B_6) ;$$

$$R_0(T) = B_0[1 - (2Q_1 + 2Q_3 + Q_5)T + q(3Q_1^2 + 3Q_3^2 + Q_5^2$$

$$\qquad + 4Q_1Q_3 + 2Q_1Q_5 + 2Q_3Q_5)T^2 + \ldots] ,$$

$$R_1(T) = R_2(T) = B_0T[Q_1 + q(Q_3Q_5 - 2Q_1^2 - Q_1Q_3 - Q_1Q_5)T + \ldots] ,$$

$$R_3(T) = R_4(T) \tag{3.79}$$

$$\qquad = B_0T[Q_3 + q(\tfrac{1}{2}Q_1^2 + Q_1Q_3 - 2Q_1Q_5 - \tfrac{3}{2}Q_3^2 - Q_3Q_5)T + \ldots] ,$$

$$R_5(T) = B_0T[Q_5 + q(2Q_1Q_3 - 2Q_1Q_5 - 2Q_3Q_5 - Q_5^2)T + \ldots] ;$$

where

$$B_4 = \exp[-3(Q_1 + Q_3)T] ,$$

$$B_5 = \exp[-(Q_1 + 3Q_3 + 2Q_5)T] ,$$

$$B_6 = \exp[-(4Q_1 + 2Q_5)T] .$$

For a general three-beam case, a similar solution to (3.4) can be found in the following way [3.6]. Suppose that in the case involving a Bragg reflection as the primary reflection 1 and a transmitted secondary reflection 2, there are three differential equations like (3.4). By differentiating these three equations twice, the following third-order differential equation is obtained:

$$\frac{d^3P_i}{dx^3} + F_a\frac{d^2P_i}{dx^2} + F_b\frac{dP_i}{dx} + F_cP_i = 0 \tag{3.80}$$

for all i, where

$$F_a = -\sum_{i=1}^{3} a_{ii} ,$$

$$F_b = \sum_{\substack{i,j \\ i \neq j}} (a_{ii}a_{jj} - a_{ij}a_{ji}) , \tag{3.81}$$

$$F_c = -\Delta .$$

The quantity Δ is the determinant of the matrix formed by a_{ij}. The solutions to (3.80) have the same form as (3.6),

$$
\begin{pmatrix} P_0 \\ P_1 \\ P_2 \end{pmatrix} = \begin{pmatrix} F_1 & F_2 & F_3 \\ F_4 & F_5 & F_6 \\ F_7 & F_8 & F_9 \end{pmatrix} \begin{pmatrix} \exp(\lambda_1 x) \\ \exp(\lambda_2 x) \\ \exp(\lambda_3 x) \end{pmatrix} . \tag{3.82}
$$

Substituting (3.82) into the boundary conditions

$$
P_0(0) \neq 0 , \quad P_1(T) = 0 , \quad P_2(0) = 0 ,
$$

$$
a_{32} \frac{dP_1(0)}{dx} - a_{22} \frac{dP_2(0)}{dx} = (a_{32}a_{21} - a_{22}a_{31}) P_0(0) ,
$$

$$
a_{32} \frac{dP_0(0)}{dx} - a_{12} \frac{dP_2(0)}{dx} = (a_{32}a_{11} - a_{12}a_{31}) P_0(0) , \tag{3.83}
$$

$$
\frac{a_{23}\dfrac{dP_0(T)}{dx} - a_{13}\dfrac{dP_1(T)}{dx}}{a_{11}a_{23} - a_{21}a_{13}} = \frac{a_{33}\dfrac{dP_1(T)}{dx} - a_{23}\dfrac{dP_2(T)}{dx}}{a_{21}a_{33} - a_{31}a_{23}} ,
$$

$$
\begin{pmatrix} \dfrac{d^2 P_0(0)}{dx^2} \\[2mm] \dfrac{d^2 P_1(0)}{dx^2} \\[2mm] \dfrac{d^2 P_2(0)}{dx^2} \end{pmatrix} = \begin{pmatrix} a_{11} & a_{12} & a_{13} \\ a_{21} & a_{22} & a_{23} \\ a_{31} & a_{32} & a_{33} \end{pmatrix} \begin{pmatrix} \dfrac{dP_0(0)}{dx} \\[2mm] \dfrac{dP_1(0)}{dx} \\[2mm] \dfrac{dP_2(0)}{dx} \end{pmatrix} ,
$$

we arrive at nine simultaneous equations which are expressed in matrix form:

$$
\underset{\sim}{M} a \underset{\sim}{F} = V a , \tag{3.84}
$$

where

$$\underset{\sim}{M}a =
\begin{pmatrix}
1 & 1 & 1 & 0 & 0 & 0 & 0 & 0 & 0 \\
0 & 0 & 0 & 0 & 0 & 0 & 1 & 1 & 1 \\
0 & 0 & 0 & a_{32}\lambda_1 & a_{32}\lambda_2 & a_{32}\lambda_3 & -a_{22}\lambda_1 & -a_{22}\lambda_2 & -a_{22}\lambda_3 \\
a_{32}-\lambda_1 & a_{32}\lambda_2 & a_{32}\lambda_3 & 0 & 0 & 0 & -a_{12}\lambda_1 & -a_{12}\lambda_2 & -a_{12}\lambda_3 \\
0 & 0 & 0 & a & b & c & 0 & 0 & 0 \\
a\lambda_1 & b\lambda_2 & c\lambda_3 & a\lambda_1 m_0 & b\lambda_2 m_0 & c\lambda_3 m_0 & a\lambda_1 n_1 & b\lambda_2 n_1 & c\lambda_3 n_1 \\
\lambda_1(a_{11}-\lambda_1) & \lambda_2(a_{11}-\lambda_2) & \lambda_3(a_{11}-\lambda_3) & \lambda_1(a_{22}-\lambda_1) & \lambda_2(a_{22}-\lambda_2) & \lambda_3(a_{22}-\lambda_3) & a_{13}\lambda_1 & a_{13}\lambda_2 & a_{13}\lambda_3 \\
a_{21}\lambda_1 & a_{21}\lambda_2 & a_{21}\lambda_3 & a_{12}\lambda_1 & a_{12}\lambda_2 & a_{12}\lambda_3 & a_{23}\lambda_1 & a_{23}\lambda_2 & a_{23}\lambda_2 \\
a_{31}\lambda_1 & a_{31}\lambda_2 & a_{31}\lambda_3 & a_{32}\lambda_1 & a_{32}\lambda_2 & a_{32}\lambda_3 & \lambda_1(a_{33}-\lambda_1) & \lambda_2(a_{33}-\lambda_2) & \lambda_3(a_{33}-\lambda_3)
\end{pmatrix}$$

(3.85)

The vectors F and Va can be expressed horizontally as

$$F = [F_1 \ F_2 \ F_3 \ F_4 \ F_5 \ F_6 \ F_7 \ F_8 \ F_9] ,$$

$$Va = [P_0(0) \ 0 \ (a_{32}a_{21} - a_{22}a_{31})P_0(0)$$

$$(a_{32}a_{11} - a_{12}a_{31})P_0(0) \ 0 \ 0 \ 0 \ 0 \ 0] , \qquad (3.86)$$

with

$$a = \exp(\lambda_1 T), \qquad b = \exp(\lambda_2 T), \qquad c = \exp(\lambda_3 T),$$

$$m_0 = \left(\frac{a_{33}}{a_{23}}\right)n_0 - \frac{a_{13}}{a_{23}}, \qquad n_1 = \frac{a_{21}a_{13} - a_{11}a_{23}}{a_{21}a_{33} - a_{31}a_{23}} . \qquad (3.87)$$

To solve (3.84), numerical calculation should be employed. In a general N-beam case, there are N^2 linear equations involved, which can be solved in the same way as described above.

3.6 Iterative Calculation for Reflection Power

An analytical expression for the reflection power of a general N-beam diffraction is difficult to obtain. Numerical calculation therefore plays an alternative role in providing values for reflection powers. On the other hand, the assumption $\mu l \ll 1$ is only marginally satisfied for x-ray diffraction. This suggests that higher-order terms are necessary in the series expansion of (3.16). Not losing generality, let us consider a general four-beam case in which L_1 and L_2 are the incident and transmitted diffractions and M_1 and M_2 are Bragg-type reflections. The assumed reflection power at a depth d_1 inside the crystal plate is a Taylor series expanded about $x = 0$, i.e.,

$$P_i(d_1) = \sum_{n=0}^{\infty} \frac{d_1^n}{n!} P_i^{(n)}(0) , \qquad (3.88)$$

for $i = L_1, L_2, M_1$ and M_2, where $P_i^{(n)}$ is the nth-order derivative with respect to x, and $P_i^{(0)}(0) = P_i(0)$. The derivatives are also functions of the powers of all participating reflections. For example,

$$P_i^{(1)}(x) = \sum_j Y_{ji}P_i(x) \qquad (3.89)$$

for the first-order derivative, where Y_{ji} are the coefficients involved in (3.4).

The nth-order derivative, which can be obtained by differentiating (3.89) $n-1$ times, has the compact form [3.8]

$$P_i^{(n)}(x) = \sum_{j_1} \sum_{j_2} \cdots \sum_{j_n} Y_{ij_1} Y_{j_1 j_2} \cdots Y_{j_{n-1}j_n} P_{j_n}(x) , \qquad (3.90)$$

where the j's can be L_1, L_2, M_1 and M_2, and

$$Y_{j_p j_q} = \frac{S_{j_q} Q_{j_p j_q}}{\gamma_{j_q}} \quad \text{for} \quad j_p \neq j_q \,,$$

$$Y_{j_p j_p} = -\frac{S_{j_p} A_{j_p}}{\gamma_{j_p}} \quad \text{for} \quad j_p = j_q \,,$$

(3.91)

with

$$A_{j_p} = \mu_0 + \sum_{j_r \neq j_p} Q_{j_p j_r} \quad \text{and}$$

(3.92)

$$S_{j_p} = \begin{cases} + & \text{for transmitted beam,} \\ - & \text{for reflected beam.} \end{cases}$$

By substituting (3.90) into (3.88), the reflection power at $x = d_1$ takes the simple form

$$P_i(d_1) = \sum_j a_{ij}(d_1) P_j(0) \,, \quad \text{where}$$

(3.93)

$$a_{ij}(d_1) = \sum_{n=0}^{\infty} \frac{d_1^n}{n!} \sum_{j_1} \sum_{j_2} \cdots \sum_{j_{n-1}} Y_{ij_1} Y_{j_1 j_2} \cdots Y_{j_{n-1} j} \,.$$

(3.94)

With the boundary conditions at the upper ($x = 0$) and lower surfaces ($x = d_1$) of the crystal plate, i.e.,

$$P_{L_1}(0) = 1 \,, \quad P_{L_2}(0) = 0 \,,$$

$$P_{M_1}(d_1) = 0 \,, \quad P_{M_2}(d_1) = 0 \,,$$

(3.95)

the approximate solution of (3.93) can be obtained as [3.9]

$$\begin{pmatrix} P_{L_1}(d_1) \\ P_{L_2}(d_1) \\ P_{M_1}(0) \\ P_{M_2}(0) \end{pmatrix} = B_{L_1}(d_1) P_{L_1}(0) \,,$$

(3.96)

where the vector

$$B_{L_1}(d_1) = \underset{\sim}{S}_{L_1}^{-1}(d_1) V_{L_1}(d_1) \,.$$

(3.97)

The 4×4 matrix $\underset{\sim}{S}$ and the vector V_{L_1} are given by

$$\underset{\sim}{S}_{L_1}(d_1) = \begin{pmatrix} 1 & 0 & -a_{L_1 M_1}(d_1) & -a_{L_1 M_2}(d_1) \\ 0 & 1 & -a_{L_2 M_1}(d_1) & -a_{L_2 M_2}(d_1) \\ 0 & 0 & -a_{M_1 M_1}(d_1) & -a_{M_1 M_2}(d_1) \\ 0 & 0 & -a_{M_2 M_1}(d_1) & -a_{M_2 M_2}(d_1) \end{pmatrix} \,,$$

$$V_{L_1}(d_1) = \begin{pmatrix} a_{L_1 L_1}(d_1) \\ a_{L_2 L_1}(d_1) \\ a_{M_1 L_1}(d_1) \\ a_{M_2 L_1}(d_1) \end{pmatrix}.$$

(3.98)

The subscript L_1 in B_{L_1}, V_{L_1} and $\underset{\sim}{S}_{L_1}$ indicates that the incident beam is L_1.

Because multiple diffraction involves transmission and reflection, it would be convenient to decompose (3.96) into two parts, one for transmission and the other for reflection:

$$P_L(d_1) = \begin{pmatrix} P_{L_1}(d_1) \\ P_{L_2}(d_1) \end{pmatrix} = T_{L_1}(d_1) P_{L_1}(0),$$

$$P_M(0) = \begin{pmatrix} P_{M_1}(0) \\ P_{M_2}(0) \end{pmatrix} = R_{L_1}(d_1) P_{L_1}(0),$$

(3.99)

where the vectors T_{L_1} and R_{L_1} are defined as

$$T_{L_1}(d_1) = \begin{pmatrix} b_{L_1 L_1}(d_1) \\ b_{L_2 L_1}(d_1) \end{pmatrix}, \qquad R_{L_1}(d_1) = \begin{pmatrix} b_{M_1 L_1}(d_1) \\ b_{M_2 L_1}(d_1) \end{pmatrix}.$$

(3.100)

The elements b in the vectors T and R can be obtained from (3.97, 98).

The calculation procedure involves only multiplication of matrices and vectors. It can be easily programmed for a computer. However, care must be taken to include appropriate Lorentz-polarization factors for the nth-order successive reflections involved in (3.90). Unfortunately, finding the correct polarization factor is difficult for successive reflections in the iterative calculation (3.90), because the polarization factor in Y_{ij} depends not only on the $(j-i)$ reflection but also on the ordering in the sequence of successive reflections, i.e., $i-j_1, j_1-j_2, j_2-j_3, \ldots$. To calculate the a_{ij} of (3.93) for i^{th}-order derivatives, a new term $Y_{j_{i-1} j_i}$ has to be included in the term a_{ij} of $(i-1)^{th}$ order, calculated in the previous iterative cycle. The polarization factor for the ith-order successive reflection is different from the polarization factor of the $(i-1)^{th}$ order (Sect. 3.3). This makes it difficult for the computer program to handle correctly and simultaneously the polarization and the iteration. As an approximation, the expression (3.60) can be used as the polarization factor $C_{ij}(j-i)$ for each Y_{ij} term in (3.90) [3.9].

3.7 Dynamical Treatment for Kinematical Reflections

The difficulty of inserting a correct polarization factor in the iterative calculation given in Sect. 3.6 arises from the fact that multiple diffraction is interpreted in terms of successive reflections. In fact, the participating reflection planes in a multiple diffraction process diffract simultaneously. Polarization factors should be introduced once in the power-transfer equation at the very beginning of the calculation. By this means, the difficulty can be eliminated. A similar way of introducing polarization factors is also employed in the dynamical theory of diffraction (Sect. 4.2). In the dynamical case, the polarization factors are associated with the electric susceptibilities appearing in the fundamental equation of wavefield. In this section, the intensity problem for kinematical reflection is treated using a dynamical formalism [3.10] so that the polarization factor is involved only in the power-transfer equation. Thus, the series expansion and the iterative procedure described in the previous sections can be omitted.

We assume that the approximate solutions of the differential equation (3.4) are exponential functions, i.e.,

$$P_i(x) = P_i \exp(-\alpha x),\tag{3.101}$$

where α is a linear attenuation coefficient. By substituting (3.101) into (3.4), the following set of linear equations is obtained:

$$(-S_i a_{ii} - \alpha)P_i - S_i \sum_{j \neq i}^{n} a_{ij}P_j = 0,\tag{3.102}$$

where

$$a_{ii} = -\frac{1}{\gamma_i}\left(\mu_0 + \sum_{j \neq i}^{N} Q_{ij}\right),\tag{3.103}$$

$$a_{ij} = \frac{Q_{ij}}{\gamma_j},$$

and S_i is given in (3.92).

Equation (3.102) can be solved as an eigenvalue problem. There are N eigenvalues and N eigenvectors for an N-beam case. The eigenvalues determine the attenuation and the eigenvectors give the ratios of the reflection powers among the N diffracted beams. In the dynamical theory, the fundamental equation of a wavefield can also be written as an eigenvalue equation. The eigenvectors yield the ratios of wavefield amplitudes among the diffracted waves, and the eigenvalues describe the way the wave propagates, the so-called mode of propagation (Sect. 4.5). Correspondingly, the state of attenuation described by α in this kinematical diffraction can be referred to as the mode of attenuation. There are N modes of attenuation. For each mode j, there exists a ratio:

$$P_{L_1}(j):P_{L_2}(j):\ldots:P_{L_{N_T}}(j):P_{M_1}(j):P_{M_2}(j):\ldots:P_{M_{N_R}}(j)$$

$$= \beta_{L_1}(j):\beta_{L_2}(j):\ldots:\beta_{L_{N_T}}(j):\beta_{M_1}(j):\beta_{M_2}(j):\ldots:\beta_{M_{N_R}}(j),\tag{3.104}$$

which corresponds to the attenuation α_j. N_T and N_R are the numbers of the transmitted L and reflected beams M, respectively. L_1 again indicates the incident beam. The proportionality constant ξ_j, defined as

$$\xi_j = \frac{P_L(j)}{\beta_L(j)} = \frac{P_M(j)}{\beta_M(j)} = \dots , \tag{3.105}$$

can be determined from the boundary conditions at $x = 0$.

The total reflected power inside the crystal is equal to that outside the crystal for each diffracted beam:

$$\sum_j P_{L_1}(j) = P_{L_1}(x = 0) = 1 \quad \text{for the incident beam } L_1 ,$$

$$\sum_j P_L(j) = 0 \quad \text{for } L = L_2, L_3, \dots L_{N_T} ,$$

$$\sum_j P_M(j) = P_M(x = 0) = P_M(0) \quad \text{for } M = M_1, M_2, \dots, M_{N_R} .$$

The summation is taken over all the permitted modes of attenuation. $P_M(0)$ is the reflection power of the Bragg-reflected beam M at $x = 0$. As pointed out by *Chang* [3.10], the number N_p of permitted modes of dynamical diffraction for infinitely thick crystals is equal to $2(N - N_R)$. Similarly, by considering the fact that $|a_{ij}| \ll |a_{ii}|$ and $\mu \gg Q_{ij}$ for highly absorbing crystals, it would be very easy to show that the number of modes in an N-beam kinematical diffraction is equal to $N - N_R$ (Sect. 4.5.2). The factor 2 in the dynamical case is due to the two polarized wavefields, E_σ and E_π, while in the kinematical case only the reflection power is considered. The α's of the permitted modes are positive. The rest of the modes, with negative α, are considered as attenuation for a back-reflected beam from the lower crystal surface.

If the crystal is infinitely thick, the modes associated with the back reflection should not be included in the calculation, since the corresponding reflection power may be greater than the incident power. This violates the law of conservation of energy. By considering only the permitted modes, the boundary conditions can be written in the following matrix form:

$$
\begin{pmatrix}
\beta_{L_1}(1) & \cdots & \beta_{L_1}(N_p) & 0 & \cdots & 0 \\
\cdot & & & & \cdot & \\
\cdot & & & & \cdot & \\
\cdot & & & \cdot & & \\
\beta_{L_{N_T}}(1) & \cdots & \beta_{L_{N_T}}(N_p) & 0 & & 0 \\
\beta_{M_1}(1) & \cdots & \beta_{M_1}(N_p) & -1 & & 0 \\
& & & 0 & -1 \; 0 \; \cdot & 0 \\
\cdot & & & & \cdot & \\
\cdot & & & \cdot & & \\
\cdot & & & & \cdot & \\
\beta_{M_{N_R}}(1) & \cdots & \beta_{M_{N_R}}(N_p) & 0 & 0 \; \cdots \; -1
\end{pmatrix}
\begin{pmatrix}
\xi_1 \\ \cdot \\ \cdot \\ \cdot \\ \xi_{N_p} \\ P_{M_1}(0) \\ P_{M_2}(0) \\ \cdot \\ \cdot \\ \cdot \\ P_{M_{N_R}}(0)
\end{pmatrix}
=
\begin{pmatrix}
1 \\ 0 \\ \cdot \\ \cdot \\ \cdot \\ \cdot \\ \cdot \\ \cdot \\ \cdot \\ \cdot \\ 0
\end{pmatrix}
\tag{3.106}
$$

The reflection powers $P_{M_1}(0), \ldots, P_{M_{N_R}}(0)$ of the reflections $M_1, \ldots, M_{N_R}$, from the upper crystal surface are then obtained. The excitations of modes, defined as (see also Sect. 4.8)

$$Ex(j) = |\xi_j| / \sum_{k=1}^{N_p} |\xi_k|, \tag{3.107}$$

are also determined.

If the crystal is thin, all the N modes are permitted in the diffraction process. They should be included in the above matrix equation. The reflection powers $P_{M_1}(0), \ldots, P_{M_{N_R}}(0)$ are obtained from the matrix equation, and the transmitted powers are equal to

$$P_L(d_1) = \sum_{j=1}^{N_p} \xi_j \beta_L(j) \exp(-\alpha_j d_1). \tag{3.108}$$

The procedure is exactly the same as that given in Sect. 3.5, except here the dynamical calculation procedure is followed.

3.8 Diffraction in Multi-Layered Crystals

We have just discussed the various methods of calculating the reflection powers in multiple diffraction from a plane-parallel crystal. The diffraction process is far more complicated when multiple diffraction occurs in a multi-layered crystal [3.9], which is composed of at least two materials. For example, Fig. 3.9 shows four-beam diffraction from a three-layered crystal sample. Suppose the lattice constants of the second layer S_2 are different from those of the first and the third layers, S_1 and S_3, such that diffraction cannot take place simultaneously for all the S_1, S_2 and S_3. Both layers S_1 and S_3 are of

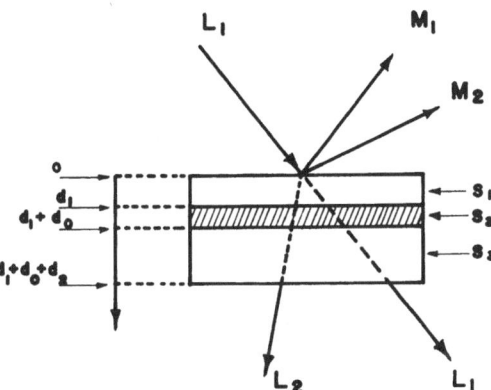

Fig. 3.9. Four-beam diffraction from a double-layered crystal plate

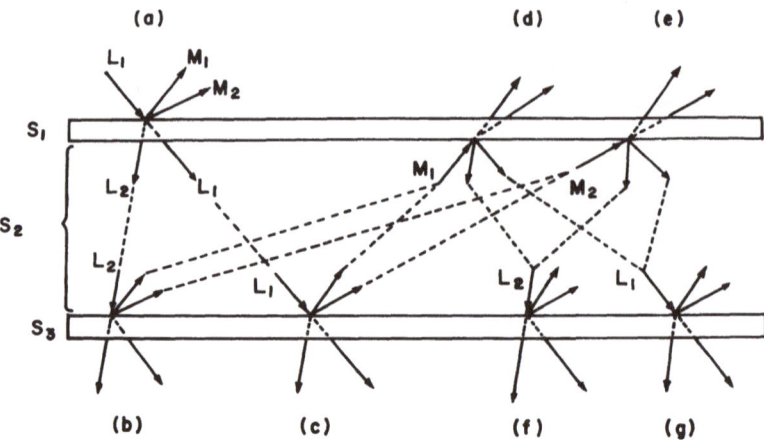

Fig. 3.10. Multiple diffraction processes from a double-layered crystal plate (after *Chang* [3.9])

the same material with the same lattice constants. Again, L_1 and L_2 are the incident and the transmitted beams, and M_1 and M_2 are the reflected beams. The thicknesses of these three layers are indicated in the figure.

When layers S_1 and S_3 are in position to diffract an incident x-ray beam, an infinite number of reflections and transmissions takes place. The diffraction process, as shown in Fig. 3.10, involves the following successive steps: (1) diffraction of the incident beam L_1 by layer S_1 (Fig. 3.10a). Beam L_1 with an incident power $P_{L_1}(0)$ generates, at $x = d_1$, two transmitted beams, L_1 and L_2, with powers equal to $P_{L_1}(d_1)$ and $P_{L_2}(d_1)$, and at $x = 0$, two zeroth-order reflected beams, M_1 and M_2, with reflection powers $P^0_{M_1}(0)$ and $P^0_{M_2}(0)$. Matrices $R_{L_1}(d_1)$ and $T_{L_1}(d_1)$ are the reflection and the transmission operators which have been defined in (3.99). (2) Absorption of the two transmitted beams, L_1 and L_2, by the S_2 layer: the transmitted powers $P_{L_1}(d_1)$ and $P_{L_2}(d_1)$ suffer ordinary absorption through the operator $U_L(d_0)$ when they traverse the layer S_2. (3) Diffraction of beams L_1 and L_2 by layer S_3 (Fig. 3.10b and c): multiple diffraction takes place for each of the incident beams L_1 and L_2. Two sets of reflected and transmitted beams are generated by the operators $R_{L_1}(d_2)$, $T_{L_1}(d_2)$ and $R_{L_2}(d_2)$, $T_{L_2}(d_2)$. The subscripts L_1 and L_2 indicate the incident beams. Considering the two transmitted beams in the direction L_1 in Fig. 3.10b, c, the sum of these two transmitted powers, denoted as $P^0_{L_2}(d_1 + d_0 + d_2)$, is the total transmitted power emerging from the lower surface of S_3, i.e., at $x = d_1 + d_0 + d_2$. Similarly, $P^0_{L_2}(d_1 + d_0 + d_2)$ is the total transmitted power for the L_2 reflection. For the reflected beams M_1 and M_2, the total powers at the upper surface of S_3, $x = d_0 + d_1$, are $P_{M_1}(d_0 + d_1)$ and $P_{M_2}(d_0 + d_1)$, which remain to participate in the next diffraction step. The block diagram, Fig. 3.11, clearly shows the sequence of the diffraction steps and the relation between the diffracted beams. (4) Absorption of the two re-

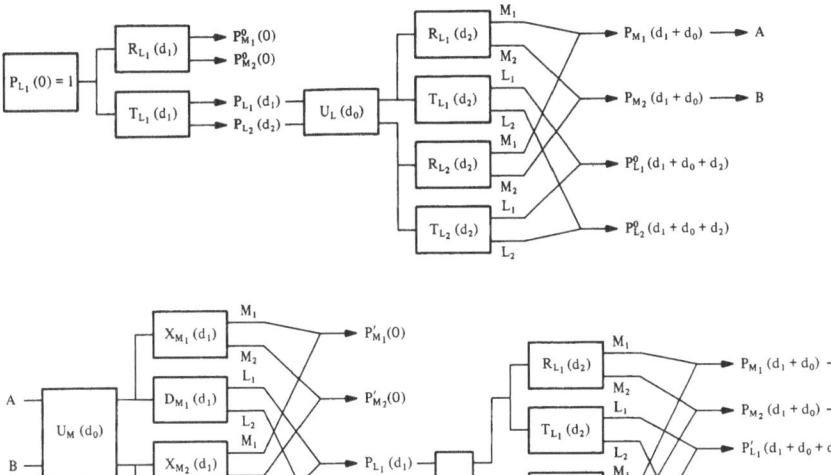

Fig. 3.11. Block diagram of Fig. 3.10 (after *Chang* [3.9])

flected beams M_1 and M_2 by the layer S_2: the two reflection powers $P_{M_1}(d_0 + d_1)$ and $P_{M_2}(d_0 + d_1)$ in step (3) suffer absorption through the operator $U_M(d_0)$ when beams M_1 and M_2 traverse the S_2 layer. (5) Backward diffractions of the reflected beams from S_1 (Fig. 3.10d, e): this step is similar to step (3) except that beams M_1 and M_2 are now the transmitted beams while L_1 and L_2 are the reflected ones. The first-order reflected powers, $P^1_{M_1}(0)$ and $P^1_{M_2}(0)$ at $x = 0$, and the transmitted powers, $P_{L_1}(d_1)$ and $P_{L_2}(d_2)$ at $x = d_1$, are generated by the transmission and reflection operators $X_{M_1}(d_1)$, $X_{M_2}(d_1)$ and $D_{M_1}(d_1)$, $D_{M_2}(d_1)$, respectively. The repetition of steps $(2-5)$ then follows to complete the diffraction process (Fig. 3.11). The labels A and B at the end of the upper row of Fig. 3.11 are the same as A and B at the beginning of the lower row. The diffraction step at A and B at the end of the lower row should be continued at A and B at the beginning of this row.

Step (1) is exactly the same as that discussed in Sect. 3.6. The reflection power $P_M(0)$ in (3.99) is replaced by $P^0_M(d_1)$. Since the steps (3) and (5) are similar to step (1) and the steps (2) and (4) are merely the attenuation of diffracted beams by the layer S_2, in the following we shall continue the consideration given in Sect. 3.6 for steps (3) and (5). For simplicity, the letters with subscripts indicate the elements of the corresponding vectors or matrices labeled with the same letters.

We only need to define the matrices T and R for step (3). These two matrices, $T_{L_2}(d_2)$ and $R_{L_2}(d_2)$, can be obtained by just substituting L_1 and d_1 by L_2 and d_2 in (3.99). Step (5) is similar to step (1). The same procedure can be taken to determine the matrices $X_{M_1}(d_1)$ and $D_{M_1}(d_1)$. Considering the boundary conditions (Fig. 3.10d)

$$P^1_{M_1}(d_1) \neq 0, \quad P^1_{M_2}(d_1) = 0,$$
$$P_{L_1}(0) = 0, \quad P_{L_2}(0) = 0, \tag{3.109}$$

a similar vector equation to (3.96) can be obtained:

$$\begin{pmatrix} P^1_{L_1}(0) \\ P^1_{L_2}(0) \\ P^1_{M_1}(d_1) \\ P^1_{M_2}(d_1) \end{pmatrix} = B_{M_1}(d_1) P^1_{M_1}(d_1), \tag{3.110}$$

where the vector

$$B_{M_1}(d_1) = S^{-1}_{M_1}(d_1) V_{M_1}(d_1). \tag{3.111}$$

The vector $V_{M_1}(d_1)$ is equivalent to $V_{L_1}(d_1)$ with M_1 replacing L_1. The matrix $S_{M_1}(d_1)$ is defined as

$$S_{M_1}(d_1) = \begin{pmatrix} -a_{L_1,L_1}(d_1) & -a_{L_1,L_2}(d_1) & 0 & 0 \\ -a_{L_2,L_1}(d_1) & -a_{L_2,L_2}(d_1) & 0 & 0 \\ -a_{M_1,L_1}(d_1) & -a_{M_1,L_2}(d_1) & 1 & 0 \\ -a_{M_2,L_1}(d_1) & -a_{M_2,L_2}(d_1) & 0 & 1 \end{pmatrix}. \tag{3.112}$$

The operators $X_{M_1}(d_1)$ and $D_{M_1}(d_1)$, similar to $T_{L_1}(d_1)$ and $R_{L_1}(d_1)$ in Sect. 3.6, are then obtained as

$$X_{M_1}(d_1) = \begin{pmatrix} b_{M_1,M_1}(d_1) \\ b_{M_2,M_1}(d_1) \end{pmatrix}, \quad D_{M_1}(d_1) = \begin{pmatrix} b_{L_1,M_1}(d_1) \\ b_{L_2,M_1}(d_1) \end{pmatrix}, \tag{3.113}$$

and similarly for $X_{M_2}(d_1)$ and $D_{M_2}(d_1)$.

The absorption operators $U_L(d_0)$ and $U_M(d_0)$ have the simple form

$$U_L(d_0) = \begin{pmatrix} u_{L_1}(d_0) \\ u_{L_2}(d_0) \end{pmatrix}, \quad U_M(d_0) = \begin{pmatrix} u_{M_1}(d_0) \\ u_{M_2}(d_0) \end{pmatrix}, \tag{3.114}$$

where $u_L(d_0) = \exp(-\mu_0 d_0/\gamma_L)$, μ_0 being the ordinary linear absorption coefficient for the S_2 layer and the particular radiation used.

Following the block diagram (Fig. 3.11), the reflected powers of M_1 and M_2 from the upper surface of layer S_1 are obtained for the first and n^{th} order of reflection:

$$P_M^0(0) = R_{L_1}(d_1) P_{L_1}(0) \,,$$

$$P_M^n(0) = X(d_1) [Ru(d_0, d_2) \quad Du(d_0, d_1)]^{n-1} Ru(d_0, d_1) Tu(d_0, d_1) \tag{3.115}$$

$$\times P_{L_1}(0) \,,$$

where

$$X(d_1) = [X_{M_1}(d_1) \quad X_{M_2}(d_1)] \,,$$

$$Ru(d_0, d_2) = \begin{pmatrix} u_{M_1}(d_0) b_{M_1, L_1}(d_2) & u_{M_1}(d_0) b_{M_1, L_2}(d_2) \\ u_{M_2}(d_0) b_{M_2, L_1}(d_2) & u_{M_2}(d_0) b_{M_2, L_2}(d_2) \end{pmatrix} \,,$$

$$Du(d_0, d_1) = \begin{pmatrix} u_{L_1}(d_0) b_{L_1, M_1}(d_1) & u_{L_1}(d_0) b_{L_1, M_2}(d_1) \\ u_{L_2}(d_0) b_{L_2, M_1}(d_2) & u_{L_2}(d_0) b_{L_2, M_2}(d_1) \end{pmatrix} \,, \tag{3.116}$$

$$Tu(d_0, d_1) = u_{L_1}(d_0) \begin{pmatrix} b_{L_1, L_1}(d_1) \\ b_{L_2, L_1}(d_1) \end{pmatrix} \,,$$

and $P_{L_1}(0) = 1$. The total reflected power is then equal to the sum

$$P_M^T(0) = \sum_{n=0}^{\infty} P_M^n(0) = R_{L_1}(d_1) + X(d_1) [I_M - F(d_2, d_1)]^{-1} G(d_2, d_1) \,, \tag{3.117}$$

where $[I_M - F]^{-1}$ is the inverse matrix of $[I_M - F]$, I_M is the unit matrix and F and G are defined as

$$F(d_2, d_1) = Ru(d_0, d_2) Du(d_0, d_1) \,,$$

$$G(d_2, d_1) = Ru(d_0, d_2) Tu(d_0, d_1) \,. \tag{3.118}$$

All the elements of F are much less than 1 for x-rays.

Similarly, the transmitted power of the nth order emerging from the lower surface of layer S_3 can be written in the form

$$P_L^n(d_0 + d_1 + d_2) = T(d_2) [Du(d_0, d_1) \quad Ru(d_0, d_2)]^n Tu(d_0, d_1) \,, \tag{3.119}$$

where $T(d_2) = [T_{L_1}(d_2) \quad T_{L_2}(d_2)]$. The total transmitted powers are

$$P_L^T(d_0 + d_1 + d_2) = \sum_{n=0}^{\infty} P_L^n(d_0 + d_1 + d_2)$$

$$= T(d_2) [I_L - E(d_1, d_2)]^{-1} Tu(d_0, d_1) \,, \tag{3.120}$$

where $E(d_1, d_2) = Du(d_0, d_1) Ru(d_0, d_2)$ and the determinant $\det[E] \ll 1$ for x-ray cases.

Equations (3.117 and 120) can be generalized for any N-beam diffraction in which N_T transmissions and N_R reflections are involved. The corresponding

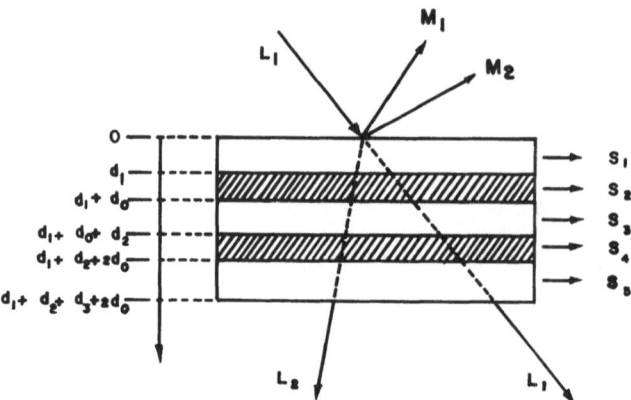

Fig. 3.12. Four-beam diffraction from a triple-layered crystal plate

dimensions of the vectors R_L, G, and Tu are $(N_R \times 1)$, $(N_R \times 1)$ and $(N_T \times 1)$. The dimensions of the matrices X, I_M and F are $(N_R \times N_R)$ and of Du, T, I_L and E, $(N_T \times N_T)$. Ru is an $(N_R \times N_T)$ matrix. A computer program for calculating the peak intensities of a general N-beam case can easily be written by taking care of the multiplication and inversion of matrices.

When a crystal is composed of five layers with three layers of one kind of material and two layers of another kind in between (Fig. 3.12), multiple diffraction in this crystal involves very many transmissions and forward and backward reflections. Because of the additional layer S_5, the combinations of the reflected M_1 and M_2 beams from layer S_5 and the beams reflected by layer S_3 should be regarded as the new incident beams at layer S_1 for the backward diffractions. Similarly, the diffracted beams, L_1 and L_2, of the backward diffractions from layers S_1 and S_3 should be summed to form the new incident beams for the forward diffractions through layer S_5. The block diagram taking these into consideration is shown in Fig. 3.13. The total power reflected from the upper surface of S_1 and the total transmitted power emerging from the lower surface of S_5 are obtained, referring to Fig. 3.13, as

$$P_M^T(0) = W_m + X(d_1)M_r[I_M - M_s]^{-1}Z_1, \quad \text{and} \tag{3.121}$$

$$P_L^T(d_1 + d_2 + d_3 + 2d_0) = W_l + T(d_3)M_t[I_L - M_u]^{-1}Z_2 \tag{3.122}$$

respectively, where

$$W_m = R_{L_1}(d_1) + X(d_1)G(d_2, d_1) + X(d_1)Z_3(d_3, d_2, d_1),$$

$$W_l = T(d_3)K(d_2, d_1) + T(d_3)Z_4(d_3, d_2, d_1),$$

$$Z_1 = \begin{pmatrix} Z_3 \\ Z_4 \end{pmatrix},$$

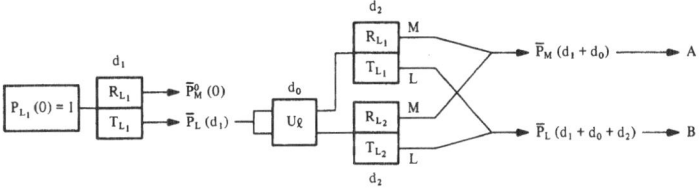

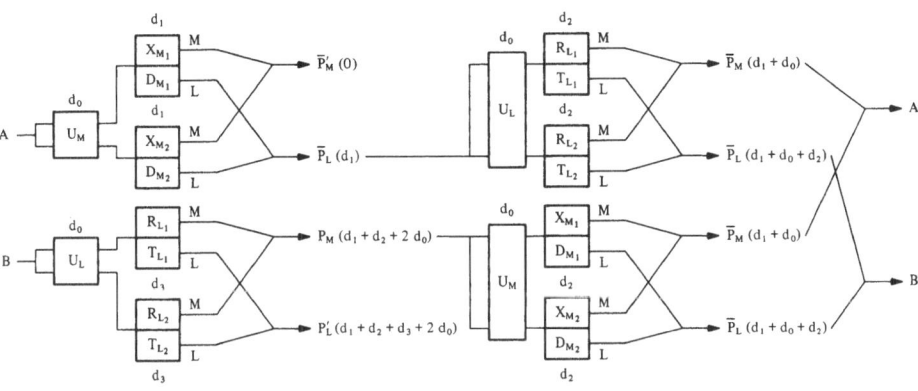

Fig. 3.13. Block diagram of Fig. 3.12 (after *Chang* [3.9])

$$Z_2 = \begin{pmatrix} Z_4 \\ Z_3 \end{pmatrix},$$

$$Z_3 = \underset{\sim}{F}(d_2, d_1)\, G(d_2, d_1) + \underset{\sim}{A}(d_2, d_3)\, K(d_2, d_1),$$

$$Z_4 = \underset{\sim}{E}(d_2, d_3)\, K(d_2, d_1) + \underset{\sim}{J}(d_2, d_1)\, G(d_2, d_1),$$

$$\underset{\sim}{M}_r = [\underset{\sim}{F}(d_2, d_1) \quad \underset{\sim}{A}(d_2, d_3)],$$

$$\underset{\sim}{M}_t = [\underset{\sim}{E}(d_2, d_3) \quad \underset{\sim}{J}(d_2, d_1)],$$

$$\underset{\sim}{M}_s = \begin{pmatrix} \underset{\sim}{F}(d_2, d_1) & \underset{\sim}{A}(d_2, d_3) \\ \underset{\sim}{J}(d_2, d_1) & \underset{\sim}{E}(d_2, d_3) \end{pmatrix},$$

$$\underset{\sim}{M}_u = \begin{pmatrix} \underset{\sim}{E}(d_2, d_3) & \underset{\sim}{J}(d_2, d_1) \\ \underset{\sim}{A}(d_2, d_3) & \underset{\sim}{F}(d_2, d_1) \end{pmatrix}.$$

(3.123)

The matrices $\underset{\sim}{A}$ and $\underset{\sim}{J}$ and the vector K are defined as

$$\underset{\sim}{A}(d_2, d_3) = \underset{\sim}{Xu}(d_0, d_2)\, \underset{\sim}{Ru}(d_0, d_3),$$

$$\underset{\sim}{J}(d_2, d_1) = \underset{\sim}{Tx}(d_0, d_2)\, \underset{\sim}{Du}(d_0, d_1),$$

$$\underset{\sim}{K}(d_2, d_1) = \underset{\sim}{Tx}(d_0, d_2)\, \underset{\sim}{Tu}(d_0, d_1), \quad \text{with}$$

(3.124)

$$\underset{\sim}{Tx}(d_0, d) = \begin{pmatrix} u_{L_1}(d_0)\, b_{L_1, M_1}(d) & u_{L_1}(d_0)\, b_{L_1, M_2}(d) \\ u_{L_2}(d_0)\, b_{L_1, M_1}(d) & u_{L_2}(d_0)\, b_{L_1, M_2}(d) \end{pmatrix}.$$

(3.125)

The matrices Xu and Ru have the same form as Tx, except that the b's need to be replaced by the corresponding elements of X and R. The dimensions, in a general N-beam case, are $(N_R \times 1)$ for W_m and Z_3, $(N_T \times 1)$ for W_l and Z_4, $(N \times 1)$ for Z_1 and Z_2, $(N \times N)$ for M_s and M_u, $(N_R \times N)$ for M_r and $(N_T \times N)$ for M_t. Calculations may be greatly simplified by including T, R and X, D in a larger matrix. Equations (3.121 and 122) for the five-layer system can be reduced to cases having single, double, triple, and quadruple layers by choosing proper values for the crystal-layer thicknesses.

3.9 Peak Width, Beam Divergence, and Mosaic Spread

The peak width of a given multiple diffraction depends on how long the secondary reciprocal lattice points take to traverse the shell of the Ewald sphere during the crystal rotation. The thickness of the shell can be affected by the effective beam divergence, which depends on the mosaic spread, the intrinsic diffraction width of a perfect crystal and the geometry of beam collimation. Similarly to the consideration of Lorentz factors (Sect. 3.3), the time for which a secondary reciprocal lattice point stays on the surface of the Ewald sphere is [3.11]

$$t = \frac{\delta_{\text{eff}} \cdot r}{v^* \cdot r}, \qquad (3.126)$$

where v^* is the speed of rotation of the secondary reciprocal lattice points with respect to the primary reflection. The quantity δ_{eff} is the effective beam divergence which contains two components, δ_v and δ_h, perpendicular and parallel to the plane of incidence and the wavelength dispersion δ_λ along the incident direction, the i_z direction:

$$\delta_{\text{eff}} = p_n \delta_h \hat{i}_x + p_n \delta_v \hat{i}_y + \delta_\lambda \hat{i}_z, \qquad (3.127)$$

Here δ_h is the smaller of the mosaic spread η and the modified horizontal beam divergence $\delta_{h,b}$ and δ_v is the width of the convolution of the mosaic spread with the modified vertical beam divergence $\delta_{v,b}$. δ_h, δ_v, and δ_λ are given by

$$\delta_h = \min(\eta, \delta_{h,b}), \quad \delta_v = \sqrt{\eta^2 + \delta_{v,b}^2}, \quad \delta_\lambda = \delta\left(\frac{1}{\lambda}\right), \qquad (3.128)$$

where

$$\delta_{h,b} = \sqrt{\delta_H^2 + \delta_{in}^2}, \quad \delta_{v,b} = \sqrt{\delta_V^2 + \delta_{in}^2}. \qquad (3.129)$$

The quantity δ_{in} is the intrinsic width of the diffraction from a perfect crystal. δ_H and δ_V are the horizontal and vertical divergences due to the geometry of the beam collimation. The unit vectors $\hat{i}_y$ and $\hat{i}_x$ are perpendicular and parallel to the plane of incidence of the primary reflection, respectively. The incident wavevector k_0 is along $\hat{i}_z$.

Referring to Figs. 3.5, 6, the linewidth ΔW of a multiple diffraction is thus expressed as

$$\Delta W \propto t = \left(\frac{\lambda}{\omega}\right) \frac{\delta_h |p_n \sin\beta \sin\theta_G| + \delta_v |p_n \cos\beta| + \delta_\lambda |\sin\beta \cos\theta_G|}{\sin\psi \cos\chi \cos\xi},$$

$$(3.130)$$

where the angles ψ, χ, ξ are defined in Sect. 3.3 and θ_G is the Bragg angle of the primary reflection. δ_λ is usually very small (10^{-5} rad). For a perfect crystal, δ_h is also very small since $\eta \simeq 0$. The linewidth only depends on the vertical divergence δ_v [3.11, 12]. Under these conditions, $\cos\beta$ is the dominant factor for a given δ_v. The smaller the angle β, the wider the linewidth.

The linewidth, besides depending on the experimental factors discussed above, also depends on whether the multiple diffraction is of the Aufhellung or Umweganregung type. For Umweg, the integrated intensity, referring to (3.48), involves the product of two Gaussians, $W_{0i}W_{i1}$. The width of this product is, of course, narrower than the width of a single Gaussian. The linewidth of Aufhellung is affected by two single Gaussians and one double Gaussian. This linewidth is therefore wider, relatively, than that of the Umweg.

4. Dynamical Theory of X-Ray Diffraction

The kinematical theory of x-ray diffraction, thus far discussed in the previous chapter, is based on assumptions which are only valid for diffraction in small crystals. When diffraction occurs in large and perfect crystals, multiple scattering results. In other words, the crystal lattice is so regular over a large volume that the reflected wave of a reflection must be further reflected back into the direction of the incident wave. The phase difference between the twice-reflected wave and the incident wave modifies considerably the amplitude of the incident wave within the crystal. If the reflection is relatively strong, the diffracted intensity should be of the same order of magnitude as that of the incident beam. Thus interaction between the incident and the scattered waves is definitely enhanced and the effect of interaction resulting from multiple scattering can no longer be neglected. Moreover, according to the optics of the visible spectrum, the phase of the scattered wave in the forward direction is retarded by about a quarter of a period from that of the incident wave. This slight modification of phase between the scattered and the incident waves happens for the reflection at each plane of atoms. The phase velocity of the resultant wave passing through the crystal is then modified. The velocity of the radiation traveling through the crystal is therefore not the velocity of light. This implies the existence of a correction to the index of refraction for x-rays in crystals. Consequently, slight modifications of the Bragg law of reflection are expected [4 1]

These consequences resulting from the interaction between the incident and the scattered waves rely mainly on the dynamical equilibrium between the resultant x-ray wavefield and the scattering atoms within the crystal. A dynamical theory of x-ray diffraction is therefore developed to account more precisely for this mechanism of diffraction, namely, the formation of wavefields in crystals. This is in contrast to the kinematical theory which takes only the reflection powers into account. In the following we shall consider this aspect and derive the fundamental equation of wavefields.

4.1 Fundamental Equation of Wavefields

The diffraction of x-rays in crystals, for both two-beam and multi-beam cases, involves electromagnetic waves propagating in a periodic array of atoms and interacting with the electron density distributed around the atomic sites. Although this phenomenon can be treated collectively as the excitation of periodically distributed dipole resonators, proposed by *Ewald* [4.2], it is easier and simpler to describe it using classical electrodynamics via Maxwell's equations. Laue's dynamical theory [4.3] of x-ray diffraction is adapted here.

The crystal is considered in Laue's treatment as a continuous distribution of negative charges with the positive charges centered at the nuclei. The scattering of an incident electromagnetic wave from the nuclei is negligibly small. Only scattering from the negative charge distribution contributes to the resultant x-ray wavefield.

Let us first recall Maxwell's equations describing the crystal field. They specify the relation between the vector fields of the electric vector E, the electric displacement D, the magnetic vector H and the magnetic induction B for a given net charge density ϱ_t and current density J_t:

$$\mathrm{curl}\,H = \frac{1}{c}\left(\frac{\partial D}{\partial t} + 4\pi J_t\right), \quad \mathrm{div}\,D = 4\pi\varrho_t,$$

$$\mathrm{curl}\,E = -\frac{1}{c}\frac{\partial B}{\partial t}, \quad \mathrm{div}\,B = 0, \tag{4.1}$$

where the Gaussian unit system is used. The quantity c is the speed of light in vacuum. The conservation of charge is described, as usual, by the continuity equation

$$\mathrm{div}\,J_t + \frac{\partial\varrho_t}{\partial t} = 0. \tag{4.2}$$

The electric and magnetic energy densities are defined as

$$W_e = \frac{1}{8\pi}|E|^2, \quad W_m = \frac{1}{8\pi}|H|^2. \tag{4.3}$$

The Poynting vector is

$$S = \frac{c}{4\pi}(E \times H). \tag{4.4}$$

The presence of external electromagnetic waves, like x-rays, causes the displacement of the negative charge with respect to the nuclei in a crystal. This leads to polarization, which can be described by

$$D = E + 4\pi P, \tag{4.5}$$

provided that the frequency of the external field is far from the natural absorption edges of the system. The vector P is the electric polarization. Similarly, the magnetic induction B can be written

$$B = H + 4\pi M , \tag{4.6}$$

where M is the magnetic polarization. In the x-ray case, D and B can be written as

$$D = \varepsilon E, \quad B = \mu H , \tag{4.7}$$

under the assumption that the conductivity is zero at x-ray frequencies. The net charge density ϱ_t and current density J_t can therefore be assumed to be zero. The electric polarization is then expressed as

$$P = \frac{\chi}{4\pi} E, \quad \text{with} \tag{4.8}$$

$$\chi = \varepsilon - 1 , \tag{4.9}$$

where $\chi/4\pi$ is the electric susceptibility for the x-ray used. The transversality of the electric displacement is preserved in the second equation of (4.1):

$$\text{div} D = 0 . \tag{4.10}$$

We assume that the field vectors E, D and B, H have the forms

$$E(r, t) = E(r) \exp(i\omega t), \quad H(r, t) = H(r) \exp(i\omega t) ,$$
$$D(r, t) = D(r) \exp(i\omega t), \quad B(r, t) = B(r) \exp(i\omega t) . \tag{4.11}$$

A simple expression for the polarization P may be obtained by considering the motion of a free electron in an external field E. The equation of motion,

$$m \frac{d^2 x}{dt^2} = -eE , \tag{4.12}$$

leads to the expression for the displacement of the electron,

$$x = -\frac{e}{m\omega^2} E . \tag{4.13}$$

The polarization P due to electric dipoles can then be written as

$$P = eN(r)x = -\frac{e^2}{m\omega^2} N(r)E , \tag{4.14}$$

where $N(r)$ is the electron density. Comparing (4.14) with (4.8), we obtain

$$\chi = -\frac{e^2\lambda^2 N(r)}{mc^2\pi}. \tag{4.15}$$

Since the classical radius of the electron $r_e = e^2/(mc^2) \simeq 2.818 \times 10^{-13}$ cm, $\lambda^2 \simeq 10^{-16}$ cm and $N \simeq 10^{23} - 10^{25}$ cm^{-3}, this approach is justified. Otherwise, higher-order multipoles should be considered.

The combination of the first and the third equations of (4.1) gives

$$\nabla^2 D(r, t) - \frac{1}{c^2}\frac{\partial^2 D(r, t)}{\partial t^2} = -\operatorname{curl}\operatorname{curl}\left(\frac{\chi}{1+\chi}D(r, t)\right), \tag{4.16}$$

where the conditions (4.5, 7–9) have been used. Since χ is of the order of magnitude of 10^{-5} for x-rays, (4.16) can be rewritten as

$$\nabla^2 D(r, t) - \frac{1}{c^2}\frac{\partial^2 D(r, t)}{\partial t^2} \simeq -\operatorname{curl}\operatorname{curl}[\chi D(r, t)]. \tag{4.17}$$

This is called the fundamental equation of wavefield in differential form. The solution of this inhomogeneous wave equation is

$$D(r, t) = D_e(r, t) + \frac{1}{4\pi}\int\frac{1}{|r-r'|}\operatorname{curl}\operatorname{curl}\left[\chi(r')D\left(r', t-\frac{|r-r'|}{c}\right)\right]dr' \tag{4.18}$$

where D_e, the wavefield for r at infinity, is the solution of the homogeneous wave equation (4.17) with $\chi = 0$. Equation (4.18) is the integral form of the wavefield equation.

Following Laue's treatment, the crystal can be treated as a medium with a periodic complex dielectric constant. Accordingly, the quantity χ in (4.15) proportional to the electric susceptibility can be expressed as a Fourier series:

$$\chi = \sum_G \chi_G\exp[-2\pi i(g_G\cdot r)], \quad\text{where} \tag{4.19}$$

$$\chi_G = -\frac{e^2\lambda^2}{mc^2\pi V}F_G. \tag{4.20}$$

A similar expression can be obtained for $N(r)$ in (4.14), by taking the Fourier transform:

$$N(r) = \frac{1}{V}\sum_G F_G e\exp[-2\pi i(g_G\cdot r)]. \tag{4.21}$$

F_G is the structure factor of the G reflection, which can also be defined as a Fourier transform:

$$F_G = \sum_j f_j\exp(2\pi i g_G\cdot r_j), \tag{4.22}$$

where g_G is the reciprocal lattice vector, f_j is the atomic scattering factor of the atom at r_j, λ and V are the wavelength and the volume of the unit cell.

The assumed solution of the fundamental equation (4.17) is a Bloch wave function which is a superposition of an infinite number of plane waves:

$$D(r, t) = \sum_G D_G \exp(i\omega t - 2\pi i K_G \cdot r), \tag{4.23}$$

where K_G is the wavevector of the G reflection satisfying Bragg's law, i.e.,

$$K_G = K_O + g_G, \tag{4.24}$$

for every G reflection involved. The term K_O is the wavevector of the incident beam O. By substituting (4.19, 23) into (4.17) and setting all the Fourier coefficients equal to zero, the fundamental equation becomes

$$(k^2 - K_G^2)D_G - \sum_L \chi_{G-L}[K_G \times (K_G \times D_L)] = 0 \tag{4.25}$$

for all L reflections. In a more compact form, according to *Laue* [4.3], this becomes

$$\frac{(K_G^2 - k^2)}{K_G^2} D_G = \sum_L \chi_{G-L} D_{L(\perp K_G)}, \tag{4.26}$$

where the vector rule $A \times (B \times C) = B(A \cdot C) - C(A \cdot B)$ has been employed. $D_{L(\perp K_G)}$ represents the vector component of D_L perpendicular to the wavevector K_G. k is the magnitude of the incident wavevector, which is equal to $1/\lambda$ in vacuum. The coefficients of the Fourier expansion for the polarization P and the electric field E are also obtained:

$$P_G = \frac{\sum_L \chi_{G-L} D_L}{4\pi}, \quad \text{and} \tag{4.27}$$

$$E_G = D_G - \sum_L \chi_{G-L} D_L. \tag{4.28}$$

Substituting the Fourier representations of H and D into the first and the third equations in (4.1), we obtain

$$D_G = \frac{K_G \times H_G}{|K_G|}, \quad H_G = \frac{K_G \times E_G}{|K_G|}. \tag{4.29}$$

The vector K_G, the magnetic field H and the electric displacement D are therefore mutually orthogonal. The electric field E lies in the plane containing D and K_G.

It should be noted that the above derivation is only an approximation. The fundamental equation (4.17) in terms of D is not in itself strictly rigorous. For instance, in Laue's treatment, the assumption

$$E \approx D \tag{4.30}$$

has been made [4.3]. However, this contradicts the fact that D is a transverse wave while E is not [4.4]. In order to avoid this contradiction and to consider the microscopic nature of x-ray diffraction, we shall use the electric field E in the fundamental equation of the wavefield. In addition, there are some merits in using E instead of D [4.5], such as the simple form for the boundary conditions and the ease in dealing with multi-beam cases. For this reason, the alternative form of the fundamental equation in terms of E is given below [4.5].

The relation between the electric field and the quantity χ can be obtained from the first and the third equations of (4.1):

$$\frac{K_G^2 E_{G(\perp K_G)} - k^2 E_G}{k^2} = \sum_L \chi_{G-L} E_L . \tag{4.31}$$

The term $E_{G(\perp K_G)}$ appears because $\mathrm{div} E \neq 0$. Because the quantity χ_{G-L} in (4.28) is of the order of 10^{-5}, the summation term involving $\chi_{G-L} D_L$ can be neglected in (4.28). The electric field E_G is then approximately parallel to D_G and accordingly perpendicular to K_G. Equation (4.31) can thus be expressed in the simple form

$$\frac{K_G^2 - k^2}{k^2} E_G = \sum_L \chi_{G-L} E_L . \tag{4.32}$$

Note that the constant k^2 stays in the denominator, and that (4.32) is simpler than (4.26). In addition, since E is directly related to the vector potential $A(r)$, (4.32) is of a more microscopic nature than (4.26). Moreover, (4.32) resembles the scalar fundamental equation for electron diffraction, in which the amplitude of a component of the electron Bloch wave replaces the E vector. The χ term for electron diffraction is the ratio of the periodic potential energy to the total energy in a crystal. In practice, (4.26) and (4.32) give essentially the same results. The difference in numerical calculation between the two is not greater than $10^{-3}\%$ [4.6].

There are N vector equations (4.32) for a general N-beam x-ray diffraction. Suppose this N-beam case consists of N_T transmitted beams, $O, L_1, \ldots,$ L_{N_T-1}, and N_R reflected beams, $M_1, \ldots, M_{N_R}$. The incident beam O is of the transmission type. These N equations are somewhat inconvenient because of their vector forms. A set of scalar equations, which is easy to deal with, can be obtained by considering each polarized component of the wavefield. We define $E_{\sigma G_i}$ and $E_{\pi G_i}$ as the σ- and π-polarized components of a given wavefield E_{G_i},

$$E_{G_i} = E_{\sigma G_i} \, \hat{\boldsymbol{\sigma}}_{G_i} + E_{\pi G_i} \, \hat{\boldsymbol{\pi}}_{G_i},$$ (4.33)

where the unit vectors $\hat{\boldsymbol{\sigma}}_{G_i}$ and $\hat{\boldsymbol{\pi}}_{G_i}$ are defined as

$$\hat{\boldsymbol{\sigma}}_{G_i} = \frac{E_{\sigma G_i}}{|E_{\sigma G_i}|}, \qquad \hat{\boldsymbol{\pi}}_{G_i} = \frac{E_{\pi G_i}}{|E_{\pi G_i}|}.$$ (4.34)

The vectors $\boldsymbol{K}_{G_i}$, $\hat{\boldsymbol{\sigma}}_{G_i}$ and $\hat{\boldsymbol{\pi}}_{G_i}$ form mutually orthogonal axes such that

$$\hat{\boldsymbol{\sigma}}_{G_i} = \hat{\boldsymbol{K}}_{G_i} \times \hat{\boldsymbol{\pi}}_{G_i}.$$ (4.35)

By multiplying (4.32) by the unit polarization vectors $\hat{\boldsymbol{\sigma}}$ and $\hat{\boldsymbol{\pi}}$ of each reflection involved, $2N$ scalar equations are obtained. They can be written in the matrix form

$$\underset{\sim}{\boldsymbol{\Phi}} E = 0,$$ (4.36)

where E is a column vector, expressed horizontally as

$$E = [E_{\sigma O} E_{\pi O} E_{\sigma G_1} E_{\pi G_1} \dots E_{\sigma G_{N-1}} E_{\pi G_{N-1}}],$$ (4.37)

and the $(2N \times 2N)$ complex matrix $\underset{\sim}{\boldsymbol{\Phi}}$ is

$$\Phi_2 = \begin{pmatrix} \chi_{O-O}-2\varepsilon_O & 0 & \chi_{O-G_1}(\hat{\sigma}_O\cdot\hat{\sigma}_{G_1}) & \chi_{O-G_1}(\hat{\sigma}_O\cdot\hat{\pi}_{G_1}) & \cdots & \chi_{O-G_{N-1}}(\hat{\sigma}_O\cdot\hat{\pi}_{G_{N-1}}) \\ 0 & \chi_{O-O}-2\varepsilon_O & \chi_{O-G_1}(\hat{\pi}_O\cdot\hat{\sigma}_{G_1}) & \chi_{O-G_1}(\hat{\pi}_O\cdot\hat{\pi}_{G_1}) & \cdots & \chi_{O-G_{N-1}}(\hat{\pi}_O\cdot\hat{\pi}_{G_{N-1}}) \\ \chi_{G_1-O}(\hat{\sigma}_{G_1}\cdot\hat{\sigma}_O) & \chi_{G_1-O}(\hat{\sigma}_{G_1}\cdot\hat{\pi}_O) & \chi_{O-O}-2\varepsilon_{G_1} & 0 & \cdots & 0 \\ \chi_{G_1-O}(\hat{\pi}_{G_1}\cdot\hat{\sigma}_O) & \chi_{G_1-O}(\hat{\pi}_{G_1}\cdot\hat{\pi}_O) & 0 & \chi_{O-O}-2\varepsilon_{G_1} & \cdots & 0 \\ \cdot & \cdot & \cdot & \cdot & \cdot & \\ \cdot & \cdot & \cdot & \cdot & & \cdot \\ \chi_{G_{N-1}-O}(\hat{\pi}_{G_{N-1}}\cdot\hat{\sigma}_O) & \chi_{G_{N-1}-O}(\hat{\pi}_{G_{N-1}}\cdot\hat{\pi}_O) & \chi_{G_{N-1}-O}(\hat{\pi}_{G_{N-1}}\cdot\hat{\sigma}_{G_1}) & \chi_{G_{N-1}-O}(\hat{\pi}_{G_{N-1}}\cdot\hat{\pi}_{G_1}) & \cdots & \chi_{O-O}-2\varepsilon_{G_{N-1}} \end{pmatrix}$$

(4.38)

with the quantities $2\varepsilon_{G_i}$ defined as

$$2\varepsilon_{G_i} = \frac{K_{G_i}^2 - k^2}{k^2}.$$

(4.39)

Because $2\varepsilon_{G_i}$ is proportional to the difference in magnitude of the wavevectors inside and outside the crystal, (4.36) is dependent on the variation of the index of refraction of x-rays in the crystal. Moreover, since $2\varepsilon_{G_i}$ is direction dependent, (4.38) exhibits the anisotropy of the index of refraction in the case of multiple diffraction.

4.2 Polarization of Wavefields

The decomposition of a wavefield into σ and π components, as discussed in the previous section, provides a proper mathematical instrument to transform the vector form of the fundamental equation into the scalar form. However, the quantities σ and π are just mathematical symbols. We need, in this section, to give them physical definition.

In a simple two-beam case (direct beam O and reflected beam G_1), the plane containing the incident and the reflected wavevectors is called the plane of incidence of the G_1 reflection. The unit polarization vectors, $\hat{\sigma}$ and $\hat{\pi}$, are usually chosen to be perpendicular and parallel to the plane of incidence, respectively, with $\hat{\pi}$ actually lying in that plane. Under these circumstances, the wavevectors K_O and K_{G_1} must be perpendicular to $\hat{\sigma}_O$, $\hat{\pi}_O$ and $\hat{\sigma}_{G_1}$, $\hat{\pi}_{G_1}$, respectively, and the following relations are obtained:

$$\hat{\sigma}_O \cdot \hat{\sigma}_{G_1} = 1, \quad \hat{\pi}_O \cdot \hat{\pi}_{G_1} = \cos 2\theta_{G_1},$$
$$\hat{\sigma}_O \cdot \hat{\pi}_{G_1} = \hat{\sigma}_{G_1} \cdot \hat{\pi}_O = 0,$$

(4.40)

where θ_{G_1} is the Bragg angle of the G_1 reflection. These relations indicate that all the σ's are perpendicular to the plane of incidence and all the π's lie in that plane. If an N-beam diffraction is wavevector coplanar (coincidental diffraction, see Sect. 2.1.2), (4.40) still holds for all the reflections.

To define the polarization vectors in the same sense as in the two-beam case for all reflections in an N-beam persistent diffraction (Sect. 2.1.2) is a difficult task, since for each reflection, there is a corresponding plane of incidence. The $\hat{\sigma}$ vector of reflection G_1 may not be perpendicular to the $\hat{\pi}$ vector of reflection G_2. It is therefore necessary to describe all the σ and π polarization vectors in terms of common coordinates so that the geometrical relation between these vectors can be easily revealed. On the other hand, there must be correlation between the σ-polarized wavefield of one reflection and the π-polarized field of the other reflections when orthogonality between all the σ's and π's cannot be achieved. The correlation becomes important in calculating diffracted intensities in N-beam cases.

The simplest way of handling the polarization vectors in N-beam cases is to treat the $\hat{\sigma}$ and $\hat{\pi}$ vectors as references for wavefields and wavevectors. In this way, one can make one's own choice of the direction of polarization, provided that (4.35) holds for all wavefields. Following *Joko* and *Fukuhara* [4.7], it is convenient to define all the σ vectors so that they lie in the plane which contains all the reciprocal lattice points involved. Figure 4.1 shows the σ vectors for a three-beam diffraction, O, G_1 and G_2, which lie in the reflection circle centered at point C. All the σ vectors are chosen to be perpendicular to and to bisect the corresponding vectors CO, CG_1 and CG_2, respectively. The wavevectors K_O, K_{G_1} and K_{G_2} within the crystal are, for the time being (without considering the dispersion effect due to the interaction between diffracted beams), defined as the vectors connecting the center W of the Ewald sphere to the corresponding reciprocal lattice points. The π vectors can then be generated, according to (4.35), from the vector product of K's and σ's. This way of defining polarization vectors can be extended to the non-coplanar coincidental multiple diffraction, though the geometrical relationships among the σ's, π's and K's are far more complicated. Detailed consideration of the choice of coordinate system to describe the vectors K, σ and π will be given in later sections for cases involving Bragg reflection and Laue transmission.

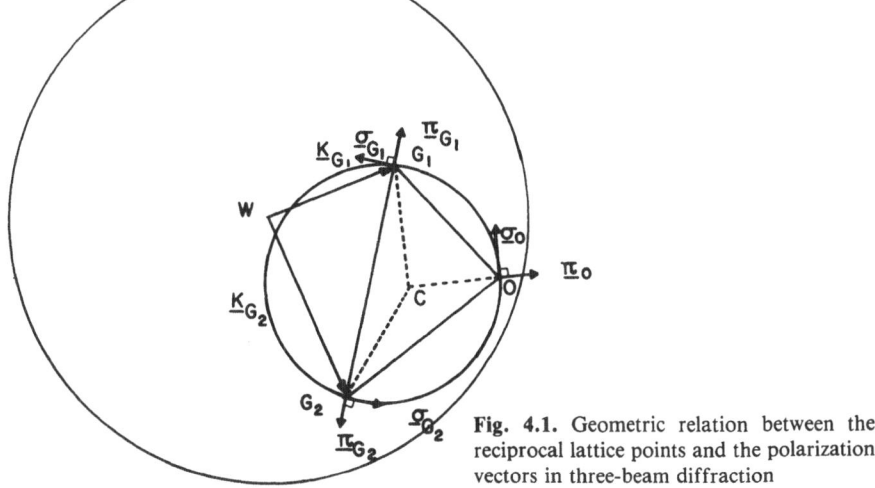

Fig. 4.1. Geometric relation between the reciprocal lattice points and the polarization vectors in three-beam diffraction

4.3 Dispersion Surface

4.3.1 Geometry of the Dispersion Surface

The fundamental equation (4.36), as can be seen from its form, is an eigen-value equation. The quantities 2ε (4.38, 39) are the eigenvalues and the column vectors E are the eigenvectors. The condition for (4.36) to have a non-trivial solution is that the determinant of the matrix must be zero:

$$\| \Phi \| = 0 . \tag{4.41}$$

This equation is called the dispersion relation, and it describes the variation of the wavevectors in N-beam diffraction with respect to the angular setting of the crystal for a given wavelength. In reciprocal space (momentum space), a surface called a dispersion surface can be constructed according to (4.41). Since the wavefields are considered to be complex variables, the eigenvalues and eigenvectors are also complex. The real part of an eigenvalue defines the coordinates of the starting point of the wavevectors within the crystal, for all reflections involved, at a given angular crystal setting. This point was called the tie point by $Ewald$ [4.2], or wavepoint. The loci of tie points when the crystal setting is varied relative to the incident beam form the dispersion surface. The imaginary part of the eigenvalue is proportional to the linear absorption coefficient. For the jth eigenvalue $2\varepsilon_{G_i}(j)$ or $K_{G_i}(j)$, the corresponding eigenvector yields the amplitude ratio of an arbitrary pair of two-component waves:

$$\frac{E_{\sigma G_l}(j)}{E_{\pi G_m}(j)} = \frac{\text{Cof}(\sigma G_l, \sigma G_l)_j}{\text{Cof}(\pi G_m, \sigma G_l)_j} , \qquad \frac{E_{\sigma G_l}(j)}{E_{\pi G_m}(j)} = \frac{\text{Cof}(\sigma G_l, \pi G_m)_j}{\text{Cof}(\pi G_m, \pi G_m)_j} , \tag{4.42}$$

where $\text{Cof}(\sigma G_l, \pi G_m)$ is the cofactor of the element $\chi_{G_m - G_l}(\pi G_m \cdot \sigma G_l)$ in the determinant $\| \Phi \|$.

Rigorously speaking, the dispersion surface described by (4.41) should be represented in a complex momentum space, since Φ is a complex matrix. However, a complex dispersion surface is difficult to visualize physically. If the imaginary part of χ is sufficiently small, which is usually the case for an x-ray far away from the absorption edge, the dispersion surface projected in real momentum space can approximately represent the actual dispersion surface. From (4.36), there are $2N$ eigenvalues, namely, $2N$ dispersion sheets for a given N-beam case. Because these $2N$ sheets are displaced in the three-dimensional reciprocal space, it is difficult to construct and to visualize the features of the N-beam dispersion surface. We shall, in the following, consider the dispersion surface first for simple one- and two-beam cases and then for the rather complicated multi-beam cases.

For simplicity, let us consider a one-beam, O, case for a linearly polarized wavefield. The matrix Φ has the dimensions (1×1). The dispersion relation takes the form

$$\chi_{O-O} - 2\varepsilon_O = 0 . \tag{4.43}$$

By combining (4.39) and (4.43), and neglecting the terms involving second- and higher-order powers of χ_{O-O}, the real and imaginary parts of the wave-vector inside the crystal are obtained as

$$K'_{O,\mathrm{r}} = k \left(1 + \frac{\chi^{\mathrm{r}}_{O-O}}{2} \right) , \qquad K'_{O,\mathrm{i}} = \frac{k}{2} \chi^{\mathrm{i}}_{O-O} , \tag{4.44}$$

and the corresponding linear absorption coefficient along K_0 is

$$\mu = -4\pi K'_{O,\mathrm{i}} = -2\pi k \chi^{\mathrm{i}}_{O-O} , \tag{4.45}$$

where χ is a very small negative quantity in comparison with k ($=1/\lambda$) in vacuum. The difference between K'_O and k is therefore due to the small correction Δ in the index of refraction n, i.e.,

$$n = \frac{K'_O}{k} = 1 - \Delta , \quad \text{where} \tag{4.46}$$

$$\Delta = -\frac{\chi_{O-O}}{2} \simeq \frac{e^2 \lambda^2}{2mc^2 \pi} \left(\frac{N_0}{V} \right) . \tag{4.47}$$

The expression for χ_{O-O} is substituted from (4.15). The term N_0/V is the electron density. The magnitude of Δ is of the order of 10^{-5} for x-ray frequencies higher than the absorption edges. The dispersion surface defined by (4.46) is a sphere with radius equal to nk.

The dispersion surface of a two-beam, O and G, diffraction can be constructed by using the dispersion sphere of the one-beam case. First we draw two dispersion spheres for beams O and G, according to (4.43), and then put these two spheres together so that the distance between point O and point G is the modulus of the reciprocal lattice vector g of the G reflection (Fig. 4.2). Interaction of beams O and G modifies further the index of refraction, $1 - \Delta$, at the intersections of these two dispersion spheres. The dispersion surface of this two-beam case is shown in Fig. 4.2. The point T on the dispersion surface is a tie point. Point L is the so-called Lorentz point (line) which is the intersection of the two one-beam spheres. The vectors K'_O ($= LO$) and K'_G ($= LG$) starting at L have magnitudes $k(1 + \Delta/2)$ or $K'^2_O = k^2(1 + \chi_{O-O}) = K'^2_G$. The wavevectors K_O and K_G from the tie point denote the modified wavevectors within the crystal which are governed by the two-beam dispersion relation derived from (4.41):

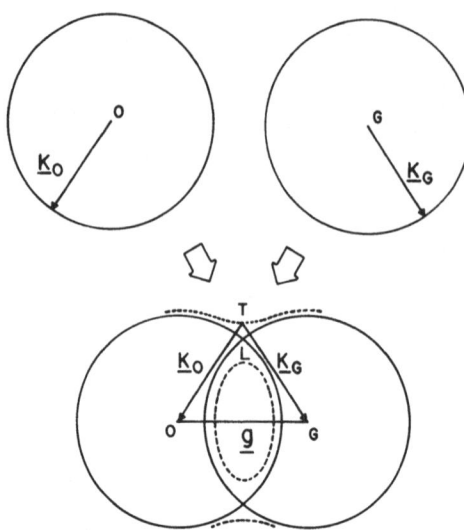

Fig. 4.2. Formation of the dispersion surface for two-beam diffraction

$$\begin{vmatrix} \chi_{O-O} - \dfrac{K_O^2 - k^2}{k^2} & 0 & \chi_{O-G}(\hat{\sigma}_O \cdot \hat{\sigma}_G) & 0 \\[2ex] 0 & \chi_{O-O} - \dfrac{K_O^2 - k^2}{k^2} & 0 & \chi_{O-G}(\hat{\pi}_O \cdot \hat{\pi}_G) \\[2ex] \chi_{G-O}(\hat{\sigma}_G \cdot \hat{\sigma}_O) & 0 & \chi_{O-O} - \dfrac{K_G^2 - k^2}{k^2} & 0 \\[2ex] 0 & \chi_{G-O}(\hat{\pi}_G \cdot \hat{\pi}_O) & 0 & \chi_{O-O} - \dfrac{K_G^2 - k^2}{k^2} \end{vmatrix} = 0,$$

$$(4.48)$$

where both σ and π polarizations are considered. This equation can obviously be represented separately for each polarization by the (2×2) determinant

$$\begin{vmatrix} K_O^2 - K_O'^2 & k^2 C \chi_{O-G} \\ k^2 C \chi_{G-O} & K_G^2 - K_G'^2 \end{vmatrix} = 0, \tag{4.49}$$

where (4.44) has been employed. The term C is the polarization factor which is defined as

$$C = \begin{cases} \hat{\sigma}_O \cdot \hat{\sigma}_G = 1, \\ \hat{\pi}_O \cdot \hat{\pi}_G = \cos 2\theta_G. \end{cases} \tag{4.50}$$

Equation (4.49) involving both χ_{O-G} and χ_{G-O} implies that the reflections from O to G and from G to O are considered. This signifies that multiple scattering is taken into account. Moreover, this equation is invariant to interchange of O and G. Hence, both beams O and G are given equal weight in considering the diffraction process. Interaction between the two beams is therefore expected to be very appreciable, especially when G is a strong reflection.

We define the quantity ΔK as the difference between K and K',

$$K_O = K'_O + \Delta K, \quad K_G = K'_G + \Delta K. \tag{4.51}$$

Equation (4.49) can be written in terms of ΔK as

$$y_O y_G = \frac{k^2 C^2}{4} \chi_{O-G} \chi_{G-O}, \quad \text{where} \tag{4.52}$$

$$y_O = \frac{K'_O \cdot \Delta K}{k}, \quad y_G = \frac{K'_G \cdot \Delta K}{k}, \tag{4.53}$$

and the second-order terms of ΔK are neglected. Since χ is very small, y_O and y_G are approximately 10^{-5} of k. The splitting of the dispersion surface near the Lorentz point is therefore very small compared with the size of the sphere. We should enlarge this portion of the dispersion surface and show its cross section in the plane of incidence (Fig. 4.3). The point La is the Laue point with $\text{La}O = \text{La}G = k$. The line La$E$ represents the wavefront of the incident wave. The vector EO, the wavevector of an incident wave, deviates from the Bragg angle θ_G by $\Delta\theta$. The distance between La and L is

$$\text{La}L = \frac{1}{2} \frac{kC\chi_{G-O}}{\cos\theta_G}. \tag{4.54}$$

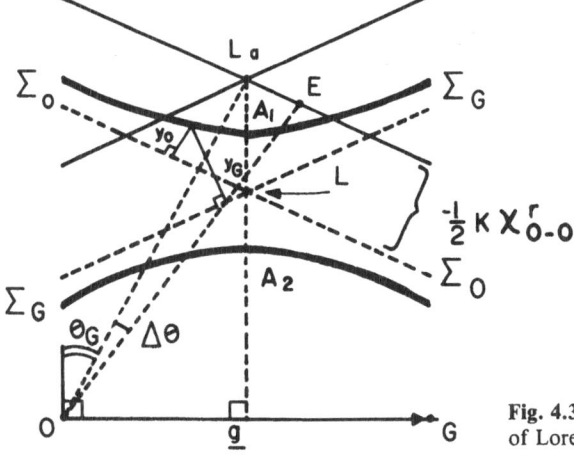

Fig. 4.3. Dispersion sheets in the vicinity of Lorentz point of Fig. 4.2

A difference in the wavevector k equal to LaL means a phase change of π. Accordingly, the difference Δk, with

$$\Delta k = 2\,|\text{La}L\,|, \tag{4.55}$$

gives a phase change of 2π. The reciprocal of Δk is usually called the extinction length, l_x, i.e.,

$$l_x = \frac{\cos\theta_G}{kC\,|\chi_{G-O}|}, \tag{4.56}$$

which is the distance between the maxima of the diffracted wave.

The quantities y_O and y_G are the distances from the tie point T to the wavefronts Σ_O and Σ_G of the waves K'_O and K'_G. When the tie point lies at A_1,

$$y_O = y_G = \frac{kC}{2}\sqrt{\text{Re}\{\chi_{O-G}\chi_{G-O}\}}, \tag{4.57}$$

where Re$\{\}$ indicates the real part of the argument. The distance A_1A_2 between the vertices of the hyperbola then takes the form

$$A_1A_2 = \frac{kC\sqrt{\text{Re}\{\chi_{O-G}\chi_{G-O}\}}}{\cos\theta_G}. \tag{4.58}$$

To obtain the dispersion surface for a three-beam, O, G_1, G_2, case, three dispersion spheres, Σ_O, Σ_{G_1} and Σ_{G_2}, for the reciprocal lattice points O, G_1 and G_2, have to be constructed. The radii of the spheres are $k(1 + \chi_{O-O}/2)$ (Fig. 4.4). These three points are connected by the three reciprocal lattice vectors $\mathbf{g}_1$, $\mathbf{g}_2$ and $\mathbf{g}_1 - \mathbf{g}_2$, of the primary G_1, the secondary G_2 and the coupling

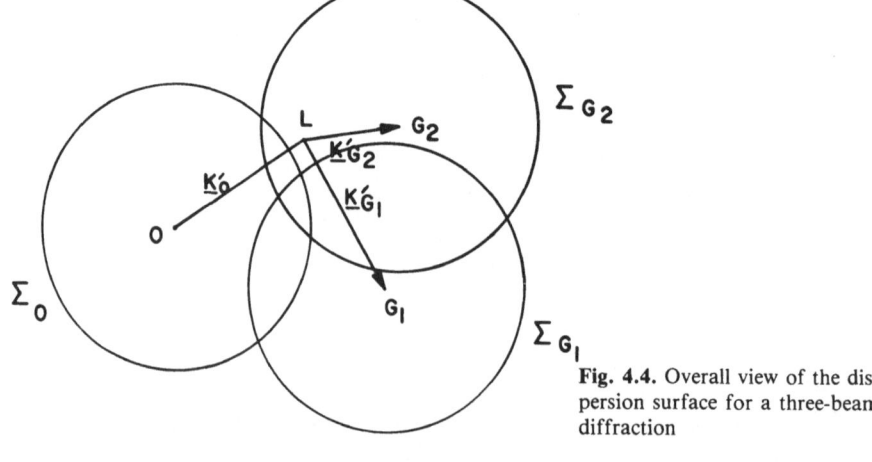

Fig. 4.4. Overall view of the dispersion surface for a three-beam diffraction

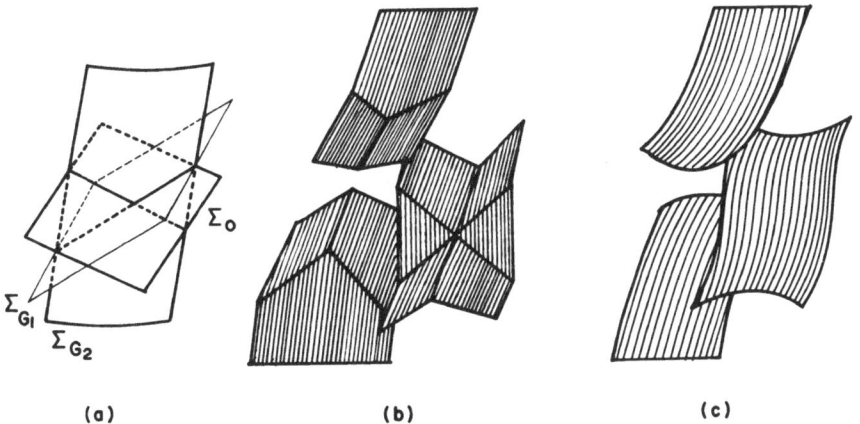

Fig. 4.5. (a) Sections of the spheres shown in Fig. 4.4 near Lorentz point; (b) the approximate planes for the dispersion sheets (c) (after *Kambe* [4.8])

reflection $G_2 - G_1$, respectively. The condition for three-beam diffraction, mentioned in Sect. 2.1.1, is fulfilled. The three spheres in the vicinity of the Lorentz point L for the three-beam diffraction are shown in Fig. 4.5a. They are represented as three planes, because they are small fractions of the surfaces of the three spheres. These planes serve as the asymptotic planes for the actual three-beam dispersion surface. The asymptotic component surfaces of Fig. 4.5a and the three actual dispersion sheets are shown in Fig. 4.5b and c [4.8]. A convenient way of visualizing the dispersion surface is to show its cross section in a two-dimensional picture. For illustration, let us cut the dispersion surface at point L with the plane C (Fig. 4.6a) perpendicular to the g_1 vector. The intersection of the dispersion surface and the plane C is shown in Fig. 4.6b. Points L and La are the three-beam Lorentz point and the Laue point, for which

$$LO = LG_1 = LG_2 = k \left(1 + \frac{\chi_{0-0}}{2} \right),$$

$$LaO = LaG_1 = LaG_2 = k .$$

(4.59)

The horizontal axis of Fig. 4.6b represents the azimuthal angle ϕ of the rotation about the vector g_1. Tie points lying in the section plane satisfy the condition $TO + g_1 = TG_1$. In other words, the Bragg reflection of G_1 occurs for all ϕ. The reciprocal lattice points O and G_1 lie in the plane passing through La and L, perpendicular to the plane of Fig. 4.6b. Point O is below and point G_1 is above the plane of this figure.

The dispersion curves in Fig. 4.6b can be described by the dispersion relation for linearly polarized three-beam diffraction:

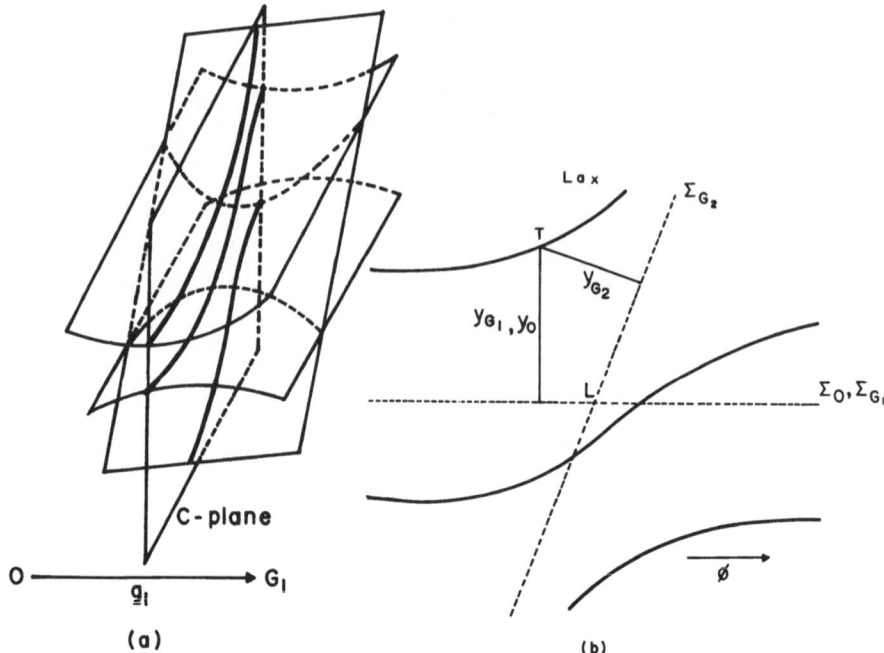

Fig. 4.6. (a) Portion of the dispersion sheets of Fig. 4.5c; (b) the projection of the dispersion sheets on the C plane perpendicular to the reciprocal lattice vector g_1

$$\begin{vmatrix} 2y_O & kC_1\chi_{O-G_1} & kC_2\chi_{O-G_2} \\ kC_1\chi_{G_1-O} & 2y_{G_1} & kC_3\chi_{G_1-G_2} \\ kC_2\chi_{G_2-O} & kC_3\chi_{G_2-G_1} & 2y_{G_2} \end{vmatrix} = 0 , \tag{4.60}$$

with $\Delta\theta = 0$. The term y_{G_2} is defined in the same way as y_O and y_{O_1} in (4.53). Only the σ-polarized wavefield is considered. The polarization factors C are

$$C_1 = \hat{\sigma}_O \cdot \hat{\sigma}_{G_1}, \quad C_2 = \hat{\sigma}_O \cdot \hat{\sigma}_{G_2}, \quad C_3 = \hat{\sigma}_{G_1} \cdot \hat{\sigma}_{G_2}. \tag{4.61}$$

In view of (4.60), the coupling reflections $G_1 - G_2$ and $G_2 - G_1$ are as important as the primary reflection G_1 and the secondary reflection G_2, provided that the structure factors involved in (4.60) for $G_1 - G_2$, $G_2 - G_1$, G_1 and G_2 have the same order of magnitude.

Alternatively, it is convenient to visualize the dispersion surface of three-beam diffractions by considering the normalized dispersion equation [4.9] for non-absorbing crystals. Let us recall (4.26). By expressing the quantity $D_{L(\perp K_G)}$ in vector form, the fundamental equation of the wavefield of (4.26) can be written as

$$\tau_G D_G + \sum_L \chi_{G-L}[D_L - \hat{u}_G(\hat{u}_G \cdot D_L)] = 0 \tag{4.62}$$

for all G, where $\hat{u}_G$ is the unit vector defined as

$$\hat{u}_G = \frac{K_G}{|K_G|}, \quad \text{and} \tag{4.63}$$

$$\tau_G = \chi_{O-O} - \frac{k^2 - K_G^2}{k^2}. \tag{4.64}$$

In the case of a three-beam, O, G_1, G_2, diffraction, there are three equations (4.62) which involve the displacements D_O, D_{G_1}, D_{G_2} and χ_{O-G_1}, χ_{O-G_2}, $\chi_{G_1-G_2}$, $\chi_{G_2-G_1}$, χ_{G_1-O} and χ_{G_2-O}. For simplicity, let us neglect absorption so that $\chi^*_{G_1-G_2} = \chi_{G_2-G_1}$. Define the triple product

$$\chi_{O-G_1}\chi_{G_1-G_2}\chi_{G_2-O} = \chi^3_{O-O}M^3 \exp(i\Psi), \tag{4.65}$$

where Ψ is the invariant phase angle of this triplet. Since the vector sum of the reciprocal lattice vectors $O-G_1$, G_1-G_2, G_2-O is null, Ψ is independent of the choice of the origin of the crystal unit cell (4.22). For normalization of the wavefield, we introduce

$$\xi_{G_1} = \frac{\tau_{G_1} m_{G_1} m^*_{G_1}}{\chi_{O-O}M^3}, \quad D'_G = n_G D_G, \quad \text{where} \tag{4.66}$$

$$m_{G_1} = \frac{\chi_{G_1-O}}{\chi_{O-O}}, \quad m^*_{G_1} = \frac{\chi_{O-G_1}}{\chi_{O-O}},$$

$$m_{G_1-G_2} = \frac{\chi_{G_1-G_2}}{\chi_{O-O}}, \quad m^*_{G_1-G_2} = \frac{\chi_{G_2-G_1}}{\chi_{O-O}}, \tag{4.67}$$

$$\frac{n_{G_1}}{n_{G_2}} = \frac{\gamma'^2 m_{G_1}}{m^*_{G_2}}, \quad \gamma' = \exp\left(-i\frac{\Psi}{3}\right).$$

Equation (4.62), after normalization, takes the form

$$\xi_O D'_O + \gamma' D'_{G_1} + \frac{D'_{G_2}}{\gamma'} - \hat{u}_O\left[\hat{u}_O \cdot \left(\gamma' D'_{G_1} + \frac{D'_{G_2}}{\gamma'}\right)\right] = 0,$$

$$\xi_{G_1} D'_{G_1} + \gamma' D'_{G_2} + \frac{D'_O}{\gamma'} - \hat{u}_{G_1}\left[\hat{u}_{G_1} \cdot \left(\gamma' D'_{G_2} + \frac{D'_O}{\gamma'}\right)\right] = 0, \tag{4.68}$$

$$\xi_{G_2} D'_{G_2} + \gamma' D'_O + \frac{D'_{G_1}}{\gamma'} - \hat{u}_{G_2}\left[\hat{u}_{G_2} \cdot \left(\gamma' D'_O + \frac{D'_{G_1}}{\gamma'}\right)\right] = 0.$$

Following *Ewald*'s treatment [4.10], (4.68) can be expressed as

$$\underset{\sim}{A} D' = Bu, \quad \text{where} \tag{4.69}$$

$$
A = \begin{pmatrix} \xi_O & \gamma' & \dfrac{1}{\gamma'} \\[2ex] \dfrac{1}{\gamma'} & \xi_{G_1} & \gamma' \\[2ex] \gamma' & \dfrac{1}{\gamma'} & \xi_{G_2} \end{pmatrix}, \quad u = \begin{pmatrix} \hat{u}_O \\[2ex] \hat{u}_{G_1} \\[2ex] \hat{u}_{G_2} \end{pmatrix}. \tag{4.70}
$$

The quantity B is a scalar.

By introducing a matrix S whose elements S_{LG} are the cofactors of the elements of A, the wavefield can be obtained as

$$
D' = \frac{BSu}{\|A\|}, \quad \text{or} \tag{4.71}
$$

$$
D'_G = \frac{\sum\limits_{L} S_{LG} B_L u_L}{|\det A_{LG}|}. \tag{4.72}
$$

Since $D_G \cdot \hat{u}_G = 0$, we obtain

$$
\sum_{L,G} S_{LG} B_L (\hat{u}_L \cdot \hat{u}_G) = 0. \tag{4.73}
$$

For B_L to have non-trivial values, the following condition must be fulfilled:

$$
\det\left[S_{LG}(\hat{u}_L \cdot \hat{u}_G)\right] = 0. \tag{4.74}
$$

By defining

$$
a = \hat{u}_{G_1} \cdot \hat{u}_{G_2}, \quad b = \hat{u}_O \cdot \hat{u}_{G_2}, \quad c = \hat{u}_O \cdot \hat{u}_{G_1}, \tag{4.75}
$$

(4.74) takes the following form as the dispersion relation for this three-beam diffraction:

$$
\|\Phi\| = \begin{vmatrix} \begin{vmatrix} \xi_{G_1} & \gamma' \\[1.5ex] \dfrac{1}{\gamma'} & \xi_{G_2} \end{vmatrix} & -c\begin{vmatrix} \gamma' & \dfrac{1}{\gamma'} \\[1.5ex] \dfrac{1}{\gamma'} & \xi_{G_2} \end{vmatrix} & b\begin{vmatrix} \gamma' & \dfrac{1}{\gamma'} \\[1.5ex] \xi_{G_1} & \gamma' \end{vmatrix} \\[5ex] -c\begin{vmatrix} \dfrac{1}{\gamma'} & \gamma' \\[1.5ex] \gamma' & \xi_{G_2} \end{vmatrix} & \begin{vmatrix} \xi_O & \dfrac{1}{\gamma'} \\[1.5ex] \gamma' & \xi_{G_2} \end{vmatrix} & -a\begin{vmatrix} \xi_O & \dfrac{1}{\gamma'} \\[1.5ex] \dfrac{1}{\gamma'} & \gamma' \end{vmatrix} \\[5ex] b\begin{vmatrix} \dfrac{1}{\gamma'} & \xi_{G_1} \\[1.5ex] \gamma' & \dfrac{1}{\gamma'} \end{vmatrix} & -a\begin{vmatrix} \xi_O & \gamma' \\[1.5ex] \gamma' & \dfrac{1}{\gamma'} \end{vmatrix} & \begin{vmatrix} \xi_O & \gamma' \\[1.5ex] \dfrac{1}{\gamma'} & \xi_{G_1} \end{vmatrix} \end{vmatrix} = 0, \tag{4.76}
$$

which can be simplified to

$$\| \Phi(\xi_O, \xi_{G_1}, \xi_{G_2}, \Psi) \|$$
$$= (\xi_O \xi_{G_1} \xi_{G_2} - \xi_O - \xi_{G_1} - \xi_{G_2} + 2\cos\Psi)(\xi_O \xi_{G_1} \xi_{G_2} - a^2 \xi_O^2$$
$$- b^2 \xi_{G_1} - c^2 \xi_{G_2} + 2abc\cos\Psi) + (1 - a^2 - b^2 - c^2 + 2abc)$$
$$\cdot (-2\xi_O \xi_{G_1} \xi_{G_2} \cos\Psi + \xi_O \xi_{G_1} + \xi_O \xi_{G_2} + \xi_{G_1} \xi_{G_2} - 1) = 0 . \qquad (4.77)$$

A similar expression in a non-normalized form has also been derived by *Lamla* [4.11].

The 'normalized' dispersion surface of a three-beam case, described by (4.77), consists of six dispersion sheets. These six sheets can be grouped into two sets, which are very similar to each other. The only difference comes from the state of polarization. We shall therefore consider only one set of dispersion sheets, i.e., three dispersion sheets. Figure 4.7 shows these three dispersion sheets in a cubic block whose sides are parallel to the axes, ξ_O, ξ_{G_1} and ξ_{G_2}, with the origin centered at $\xi_O = \xi_{G_1} = \xi_{G_2} = 0$. The sides of the cube represent the situations for $\xi_O = 0$, $\xi_{G_1} = 0$ and $\xi_{G_2} = 0$, as the corresponding ξ_{G_1}, ξ_{G_2}; ξ_{G_2}, ξ_O; ξ_O, ξ_{G_1} approach infinity respectively. Two limiting cases are associated with two-beam diffraction:

$$\xi_j \xi_k = 1 , \qquad \xi_j \xi_k = (\hat{u}_j \cdot \hat{u}_k)^2 , \qquad (4.78)$$

where j, k can be any one of O, G_1 and G_2, $j \neq k$.

Three sheets of the dispersion surface are drawn in the faces of the cube. Sheet I intersects three adjacent faces through the closed curve I, which consists of the branch corresponding to $\xi_j \cdot \xi_k = 1$ for positive ξ_j. Sheet I is concave-up. Sheet III intersects the cube at the corner near the origin. Curve III, the intersection of sheet III with the cubic faces, consists of the branches of the hyperbolae $\xi_j \cdot \xi_k = (\hat{u}_j \cdot \hat{u}_k)^2$ for $\xi_j < 0$. Sheet III is convex-down. The

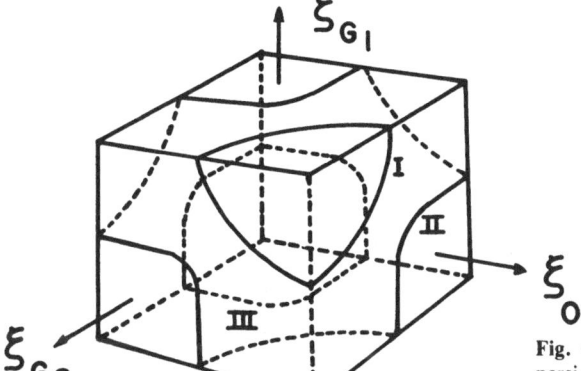

Fig. 4.7. Intersections of three dispersion sheets of the normalized dispersion surface in a three-beam case

warped sheet intersecting the faces of the cube through curve II is sheet II. Curve II consists alternately of branches of the hyperbolae $\xi_j \cdot \xi_k = 1$, the dashed line, and $\xi_j \xi_k = (\hat{u}_j \cdot \hat{u}_k)^2$, the solid line. Sheet II lies in between sheets I and III. These three dispersion sheets do not, in general, intersect each other.

For a centrosymmetric crystal which possesses a center of inversion, the phase angle Ψ is equal to zero, and M is either positive or negative according to (4.22). All the normalized parameters given in (4.66, 67) are real provided that the origin is at the inversion center. Under this condition, the dispersion equation becomes

$$\| \Phi \| = (\xi_0 \xi_{G_1} \xi_{G_2} - \xi_0 - \xi_{G_1} - \xi_{G_2} + 2)(\xi_0 \xi_{G_1} \xi_{G_2} - a^2 \xi_0 - b^2 \xi_{G_1}$$
$$- c^2 \xi_{G_2} + 2abc) + V_0^2(-2\xi_0 \xi_{G_1} \xi_{G_2} + \xi_0 \xi_{G_1} + \xi_0 \xi_{G_2} + \xi_{G_1} \xi_{G_2} - 1)$$
$$= 0, \tag{4.79}$$

where the factor $V_0^2 = 1 - a^2 - b^2 - c^2 + 2abc$ represents the volume $\hat{u}_0 \cdot (\hat{u}_{G_1} \times \hat{u}_{G_2})$. There exist the following relations which are deduced from (4.79):

$$\xi_0 = \xi_{G_1} = 1, \quad \xi_0 = \xi_{G_2} = 1, \quad \xi_{G_1} = \xi_{G_2} = 1, \tag{4.80}$$

for arbitrary ξ_{G_2}, ξ_{G_1} and ξ_0, respectively. Equation (4.80) stands for three straight lines on the dispersion surface. The intersection of these three lines lies at $\xi_0 = \xi_{G_1} = \xi_{G_2} = 1$, which indicates the triplet degenerate point. This degeneracy is often encountered in symmetric transmission cases. We shall discuss this fact in detail in Sect. 4.5. Discussions about the influence of the triplet invariant phase Ψ on the dispersion surface for a non-centrosymmetric crystal will be postponed until the x -ray phase problem is mentioned in Chap. 7.

It should be noted that in the case of coplanar coincident diffraction, $V_0^0 = 0$ because $\hat{u}_0 \cdot \hat{u}_{G_1} \times \hat{u}_{G_2} = 0$ (Sect. 2.1.2). The dispersion relation becomes

$$\| \Phi \| = (\xi_0 \xi_{G_1} \xi_{G_2} - \xi_0 - \xi_{G_1} - \xi_{G_2} + 2)(\xi_0 \xi_{G_1} \xi_{G_2} - a^2 \xi_0$$
$$- b^2 \xi_{G_1} - c^2 \xi_{G_2} + 2abc) = 0. \tag{4.81}$$

For higher-order multiple diffraction, the dispersion surface can be constructed in a way similar to that described for three-beam cases, and can be visualized easily in various section planes.

The tie points on the surface should be described by at least three parameters, since the dispersion surface is a three-dimensional entity. From the discussion above, the position of a tie point may be defined by the deviation angle $\Delta\theta$ from the angle θ_{G_1} of the G_1 reflection, the azimuth angle ϕ of the rotation around g_1, and an additional parameter which must be proportional to the distance from a tie point to a point, say an entrance point, lying on the incident wavefront. This distance, after *Ewald*, is called 'accommodation' (Anpassung) [4.2], and it is related to the excitation of the dispersion surface by the incident wave.

4.3.2 Excitation of the Dispersion Surface

The tie point on a dispersion surface was more precisely called the excitation point by *Laue* [4.3] since it indicates the position at which the dispersion surface is excited by an incident wave. This position is thus determined, as discussed in the previous section, by the angles $\Delta\theta$ and ϕ, which specify the relative position of the incident beam with respect to the crystal, and by the accommodation [4.2], which will be discussed here.

The incident wave is usually represented by the wavevector *EO* in reciprocal space. The point E is the entrance point on the wavefront of the O wave. The way in which the tie point is excited is governed by the conservation of momentum, i.e., the continuity of the tangential components of the wavevectors inside and outside the crystal at the crystal boundary. Figure 4.8 shows the excitation of the dispersion surface of a three-beam diffraction at the tie points T_1, T_2 and T_3 by the incident wave *EO*. *LaE* stands for the wavefront of the incident wave. To fulfill the boundary condition for wavevectors,

$$(EO)_t = (T_jO)_t , \tag{4.82}$$

for $j = 1$, 2 and 3. The subscript t indicates the tangential component. Assume that the angles between *EO*, T_jO and the vector ET_j are α_0 and α. The magnitudes of these vectors are

$$EO = n_0k, \quad T_jO = nk , \tag{4.83}$$

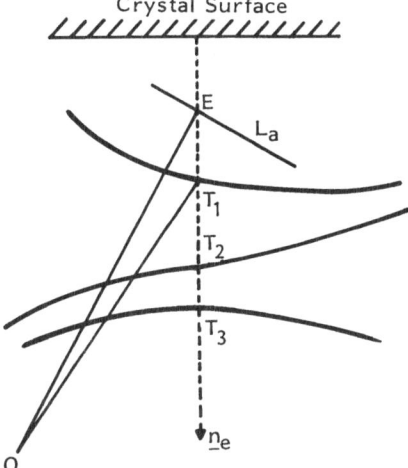

Crystal Surface

Fig. 4.8. Definition of the tie points

where $n_0 = 1$ is the index of refraction of x-rays in vacuum. The boundary condition becomes

$$n_0 \sin \alpha_0 = n \sin \alpha . \tag{4.84}$$

This is the familiar Snell's law. Equation (4.84) also implies that ET_j is perpendicular to the crystal surface. For a given entrance point E on the incident wavefront, the excited tie points T_j can be found by drawing a line perpendicular to the crystal surface. The intersections of this line with the dispersion surface are the tie points T_j. The distance between point E and the tie point T_j is proportional to the accommodation $\delta(j)$, defined as

$$ET_j = k\delta(j) . \tag{4.85}$$

The corresponding wavevectors of the excited waves at this tie point are

$$K_O(j) = k_O - k\delta(j)\hat{n}_e , \quad K_{G_1}(j) = k_{G_i} - k\delta(j)\hat{n}_e , \tag{4.86}$$

where $\hat{n}_e$ is the inward crystal surface normal. $\delta(j)$ is the jth accommodation determined by the jth eigenvalue of (4.36). Since k_O and k_G depend also on the position of the entrance point E, specified by $\Delta\theta$ and ϕ, the difference between k_G^2 and the quantity k^2 involved in (4.39) can be expressed as

$$k_G^2 - k^2 = k^2 \alpha_G , \tag{4.87a}$$

where α_G is a function of $\Delta\theta$ and ϕ. One can easily obtain α_G when a proper coordinate system is chosen for the wavevectors [see, for example, (5.115) in Sect. 5.2]. With the aid of (4.87a), the corresponding 2ε's therefore take the form

$$2\varepsilon_G(j) = -2\delta(j)\gamma_G + \alpha_G , \tag{4.87b}$$

where $\gamma_G = \hat{k}_G \cdot \hat{n}_e$.

The accommodation $k\delta(j)$ depends very much on the crystal surface normal n_e, which is different in the Bragg and the Laue cases. The excitation of the dispersion surface differs accordingly when changing from Bragg-type reflection to the Laue type. Since multiple diffraction involves either Bragg- or Laue-type diffraction, we shall discuss the excitation of the dispersion surface in the two most common cases, i.e., symmetric Bragg and symmetric Laue.

Figure 4.9a shows the dispersion surface of a two-beam diffraction. The points O and G are reciprocal lattice points. The vector OG is the reciprocal lattice vector g of the reflection G. For a symmetric Bragg reflection, the crystal surface represented by MR is perpendicular to g. The incident wave at point E excites the dispersion surface at T_1 and T_2 of the same branch (sheet) of the dispersion surface. If the entrance point moves to E', the dispersion surface will not be excited by the incident wave. This diffraction is governed

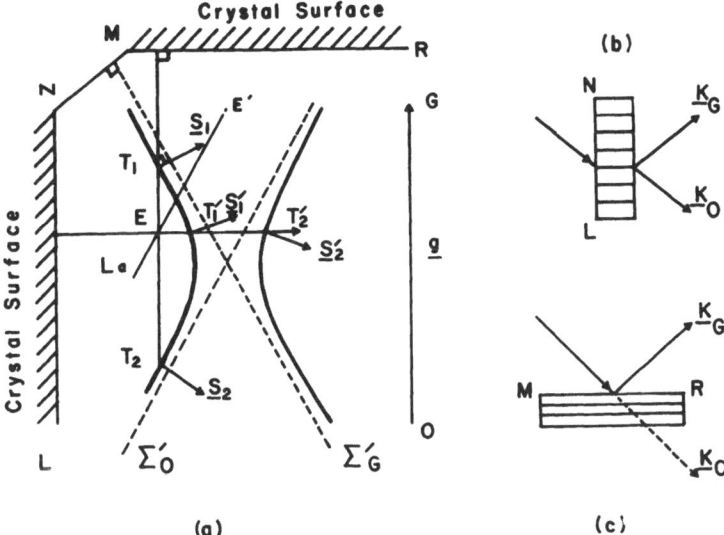

Fig. 4.9a – c. Relation between the dispersion surface (**a**) and the diffraction geometries (**b**) and (**c**)

by the total reflection. For a symmetric Laue case, the crystal surface NL is parallel to g. The points T_1' and T_2', excited by the incident wave EO, lie on the two different branches of the dispersion surfaces. If the crystal is cut so that the surface is parallel to MN which is perpendicular to the asymptotic surface Σ_G', the reflection vector EG is parallel to the crystal surface MN. This is a surface diffraction case. This kind of diffraction is often considered as an extremely asymmetric case. We shall discuss this special situation in Sect. 5.4.1.

4.4 Energy Flow

4.4.1 Poynting Vectors and the Dispersion Surface

Equation (4.37), obtained from the eigenvectors, gives the amplitude ratios of the component wavefields. If any one of the wavefield amplitudes, say $E_{G_1}(j)$, is known, the rest of the amplitudes for a given j^{th} eigenvector are determined. The amplitude E_{G_1} can be obtained by considering the boundary conditions at the entrance surface of the crystal. The resultant wavefields are therefore the superposition of all the component waves, i.e.,

$$E(r, t) = \exp(\mathrm{i}\omega t) \sum_j \sum_{\sigma G} E_{\sigma G}(j) \exp[-2\pi \mathrm{i} K_G(j) \cdot r] \, ,$$

$$H(r, t) = \exp(\mathrm{i}\omega t) \sum_j \sum_{\sigma G} H_{\sigma G}(j) \exp[-2\pi \mathrm{i} K_G(j) \cdot r] \, ,$$

$$(4.88)$$

where the summations are taken over all the jth wavefield and over all the polarized wavefield components of all reflections G, and where σG includes all σ's and π's.

The Poynting vector which describes the energy flow is defined in (4.4). Consider the wavevector as a complex variable,

$$K_G = K_{G,r} - iK_{G,i}, \tag{4.89}$$

where $K_{G,r}$ and $K_{G,i}$ are the real and imaginary parts of the wavevector. The imaginary part can be expressed in terms of the imaginary part of the eigenvalue. For the jth eigenvalue,

$$K_{G,i}(j) = k\delta_i(j)\hat{n}_e = K_i(j), \tag{4.90}$$

which is independent of the reflection directions and is normal to the crystal surface, i.e., along the vector n_e.

For simplicity, let us consider here the Poynting vector $S(j)$ associated with the jth dispersion sheet. Substituting (4.89, 90) into (4.4), the jth Poynting vector becomes

$$S(j) = \frac{c}{4\pi} \exp(-4\pi K_i(j) \cdot r) \sum_{\sigma G} \sum_{\sigma G'} [E_{\sigma G}(j) \times H_{\sigma G'}(j)]$$

$$\times \exp[2\pi i(K_{G,r}(j) - (K_{G',r}(j)) \cdot r], \tag{4.91}$$

where

$$E_{\sigma G}(j) \times H_{\sigma G'}(j) = E_{\sigma G}(j) \times [\hat{K}_{G'}(j) \times E_{\sigma G'}(j)]. \tag{4.92}$$

Since the terms involved will have to be compared with experimental observation, the temporal and spatial average of the Poynting vector is of physical interest. The averaged Poynting vector has the simple form

$$\langle S(j)\rangle = \frac{c}{8\pi} \exp(-4\pi K_i(j) \cdot r) \sum_{\sigma G} \hat{K}_G (E_{\sigma G} E_{\sigma G}^*), \tag{4.93}$$

where the cross terms, $G' \neq G$, are singled out because of the spatial averaging of the term $\exp[2\pi i(K_{G,r}(j) - K_{G',r}(j)) \cdot r]$. The implication of (4.93) is that the Poynting vector of a Bloch wave is the vector sum of the Poynting vectors of the component waves.

The term $E_{\sigma G} E_{\sigma G}^*$, referring to (4.42), can be written in terms of the cofactors. Assume the crystal is non-absorbing, such that

$$\chi_{G_l, G_m} = \chi_{G_m, G_l}^*, \tag{4.94}$$

where the asterisk means complex conjugate. The corresponding cofactors have the same relation:

$$\text{Cof}(\sigma G_l, \pi G_m)_j = [\text{Cof}(\pi G_m, \sigma G_l)_j]^* \,. \tag{4.95}$$

Taking the complex conjugate of one expression in (4.42) and multiplying it by the other expression, we obtain

$$\frac{|E_{\sigma G_l}(j)|^2}{|E_{\pi G_m}(j)|^2} = \frac{\text{Cof}(\sigma G_l, \sigma G_l)_j}{\text{Cof}(\pi G_m, \pi G_m)_j}, \quad \text{or} \tag{4.96}$$

$$\frac{|E_{\sigma G_l}(j)|^2}{\text{Cof}(\sigma G_l, \sigma G_l)_j} = \frac{|E_{\pi G_m}(j)|^2}{\text{Cof}(\pi G_m, \pi G_m)_j} = \ldots = \text{const} \,. \tag{4.97}$$

By combining (4.97) and (4.93), the Poynting vector can be written as

$$\langle S(j) \rangle \propto \frac{c}{8\pi} \exp[-4\pi K_i(j) \cdot r] \sum_{\sigma G} \hat{K}_G \text{Cof}(\sigma G, \sigma G)_j \,. \tag{4.98}$$

Let us now expand the determinant $\| \Phi \|$ of (4.38) for a given mode j in terms of the elements of the σG column and the cofactors:

$$\| \Phi \| = (\chi_{0-0} - 2\varepsilon_G) \text{Cof}(\sigma G, \sigma G) + \sum_{\sigma M} \chi_{G-M} (\hat{\sigma}_M \cdot \hat{\sigma}_G) \text{Cof}(\sigma G, \sigma M) \,, \tag{4.99}$$

where the sum is over all σ's and π's. The subscript j is left out. We introduce an infinitesimal variation $\Delta \tau$ to the wavevectors along the tangent of the dispersion surface at a given tie point T (Fig. 4.10) such that the vectors $K_O + \Delta \tau$ and $K_G + \Delta \tau$ for all G's satisfy the dispersion relation (4.41), i.e.,

$$\Delta^k \| \Phi \| = 0 \,. \tag{4.100}$$

Δ^k is the variational operator with respect to k. Differentiating the expanded form of (4.100) with respect to K_G yields the expression

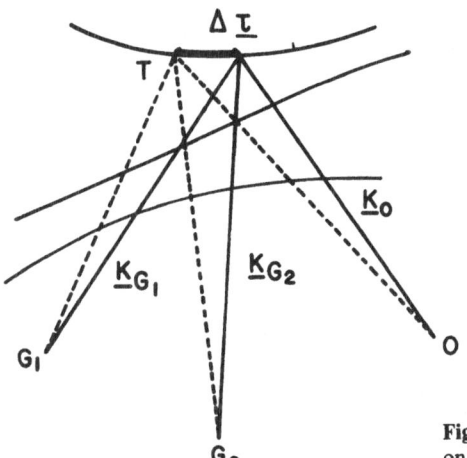

Fig. 4.10. Definition of the vector variation $\Delta \tau$ on a dispersion sheet

$$\Delta^k \| \Phi \| = -(\Delta\tau \cdot K_{G_1}) \frac{2\,\text{Cof}(\sigma G_1, \sigma G_1)}{k^2} + (\chi_{O-O} - 2\varepsilon_{G_1})$$

$$\Delta^k [\text{Cof}(\sigma G_1, \sigma G_1)] + \sum_{\sigma M} \chi_{G_1-M}(\hat{\sigma}_M \cdot \hat{\sigma}_{G_1}) \Delta^k [\text{Cof}(\sigma G_1, \sigma M)]$$

$$= -(\Delta\tau \cdot K_{G_1}) \frac{2\,\text{Cof}(\sigma G_1, \sigma G_1)}{k^2} + \Delta^k_{\sigma G_1} \| \Psi \|, \tag{4.101}$$

where $\Delta^k_{\sigma G_1}$ is the variational operator operating on all elements except the elements of the σG_1 column and σG_1 row. This operator generates

$$\Delta^k_{\sigma G_1} \| \Phi \| = -(\Delta\tau \cdot K_{G_2}) \frac{2\,\text{Cof}(\pi G_2, \pi G_2)}{k^2} + \Delta^k_{\sigma G_1, \pi G_2} \| \Phi \|. \tag{4.102}$$

Continuing further this variation operation, we obtain, finally,

$$\Delta^k \| \Phi \| = -\Delta\tau \cdot \sum_{\sigma M} K_M \frac{2\,\text{Cof}(\sigma M \cdot \sigma M)}{k^2} = 0, \tag{4.103}$$

where the relation

$$\Delta^k_{\sigma O, \pi O, \sigma G_1, \pi G_1, \ldots, \sigma G_{N-1}, \pi G_{N-1}} \| \Phi \| = 0 \tag{4.104}$$

has been employed. By comparing (4.103) with (4.98), the conclusion

$$\langle S(j) \rangle \cdot \Delta\tau = 0 \tag{4.105}$$

is obtained. The direction of the energy flow associated with each dispersion sheet is normal to the tangential plane of the dispersion sheet. The same conclusion can be obtained from the consideration of a wave packet [4.12]. This result is analogous to the similar phenomenon in the crystal optics of visible spectra, in which the wave propagation is normal to the index surface [4.13].

4.4.2 Group Velocity and Energy Flow

If the dispersion surface slightly deviates from the surface of the corresponding constant frequency (energy) ω_0 by ΔK_v, the actual frequency ω can be written as

$$\omega = \omega_0 + \left(\frac{\partial\omega}{\partial K_v} \right)_\tau (\Delta K_v), \tag{4.106}$$

where the second term is related to the group velocity v_g,

$$v_g = \left(\frac{\partial\omega}{\partial K_v} \right)_\tau \hat{v} = \text{grad}_{K_v}\,\omega. \tag{4.107}$$

The derivative is independent of the direction τ since the variation ΔK_ν is assumed to be very small. The relation between the group velocity and the Poynting vector can be derived from the dispersion relation (4.41) in a way similar to the derivation of (4.105).

Let us take into account the variation of the dispersion surface with respect to both $\Delta\omega$ and ΔK_ν. This yields

$$\Delta \|\Phi\| = \frac{\partial\|\Phi\|}{\partial K_\nu}\Delta K_\nu + \frac{\partial\|\Phi\|}{\partial\omega}\Delta\omega = 0 , \tag{4.108}$$

where the derivative with respect to K_ν can be obtained in the same way as for $\Delta^k\|\Phi\|$ of (4.101):

$$\frac{\partial\|\Phi\|}{\partial K_\nu} = -\hat{\nu}\cdot\sum_{\sigma M} K_M\frac{2\,\mathrm{Cof}(\sigma M,\sigma M)}{k^2} . \tag{4.109}$$

The second derivative in (4.108) can be performed by replacing k^2 by ω^2/c^2 in (4.38, 39, 41), and then by following the same procedure as for $\partial\|\Phi\|/\partial K_\nu$. A similar result to (4.109) is obtained:

$$\frac{\partial\|\Phi\|}{\partial\omega} = \frac{2}{\omega}\sum_{\sigma M}\left(\frac{K_M}{k}\right)^2\mathrm{Cof}(\sigma M,\sigma M)$$

$$\simeq \frac{\omega}{c^2}\sum_{\sigma M}\frac{2\,\mathrm{Cof}(\sigma M,\sigma M)}{k^2} , \tag{4.110}$$

where $K_M^2 \simeq (\omega/c)^2$. Substituting (4.109, 110) into (4.108), we obtain

$$\frac{1}{c}\left(\frac{\partial\omega}{\partial K_\nu}\right)_\tau\sum_{\sigma M}\mathrm{Cof}(\sigma M,\sigma M) = \hat{\nu}\cdot\sum_{\sigma M}\hat{K}_M\mathrm{Cof}(\sigma M,\sigma M) . \tag{4.111}$$

Since the average electric and magnetic energy densities associated with each dispersion sheet can be written as

$$\langle W_e\rangle = \langle W_m\rangle$$

$$\propto \frac{1}{16\pi}\exp(-4\pi K_i\cdot r)\sum_{\sigma M}E_{\sigma M}E_{\sigma M}^*$$

$$= \frac{1}{16\pi}\exp(-4\pi K_i\cdot r)\sum_{\sigma M}\mathrm{Cof}(\sigma M,\sigma M) , \tag{4.112}$$

the relation between the group velocity and the Poynting vector is therefore obtained by combining (4.108, 111, 112, 98):

$$(\langle W_e\rangle + \langle W_m\rangle)\,v_g = \langle S\rangle . \tag{4.113}$$

In general, a velocity of energy transport, v_e, may be defined as

$$v_e = \frac{c \sum_{\sigma G} \hat{K}_G |E_{\sigma G}|^2}{\sum_{\sigma G} |E_{\sigma G}|^2} .$$

(4.114)

4.5 Modes of Wave Propagation

4.5.1 Wavefields of Modes

The eigenvalues $K_G(j)$ specify the type of wave propagation inside the crystal. This type is called the mode of propagation [4.14], analogous to the normal modes in the classical vibration of continuum matter. As a matter of fact, the large number of x-ray wavefields within the crystal can be described as the normal modes of vibration of the electric-field vectors, since each wavefield of each mode not only travels in the direction of propagation but also forms a standing wave normal to that direction.

To illustrate the normal modes of the x-ray wavefield, let us consider a simple two-beam transmission case at the exact two-beam diffraction position. We rewrite the fundamental equation of wavefields in the form

$$\begin{pmatrix} 2y_O & kC\chi_{O-G} \\ kC\chi_{G-O} & 2y_G \end{pmatrix}\begin{pmatrix} E_O \\ E_G \end{pmatrix} = 0 ,$$

(4.115)

where $C = 1$ for $E_O = E_{\sigma O}$ and $E_G = E_{\sigma G}$, and $C = \cos 2\theta_G$ for $E_O = E_{\pi O}$ and $E_G = E_{\pi G}$. The wavefield-amplitude ratio is

$$\frac{E_G}{E_O} = -\frac{kC\chi_{G-O}}{2y_G} = -\frac{2y_O}{kC\chi_{O-G}} .$$

(4.116)

At the exact two-beam point, according to (4.57),

$$y_O = y_G = y = \pm \frac{kC}{2}\sqrt{\chi_{O-G}\chi_{G-O}} .$$

(4.117)

Let us define the four eigenvalues, or the modes, as

$$\left.\begin{matrix} y(1) \\ y(4) \end{matrix}\right\} = \pm \frac{k}{2}\sqrt{\chi_{O-G}\chi_{G-O}}$$

(4.118)

for the σ polarization and

$$\left.\begin{matrix} y(2) \\ y(3) \end{matrix}\right\} = \pm \frac{k}{2}\sqrt{\chi_{O-G}\chi_{G-O}} \cos 2\theta_G$$

(4.119)

for the π polarization. The corresponding wavefield-amplitude ratios, the eigenvectors of (4.115), are

$$\frac{E_O(1)}{E_G(1)} = \frac{E_O(2)}{E_G(2)} = -\frac{\chi_{O-G}}{\sqrt{\chi_{O-G}\chi_{G-O}}},$$

$$\frac{E_O(3)}{E_G(3)} = \frac{E_O(4)}{E_G(4)} = \frac{\chi_{O-G}}{\sqrt{\chi_{O-G}\chi_{G-O}}}.$$
(4.120)

If the crystal is non-absorbing, the amplitude ratios are $+1$ and -1:

$$\frac{E_O(1)}{E_G(1)} = \frac{E_O(2)}{E_G(2)} = 1, \quad \frac{E_O(3)}{E_G(3)} = \frac{E_O(4)}{E_G(4)} = -1.$$
(4.121)

Since χ is a negative quantity and since (4.35) must be satisfied for the σ and π components of the wavefields, $E_O(1)$, parallel to $E_O(4)$, and $E_G(1)$, antiparallel to $E_G(4)$, represent the σ components of the direct and the reflected waves, respectively. The quantities $E_O(2)$, $E_G(2)$ and $E_O(3)$, $E_G(3)$ with the amplitude ratios given in (4.121), lying in the same plane of incidence, stand for the π components. The four modes of propagation are shown schematically in Fig. 4.11.

If the incident wave is assumed to be σ polarized, only modes (1) and (4) are excited. In both modes, the wavefields E_O and E_G are perpendicular to the plane of incidence and are parallel to the G-reflecting planes. The geometric relations between the incident and the diffracted beams and the reflection

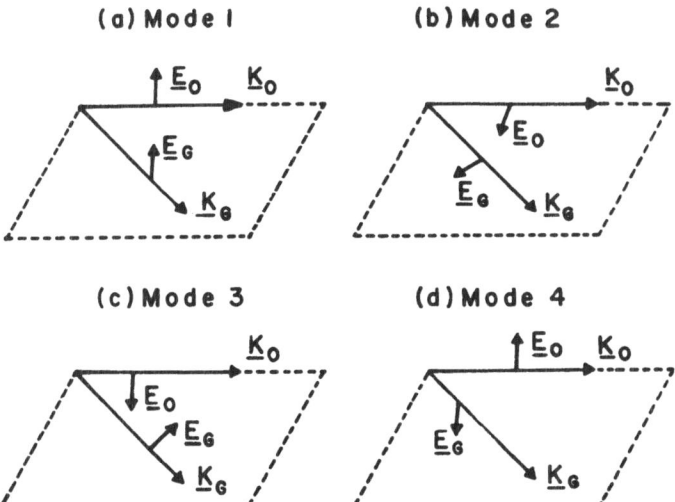

Fig. 4.11 a – d. Definition of modes of wave propagation in a two-beam case (after *Saccocio* and *Zajac* [4.14])

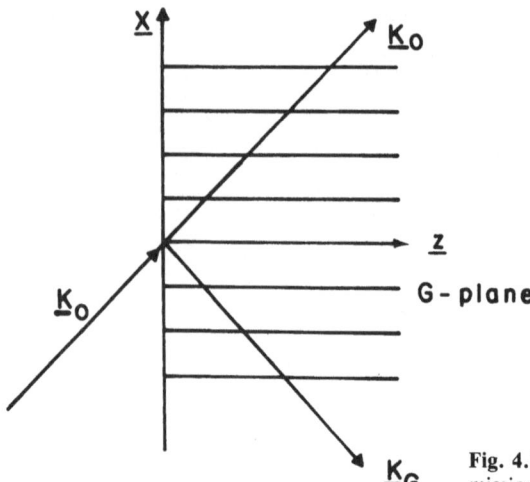

Fig. 4.12. Coordinates for a two-beam transmission diffraction

planes are shown in Fig. 4.12. If the incident angle is the Bragg angle of the G reflection, the interaction of E_O and E_G forms standing waves for both modes in the x direction and traveling waves in the positive z direction. Since $E_O(1)$ and $E_G(1)$ are parallel to each other, the standing wave thus formed has its amplitude nodes in the reflection planes. The other mode, mode 4, with $E_O(4)$ opposite to $E_G(4)$, has its antinodes in the reflection planes. These are shown in Fig. 4.13. These two modes can be described by the mathematical expressions

$$E(1) = E_1 \sin\left(\frac{\pi x}{d}\right) \exp[-2\pi i K(1)z],$$

$$E(4) = E_4 \cos\left(\frac{\pi x}{d}\right) \exp[-2\pi i K(4)z].$$

$$(4.122)$$

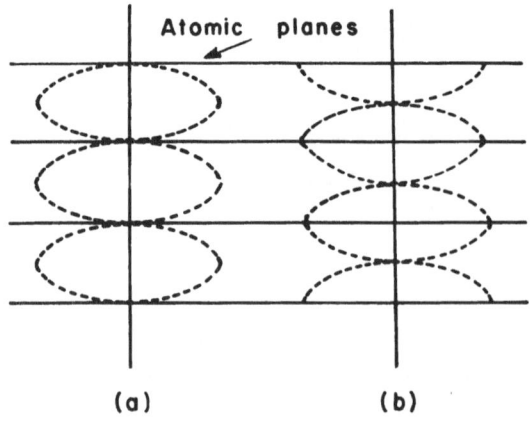

(a)   (b)

Fig. 4.13a, b. Standing-wave patterns of (a) mode 1 and (b) mode 4 for a σ-polarized two-beam transmission case

The interaction between the wavefield amplitude of mode 1 and the scattering electrons located at the reflecting planes is much weaker than the interaction between mode 4 and the electrons, because the amplitude of mode 1 at the reflecting planes is very small. In other words, mode 4 is more absorbed than mode 1. Moreover, mode 1 moves faster than mode 4. The difference in wavevectors between the two modes is

$$|\Delta K| = |K(1) - K(4)|$$
$$= y(1) - y(4)$$
$$= kC\sqrt{\chi_{O-G}\chi_{G-O}} \,. \tag{4.123}$$

The energy flow is along the direction of wave propagation, i.e., along the z direction (the reflecting planes). No energy flow propagates along the x direction since both modes have a standing-wave character in this direction. It should be noted that modes 2 and 3 do not form standing waves in the crystal because of the geometric relation between the direction of the wavefield and the reflecting planes.

It is difficult to find exact solutions of the dispersion equation (4.41) for general three-beam and higher-order diffractions. In order to reveal the modes of propagation in these cases, we shall consider symmetric three-beam transmission cases at the exact three-beam diffraction position. Suppose the three-beam, O, G_1 and G_2, case is symmetric, such that the reciprocal lattice vector magnitudes

$$g_1 = g_2 = |g_2 - g_1|, \quad \text{and} \tag{4.124}$$

$$\chi_{O-G_1} = \chi_{O-G_2} = -\chi_{G_1-G_2} = \chi_{G_2-G_1} = \chi_{G_1-O} = \chi_{G_2-O} = a \,. \tag{4.125}$$

The polarization vectors σ and π for the reflections O, G_1 and G_2 are defined in Fig. 4.14, where $\hat{\sigma}_O$ and $\hat{\sigma}_{G_1}$ are perpendicular to the plane of incidence of

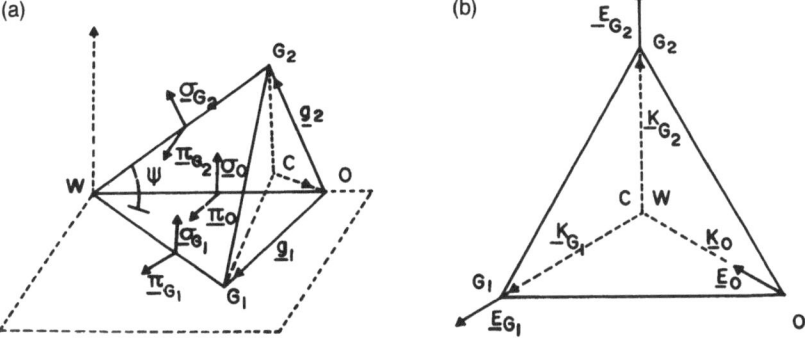

Fig. 4.14. (a) Definition of polarization vectors for a symmetric three-beam diffraction; (b) wavefields of the nondegenerate mode

the G_1 reflection, which contains K_O and K_{G_1}, and $\hat{\pi}_O$ and $\hat{\pi}_{G_1}$. The angle between K_{G_2} and this G_1 plane of incidence is ψ. The point C is the center of the circle circumscribing O, G_1 and G_2. W is the center of the Ewald sphere. The vector σ_{G_2} is chosen to be perpendicular to $\overline{WG_2}$ and g_1. All the σ's, π's and K's satisfy the condition (4.35). The polarization factors are obtained accordingly [4.14]:

$$\hat{\sigma}_O \cdot \hat{\pi}_{G_1} = \hat{\sigma}_{G_1} \cdot \hat{\pi}_O = \hat{\sigma}_O \cdot \hat{\pi}_{G_2} = \hat{\sigma}_{G_1} \cdot \hat{\pi}_{G_2},$$

$$\hat{\sigma}_{G_2} \cdot \hat{\pi}_O = -\hat{\sigma}_{G_2} \cdot \hat{\pi}_{G_1} = \sin\theta_{G_1} \sin\psi = C_1,$$

$$\hat{\sigma}_O \cdot \hat{\sigma}_{G_2} = \hat{\sigma}_{G_1} \cdot \hat{\sigma}_{G_2} = \cos\psi = C_2, \qquad (4.126)$$

$$\hat{\pi}_O \cdot \hat{\pi}_{G_2} = \hat{\pi}_{G_1} \cdot \hat{\pi}_{G_2} = \cos\theta_{G_1} = C_3,$$

$$\hat{\pi}_O \cdot \hat{\pi}_{G_1} = \cos 2\theta_{G_1} = C.$$

It is readily seen that C_1, C_2 and C_3 are functions of C, i.e.,

$$C_1^2 = \frac{(1+2C)(1-C)^2}{2(1+C)}, \qquad C_2^2 = \frac{2C^2}{1+C}, \qquad C_3^2 = \frac{1+C}{2}. \qquad (4.127)$$

The fundamental equation of the wavefield (4.36) can be simplified to

$$\begin{pmatrix} y_O' & 0 & a & 0 & C_2a & 0 \\ 0 & y_O' & 0 & Ca & C_1a & C_3a \\ a & 0 & y_{G_1}' & 0 & -C_2a & 0 \\ 0 & Ca & 0 & y_{G_1}' & C_1a & -C_3a \\ C_2a & C_1a & -C_2a & C_1a & y_{G_2}' & 0 \\ 0 & C_3a & 0 & -C_3a & 0 & y_{G_2}' \end{pmatrix} \begin{pmatrix} E_{\sigma O} \\ E_{\pi O} \\ E_{\sigma G_1} \\ E_{\pi G_1} \\ E_{\sigma G_2} \\ E_{\pi G_2} \end{pmatrix} = 0, \qquad (4.128)$$

where $y' = 2y/k$, and y is defined in (4.53). At the exact three-beam position, $y_O' = y_{G_1}' = y_{G_2}' = y'$, and (4.41) and (4.128) can be used to show

$$y'(1) = 1 - 2C,$$

$$y'(2) = y'(3) = 1 + C, \qquad (4.129)$$

$$y'(4) = y'(5) = y'(6) = -1.$$

This means that $y'(1)$ is the unique mode, $y'(2)$ and $y'(3)$ are doubly degenerate modes and $y'(4)$, $y'(5)$ and $y'(6)$ are triply degenerate modes.

The wavefield-amplitude ratios for mode 1 corresponding to $y(1)$ can be obtained from (4.128) by setting $y'(1) = 1 - 2C$, i.e.,

$$E_{\sigma O}(1):E_{\pi O}(1):E_{\sigma G_1}(1):E_{\pi G_1}(1):E_{\sigma G_2}(1)$$

$$= -\frac{1}{\beta}:1:\frac{1}{\beta}:1:-\frac{\sqrt{2(1+C)}}{\beta},$$

$$E_{\pi G_2}=0,$$ (4.130)

where $\beta = \sqrt{1+2C}$. This implies:

$$|E_O|=|E_{G_1}|=|E_{G_2}|=\sqrt{\frac{2(1+C)}{1+2C}}\,|E_{\pi O}|.$$ (4.131)

The wavefields are represented in Fig. 4.14b, which is the side view of Fig. 4.14a on the plane OG_1G_2. E_O, E_{G_1}, E_{G_2} lie respectively in the planes, CWO, CWG_1 and CWG_2. Since OC, G_1C and G_2C are perpendicular to G_1G_2, OG_2 and OG_1, respectively, the vectors E_O, E_{G_1} and E_{G_2} lie in the atomic planes of G_1-G_2, G_2 and G_1, respectively.

For the triply degenerate case, the wavefield equation can be obtained by setting $y'(4) = -1$:

$$E_{\sigma O}-E_{\sigma G_1}=\frac{C_1C_2}{1-C_2^2}(E_{\pi O}+E_{\pi G_1}),$$

$$E_{\sigma O}-E_{\sigma G_1}=\frac{1-C_1^2-C_3^2}{C_1C_2}E_{\pi O}+\frac{C+C_1^2-C_3^2}{C_1C_2}E_{\pi G_1},$$

$$E_{\sigma G_2}=C_2(E_{\sigma O}-E_{\sigma G_1})+C_1(E_{\pi O}+E_{\pi G_1}),$$ (4.132)

$$E_{\pi G_2}=C_3(E_{\pi O}-E_{\pi G_1}),$$

$$E_{\pi G_1}=0.$$

Comparing the coefficients of $E_{\pi O}$ in the first two equations,

$$\frac{C_1C_2}{1-C_2^2}\neq\frac{1-C_1^2-C_3^2}{C_1C_2},$$ (4.133)

we conclude that $E_{\pi O}=0$. This leads to

$$E_{\sigma O}=E_{\sigma G_1},\qquad E_{\sigma G_2}=E_{\pi G_2}=0.$$ (4.134)

Fig. 4.15a shows this mode, mode 4, where E_O and E_{G_1} are perpendicular to the OWG_1 plane. Because of the symmetry between the reciprocal lattice points O, G_1 and G_2, the other two modes associated with $y'(5)$ and $y'(6)$ can be determined easily by permuting O, G_1 and G_2 in (4.132). This is equivalent to redefining the polarization vectors in these two modes. In mode 5, the $\hat{\sigma}_O$ and $\hat{\sigma}_{G_2}$ are redefined to be perpendicular to the plane OWG_2. Similarly, $\hat{\sigma}_{G_1}$

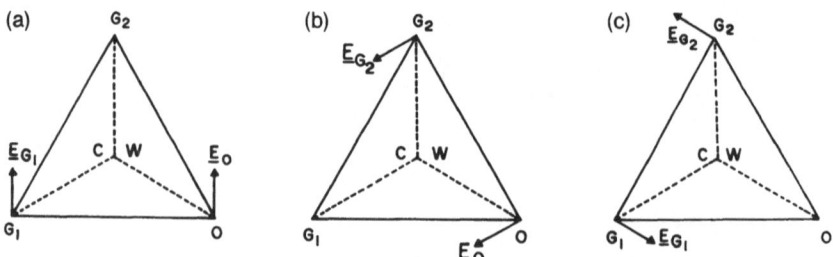

Fig. 4.15 a–c. Wavefields of the three-fold degenerate mode for a symmetric three-beam case

and $\hat{\sigma}_{G_2}$ are redefined for mode 6 to be perpendicular to the plane WG_1G_2. The corresponding wavefield-amplitude relations are

$$E_{\pi O} = E_{\pi G_1} = E_{\pi G_2} = E_{\sigma G_1} = 0 , \qquad E_{\sigma O} = E_{\sigma G_2} , \tag{4.135}$$

and

$$E_{\pi O} = E_{\pi G_1} = E_{\pi G_2} = E_{\sigma O} = 0 , \qquad E_{\sigma G_1} = -E_{\sigma G_2} , \tag{4.136}$$

for modes 5 and 6, respectively. These two modes are also shown schematically in Figs. 4.15b and c. The direction of E_{G_1} is opposite to that of E_{G_2} because $\chi_{G_1-G_2} = -\chi_{G_2-G_1}$ in (4.125). In comparison with (4.121) for the two-beam case, (4.135, 136) show that the triply degenerate modes are σ-polarized two-beam modes, which are not influenced by the presence of the third field. In other words, the situation is equivalent to a three-beam diffraction of a σ-polarized incident wave taking place when the crystal undergoes a rotation around the reciprocal lattice vector parallel to $\hat{\sigma}_O$.

In the doubly degenerate case, $y'(2) = y'(3) = 1 + C$. The wavefield amplitudes are governed by the five equations

$$E_{\sigma O} + E_{\sigma G_1} = 0 ,$$

$$E_{\pi O} + E_{\pi G_1} = -\frac{2C_1}{1+2C} E_{\sigma G_2} ,$$

$$E_{\pi O} - E_{\pi G_1} = -2C_3 E_{\pi G_2} , \tag{4.137}$$

$$E_{\sigma G_2} = -\frac{C}{2C_2}(E_{\sigma O} - E_{\sigma G_1}) ,$$

$$E_{\pi G_2} = \frac{C_3}{1+C}(E_{\pi O} - E_{\pi G_1}) .$$

Considering the third and the fifth equations, we obtain $E_{\pi G_2} = 0$. The ratios between the wavefield amplitudes are

$$E_{\sigma O}:E_{\pi O}:E_{\sigma G_1}:E_{\pi G_1}:E_{\sigma G_2}=1:\frac{CC_1}{C_2(1+2C)}:-1:\frac{CC_1}{C_2(1+2C)}:-\frac{C}{C_2},$$

(4.138)

$$E_{\pi G_2}=0\ .$$

The resultant fields are then

$$|E_{G_1}|=|E_O|,\quad |E_{G_2}|=\sqrt{\frac{2(1+2C)}{5+C}}\,|E_O|.$$

(4.139)

The angles between $E_{\pi O}$ and $E_{\sigma O}$, and between $E_{\pi G_1}$ and $E_{\sigma G_1}$, are α and $-\alpha$ where

$$\tan\alpha=-\frac{1-C}{2\sqrt{1+2C}}\ .$$

(4.140)

This mode is shown in Fig. 4.16a. By permuting O, G_1 and G_2, two additional modes are obtained which are shown in Fig. 4.16b and c. However, the last one (Fig. 4.16c) is not a normal mode, because it can be represented as a linear combination of mode 1 in Fig. 4.14b and mode 6 in Fig. 4.15c. In this particular symmetric three-beam transmission case, we have therefore six normal modes. The resultant wavefield for a given reflection G is equal to the wavefield sum of all the modes, i.e.,

$$E_G=\sum_j[\,\hat{\sigma}_G E_{\sigma G}(j)+\hat{\pi}_G E_{\pi G}(j)]\ .$$

(4.141)

Modes of propagation can also be discussed analytically for three-beam Borrmann diffraction with a zero-coupling between the primary and the secondary reflections at the exact three-beam position. Assuming $\chi_{G_1-G_2}=0$ and keeping the directions of the polarization vectors as defined in Fig. 4.14, we obtain the wavefield equation

$$\begin{pmatrix} y' & 0 & C_1'\chi_{O-G_1} & 0 & C_2'\chi_{O-G_2} & 0 \\ 0 & y' & 0 & C_3'\chi_{O-G_1} & C_4'\chi_{O-G_2} & C_5'\chi_{O-G_2} \\ C_1'\chi_{G_1-O} & 0 & y' & 0 & 0 & 0 \\ 0 & C_3'\chi_{G_1-O} & 0 & y' & 0 & 0 \\ C_2'\chi_{G_2-O} & C_4'\chi_{G_2-O} & 0 & 0 & y' & 0 \\ 0 & C_5'\chi_{G_2-O} & 0 & 0 & 0 & y' \end{pmatrix} \begin{pmatrix} E_{\sigma O} \\ E_{\pi O} \\ E_{\sigma G_1} \\ E_{\pi G_1} \\ E_{\sigma G_2} \\ E_{\pi G_2} \end{pmatrix}$$

$$=0\ ,$$

(4.142)

where

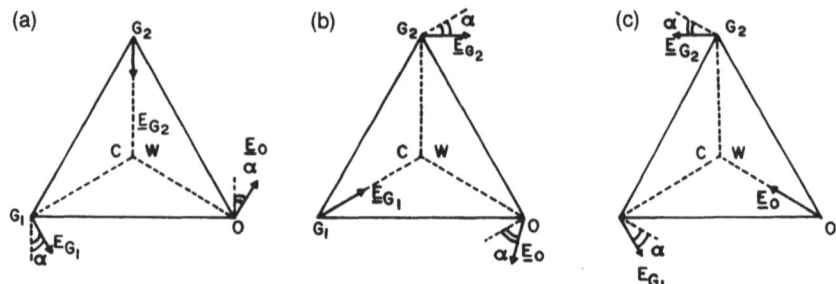

Fig. 4.16a – c. Wavefields of the two-fold degenerate mode for a symmetric three-beam case

$$C_1' = (\hat{\sigma}_O \cdot \hat{\sigma}_{G_1}) = 1, \quad C_2' = (\hat{\sigma}_O \cdot \hat{\sigma}_{G_2}),$$

$$C_3' = (\hat{\pi}_O \cdot \hat{\pi}_{G_1}), \quad C_4' = (\hat{\pi}_O \cdot \hat{\sigma}_{G_2}),$$

$$C_5' = (\hat{\pi}_O \cdot \hat{\pi}_{G_2}), \quad \text{and}$$

$$y' = \chi_{O-O} + 2\gamma\delta, \tag{4.143}$$

with $\gamma_O = \gamma_{G_1} = \gamma_{G_2} = \gamma$, which can be derived from (4.51, 53, 86, 87). From the last four equations in (4.142), $E_{\sigma G_1}$, $E_{\pi G_1}$, $E_{\sigma G_2}$ and $E_{\pi G_2}$ take the following explicit forms, for $y' \neq 0$:

$$E_{\sigma G_1} = -\frac{C_1' \chi_{G_1-O}}{y'} E_{\sigma O}, \quad E_{\pi G_1} = -\frac{C_3' \chi_{G_1-O}}{y'} E_{\pi O},$$

$$E_{\sigma G_2} = -(C_2' E_{\sigma O} + C_4' E_{\pi O}) \frac{\chi_{G_2-O}}{y'}, \quad E_{\pi G_2} = -\frac{C_5' \chi_{G_2-O}}{y'} E_{\pi O}. \tag{4.144}$$

Substituting (4.144) into the first two equations of (4.142) we obtain

$$\frac{E_{\sigma O}}{E_{\pi O}} = \frac{C_2' C_4' \chi_{O-G_2} \chi_{G_2-O}/y'}{y' - (C_1'^2 \chi_{O-G_1} \chi_{G_1-O} + C_2'^2 \chi_{O-G_2} \chi_{G_2-O})/y'}$$

$$= \frac{y' - (C_3'^2 \chi_{O-G_1} \chi_{G_1-O} + C_4'^2 \chi_{O-G_2} \chi_{G_2-O})/y'}{C_2' C_4' \chi_{O-G_2} \chi_{G_2-O}/y'}. \tag{4.145}$$

Equation (4.145) leads to the dispersion relation

$$y'^2(y'^4 - 2\zeta^2 y'^2 + \xi\zeta^4) = 0, \tag{4.146}$$

so that

$$y'^2 = 0, \quad y'^2 = \zeta^2(1 \pm \sqrt{1-\xi}), \quad \text{where} \tag{4.147}$$

$$\zeta^2 = \tfrac{1}{2} [(C_1'^2 + C_3'^2) \chi_{O-G_1} \chi_{G_1-O} + (C_2'^2 + C_4'^2) \chi_{O-G_2} \chi_{G_2-O}],$$

$$\xi = \frac{1}{\zeta^4} [C_1'^2 C_3'^2 (\chi_{O-G_1} \chi_{G_1-O})^2$$
$$+ (C_1'^2 C_4'^2 + C_2'^2 C_3'^2) \chi_{O-G_2} \chi_{G_2-O} \chi_{O-G_1} \chi_{G_1-O}] \, . \tag{4.148}$$

The relation $y'^2 = 0$ implies there exist two-fold degenerate modes whose wavefield amplitudes satisfy the conditions

$$E_{\sigma O} = E_{\pi O} = 0 \, ,$$

$$\frac{E_{\sigma G_1}}{E_{\sigma G_2}} = -\frac{C_2' \chi_{O-G_2}}{C_1' \chi_{O-G_1}} \, , \tag{4.149}$$

$$E_{\pi G_2} = -\frac{C_4'}{C_5'} E_{\sigma G_2} - \left(\frac{C_3' \chi_{O-G_1}}{C_5' \chi_{O-G_2}} \right) E_{\pi G_1} \, .$$

Setting $E_{\pi G_1} = 0$, we obtain the wavefield-amplitude ratio:

$$E_{\sigma G_1} : E_{\sigma G_2} : E_{\pi G_2} = -\frac{C_4' \chi_{O-G_2}}{C_1' \chi_{O-G_1}} : 1 : -\frac{C_4'}{C_5'} \, ,$$
$$\tag{4.150}$$
$$E_{\sigma O} = E_{\pi O} = E_{\pi G_1} = 0 \, ,$$

for one of the degenerate modes. The wavefield-amplitude ratio for the other degenerate mode can be found to be

$$E_{\sigma G_1} : E_{\pi G_1} : E_{\sigma G_2} : E_{\pi G_2}$$

$$= -\frac{C_4' \chi_{O-G_2}}{C_1' \chi_{O-G_1}} : -\frac{C_5'^2 b_0}{C_3' C_4'} : 1 : \left(-\frac{C_4'}{C_5'} + \frac{C_5' \chi_{O-G_1} b_0}{C_4' \chi_{O-G_2}} \right) \, ,$$
$$\tag{4.151}$$

$$E_{\sigma O} = E_{\pi O} = 0, \quad \text{where}$$

$$b_0 = \left(\frac{C_4' \chi_{O-G_2}}{C_5' \chi_{O-G_1}} \right)^2 + \left(\frac{C_4'}{C_5'} \right)^2 + 1 \, . \tag{4.152}$$

The degeneracy due to $y'^2 = 0$ indicates that the corresponding two dispersion sheets intersect at the exact three-beam diffraction position. Since $y' = 0$ implies $\delta = -\chi_{O-O}/(2\gamma)$, the intersection lies at the Lorentz point in reciprocal space.

The second relation in (4.147) leads to four solutions for y',

$$y'(3), y'(4) = \zeta \sqrt{1 \pm \sqrt{1-\xi}} \, , \tag{4.153}$$
$$y'(5), y'(6) = -\zeta \sqrt{1 \pm \sqrt{1-\xi}} \, .$$

The corresponding wavefield-amplitude ratio according to (4.142) is thus

$$E_{\sigma O} : E_{\pi O} : E_{\sigma G_1} : E_{\pi G_1} : E_{\sigma G_2} : E_{\pi G_2}$$

$$= b_1 : 1 : -\frac{C_1' \chi_{G_1 - O} b_1}{y'(j)} : -\frac{C_3' \chi_{G_1 - O}}{y'(j)} : b_2 : -\frac{C_5' \chi_{G_2 - O}}{y'(j)} , \qquad (4.154)$$

where

$$b_1 = \frac{y'^2(j) - (C_3'^2 \chi_{O - G_1} \chi_{G_1 - O} + C_4'^2 \chi_{O - G_2} \chi_{G_2 - O})}{C_2' C_4' \chi_{O - G_2} \chi_{G_2 - O}} ,$$

$$b_2 = -\frac{\chi_{G_2 - O}}{y'(j)} (C_2' b_1 + C_4') . \qquad (4.155)$$

It is worthwhile to note that although the dispersion equation (4.41) is independent of the choice of the polarization vectors σ and π, the modes of propagation do depend on the polarization vectors chosen. *Joko* and *Fukuhara* [4.7] defined the polarization vectors in plane $OG_1 G_2$ (Fig. 4.17a) for the symmetric three-beam Laue diffraction just mentioned. The polarization factors are

$$\hat{\sigma}_O \cdot \hat{\sigma}_{G_1} = -\tfrac{1}{2} ,$$

$$\hat{\pi}_O \cdot \hat{\pi}_{G_1} = \tfrac{1}{2} - C ,$$

$$\hat{\pi}_O \cdot \hat{\sigma}_{G_1} = -\hat{\pi}_O \cdot \hat{\sigma}_{G_2} = -\frac{\sqrt{1 + 2C}}{3} , \qquad (4.156)$$

where C is defined in (4.126). The other polarization factors can be obtained by permuting O, G_1 and G_2. The eigenvalues are the same as in (4.129). The corresponding wavefield amplitudes are listed in Table 4.1 for the six modes of propagation. In the table,

$$a_1 = \sqrt{\frac{3}{2(2 + C)}} , \qquad a_2 = \sqrt{\frac{1 + 2C}{2(2 + C)}} . \qquad (4.157)$$

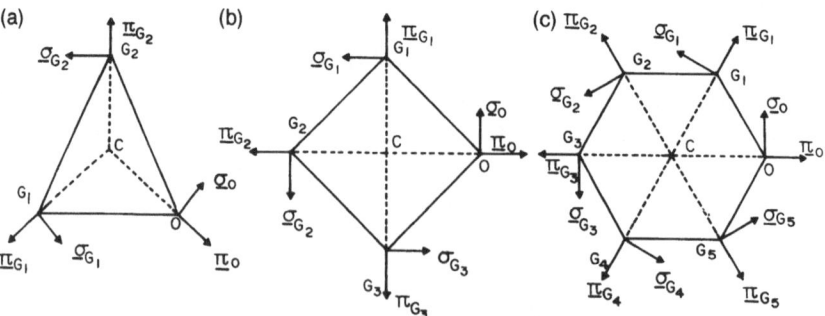

Fig. 4.17a – c. Definition of the polarization vectors for (**a**) three-beam, (**b**) four-beam and (**c**) six-beam symmetric diffractions (after *Joko* and *Fukuhara* [4.7])

Table 4.1. Ratios of the wavefield amplitudes of the symmetric three-beam transmission case

Mode	$E_{\sigma O}$	$E_{\sigma G_1}$	$E_{\sigma G_2}$	$E_{\pi O}$	$E_{\pi G_1}$	$E_{\pi G_2}$
1	0	0	0	$\dfrac{1}{\sqrt{3}}$	$\dfrac{1}{\sqrt{3}}$	$\dfrac{1}{\sqrt{3}}$
2	$\sqrt{\dfrac{2}{3}}\,a_1$	$-\sqrt{\dfrac{1}{6}}\,a_1$	$-\sqrt{\dfrac{1}{6}}\,a_1$	0	$\dfrac{a_2}{\sqrt{2}}$	$-\dfrac{a_2}{\sqrt{2}}$
3	0	$-\dfrac{a_1}{\sqrt{2}}$	$\dfrac{a_1}{\sqrt{2}}$	$-\dfrac{a_2}{\sqrt{2}}$	$\dfrac{a_2}{\sqrt{2}}$	$-\dfrac{a_2}{\sqrt{6}}$
4	$\dfrac{1}{\sqrt{3}}$	$\dfrac{1}{\sqrt{3}}$	$\dfrac{1}{\sqrt{3}}$	0	0	0
5	$-\sqrt{\dfrac{2}{3}}\,a_2$	$\dfrac{a_2}{\sqrt{6}}$	$\dfrac{a_2}{\sqrt{6}}$	0	$\dfrac{a_1}{\sqrt{2}}$	$-\dfrac{a_1}{\sqrt{2}}$
6	0	$\dfrac{a_2}{\sqrt{2}}$	$-\dfrac{a_2}{\sqrt{2}}$	$\sqrt{\dfrac{2}{3}}\,a_1$	$-\dfrac{a_1}{\sqrt{6}}$	$-\dfrac{a_1}{\sqrt{6}}$

Although the wavefield amplitudes are different from the ones previously obtained, it can still be seen that, for example, mode 1 in both polarization vector systems shows the same characteristics. If (4.33) is used to calculate the resultant wavefield in one direction, it can be proved easily that the two polarization vector systems eventually give the same results. It is also clear that the polarization vector system defined first connects the three-beam case better with two-beam diffraction because at least four polarization vectors are referred to the plane of incidence of the G_1 reflection. The latter polarization system defined above provides a mathematically and geometrically simple way of finding the eigenvectors for degenerate eigenvalues and of presenting the wavefields in a two-dimensional plane, i.e., the OG_1G_2 plane. We shall use the second polarization vector system for symmetric four-beam and six-beam transmission cases (Fig. 4.17b, c). Without showing the detailed calculation procedures for the eigenvalues and eigenvectors, results of the four-beam and six-beam diffraction cases are summarized as follows [4.7]:

i) *Four-beam case*, O, G_1, G_2, G_3, with $\chi_{G_1-O} = -\chi_{G_3-O} = -\chi_{G_2-G_1} = -\chi_{G_1-G_2}$, $\chi_{G_1-G_3} = \chi_{G_2-O}$. The polarization factors are

$$\hat{\sigma}_O \cdot \hat{\pi}_{G_2} = \hat{\sigma}_O \cdot \hat{\sigma}_{G_1} = 0 \, ,$$

$$\hat{\sigma}_O \cdot \hat{\sigma}_{G_2} = -1 \, ,$$

$$\hat{\pi}_O \cdot \hat{\pi}_{G_1} = 1 - C \, , \tag{4.158}$$

$$\hat{\pi}_O \cdot \hat{\pi}_{G_2} = 1 - 2C \, ,$$

$$\hat{\sigma}_O \cdot \hat{\pi}_{G_1} = -\hat{\sigma}_O \cdot \hat{\sigma}_{G_3} = \sqrt{C} \, ,$$

where $C = \cos 2\theta_{G_1}$. The eigenvalues are

$$y'(1), y'(2) = \pm 2(1-C)\chi_{G_1-O} + (1-2C)\chi_{G_2-O},$$

$$y'(3) = y'(4) = -\chi_{G_2-O},$$

$$y'(5) = y'(6) = \chi_{G_2-O} + 2\sqrt{C}\,\chi_{G_1-O}(\xi + \sqrt{1+\xi^2}),$$

$$y'(7) = y'(8) = \chi_{G_2-O} + 2\sqrt{C}\,\chi_{G_1-O}(\xi - \sqrt{1+\xi^2}),$$

(4.159)

where

$$\xi = -\frac{(1-C)\chi_{G_2-O}}{2\sqrt{C}\,\chi_{G_1-O}}.$$

(4.160)

The wavefield amplitudes of the eight modes are given in Table 4.2, where the term

$$\xi_\pm = \frac{1}{2}\sqrt{1 \pm \frac{\xi}{\sqrt{1+\xi^2}}}.$$

(4.161)

has been introduced.

ii) *Six-beam case*, O, G_1, G_2, G_3, G_4 and G_5, with $\chi_{G_1-O} = -\chi_{G_5-O} = \chi_{G_2-G_1} = -\chi_{G_3-G_2} = \chi_{G_4-G_3} = \chi_{G_5-G_4}$, $\chi_{G_2-O} = \chi_{G_3-G_1} = -\chi_{G_4-G_2} = \chi_{G_5-G_3}$, $\chi_{G_3-O} = \chi_{G_4-G_1} = \chi_{G_5-G_2}$. The products of the polarization vectors are (Fig. 4.17c)

Table 4.2. Ratios of the wavefield amplitudes of the symmetric four-beam transmission

Mode	$E_{\sigma O}$	$E_{\sigma G_1}$	$E_{\sigma G_2}$	$E_{\sigma G_3}$	$E_{\pi O}$	$E_{\pi G_1}$	$E_{\pi G_2}$	$E_{\pi G_3}$
1	0	0	0	0	$\frac{1}{2}$	$\frac{1}{2}$	$\frac{1}{2}$	$\frac{1}{2}$
2	0	0	0	0	$\frac{1}{2}$	$-\frac{1}{2}$	$\frac{1}{2}$	$-\frac{1}{2}$
3	$\frac{1}{\sqrt{2}}$	0	$\frac{1}{\sqrt{2}}$	0	0	0	0	0
4	0	$\frac{1}{\sqrt{2}}$	0	$\frac{1}{\sqrt{2}}$	0	0	0	0
5	0	$-\xi_-$	0	ξ_-	ξ_+	0	$-\xi_+$	0
6	ξ_-	0	$-\xi_-$	0	0	ξ_+	0	$-\xi_+$
7	0	ξ_+	0	$-\xi_+$	ξ_-	0	$-\xi_-$	0
8	$-\xi_+$	0	ξ_+	0	0	ξ_-	0	$-\xi_-$

$$\hat{\sigma}_O \cdot \hat{\sigma}_{G_3} = -1 \, ,$$

$$\hat{\sigma}_O \cdot \hat{\sigma}_{G_1} = \hat{\sigma}_O \cdot \hat{\sigma}_{G_5} = -(\hat{\sigma}_O \cdot \hat{\sigma}_{G_2}) = -(\hat{\sigma}_O \cdot \hat{\sigma}_{G_4}) = \tfrac{1}{2} \, ,$$

$$\hat{\sigma}_O \cdot \hat{\pi}_{G_3} = 0 \, ,$$

$$\hat{\pi}_O \cdot \hat{\pi}_{G_1} = \hat{\pi}_O \cdot \hat{\pi}_{G_5} = \tfrac{3}{2} - C = a_3 \, ,$$

$$\hat{\pi}_O \cdot \hat{\pi}_{G_2} = \hat{\pi}_O \cdot \hat{\pi}_{G_4} = \tfrac{5}{2} - 3C = a_4 \, ,$$

$$\hat{\pi}_O \cdot \hat{\pi}_{G_3} = 3 - 4C = a_5 \, ,$$

$$\hat{\sigma}_O \cdot \hat{\pi}_{G_1} = \hat{\sigma}_O \cdot \hat{\pi}_{G_3} = -(\hat{\sigma}_O \cdot \hat{\pi}_{G_4}) = -(\hat{\sigma}_O \cdot \hat{\pi}_{G_5})$$
$$= \tfrac{1}{2}\sqrt{3(2C-1)} = a_6 \, ,$$

(4.162)

where $C = \cos 2\theta_{G_1}$. Define

$$\xi = \frac{(a_3 - \tfrac{1}{2})\chi_{G_1 - O} + (a_4 + \tfrac{1}{2})\chi_{G_2 - O} - (a_5 + 1)\chi_{G_3 - O}}{2a_6(\chi_{G_1 - O} - \chi_{G_2 - O})} \, ,$$

$$\eta = \frac{(a_3 - \tfrac{1}{2})\chi_{G_1 - O} - (a_4 + \tfrac{1}{2})\chi_{G_2 - O} - (a_5 + 1)\chi_{G_3 - O}}{2a_6(\chi_{G_1 - O} + \chi_{G_2 - O})} \, ,$$

(4.163)

$$\xi_{\pm} = \sqrt{\frac{1}{8}\left(1 \pm \frac{\xi}{\sqrt{3 + \xi^2}}\right)} \, , \qquad \eta_{\pm} = \sqrt{\frac{1}{8}\left(1 \pm \frac{\eta}{\sqrt{3 + \eta^2}}\right)} \, .$$

The eigenvalues, in terms of the polarization factors ξ and η, are expressed as

$$y'(1), y'(2) = \pm 2a_3 \chi_{G_1 - O} + 2a_4 \chi_{G_2 - O} \pm a_5 \chi_{G_3 - O} \, ,$$

$$y'(3), y'(4) = \pm \chi_{G_1 - O} - \chi_{G_2 - O} \mp \chi_{G_3 - O} \, ,$$

$$\left.\begin{array}{c} y'(5) \\ y'(7) \end{array}\right\} = \left.\begin{array}{c} y'(6) \\ y'(8) \end{array}\right\} = -a_3 \chi_{G_1 - O} - a_4 \chi_{G_2 - O} + a_5 \chi_{G_3 - O}$$
$$+ a_6(\chi_{G_1 - O} - \chi_{G_2 - O})(\xi \pm \sqrt{3 + \xi^2}) \, ,$$

(4.164)

$$\left.\begin{array}{c} y'(9) \\ y'(11) \end{array}\right\} = \left.\begin{array}{c} y'(10) \\ y'(12) \end{array}\right\} = a_3 \chi_{G_1 - O} - a_4 \chi_{G_2 - O} - a_5 \chi_{G_3 - O}$$
$$+ a_6(\chi_{G_1 - O} + \chi_{G_2 - O})(-\eta \pm \sqrt{3 + \eta^2}) \, .$$

The wavefield amplitudes for the twelve modes are listed in Table 4.3.

Suppose that the three-beam, four-beam and six-beam cases discussed above involve the same reflection G_1. There are interesting facts which can be seen by comparing the modes of propagation in these three symmetric cases. For example, in the completely π-polarized modes, the amplitude is proportional to $1/N$ where N is the number of beams. Additionally, the values of the amplitude ratios in Tables 4.1 – 3 seem to be related to the terms $\sin(2\pi n/N)$ and $\cos(2\pi n/N)$. This is not surprising due to the fact that the wavefields

Table 4.3. Ratios of the wavefield amplitudes of the symmetric six-beam transmission case

Mode	$E_{\sigma O}$	$E_{\sigma G_1}$	$E_{\sigma G_2}$	$E_{\pi G_3}$	$E_{\pi G_4}$	$E_{\pi G_5}$
1	0	0	0	0	0	0
2	0	0	0	0	0	0
3	$\dfrac{1}{\sqrt6}$	$\dfrac{1}{\sqrt6}$	$\dfrac{1}{\sqrt6}$	$\dfrac{1}{\sqrt6}$	$\dfrac{1}{\sqrt6}$	$\dfrac{1}{\sqrt6}$
4	$\dfrac{1}{\sqrt6}$	$-\dfrac{1}{\sqrt6}$	$\dfrac{1}{\sqrt6}$	$-\dfrac{1}{\sqrt6}$	$\dfrac{1}{\sqrt6}$	$-\dfrac{1}{\sqrt6}$
5	$\dfrac{2\xi_+}{\sqrt3}$	$-\dfrac{\xi_+}{\sqrt3}$	$-\dfrac{\xi_+}{\sqrt3}$	$\dfrac{2\xi_+}{\sqrt3}$	$-\dfrac{\xi_+}{\sqrt3}$	$-\dfrac{\xi_+}{\sqrt3}$
6	0	$-\xi_+$	ξ_+	0	$-\xi_+$	ξ_+
7	$-\dfrac{2\xi_-}{\sqrt3}$	$\dfrac{\xi_-}{\sqrt3}$	$\dfrac{\xi_-}{\sqrt3}$	$-\dfrac{2\xi_-}{\sqrt3}$	$\dfrac{\xi_-}{\sqrt3}$	$\dfrac{\xi_-}{\sqrt3}$
8	0	ξ_-	$-\xi_-$	0	ξ_-	$-\xi_-$
9	$\dfrac{2\eta_-}{\sqrt3}$	$\dfrac{\eta_-}{\sqrt3}$	$-\dfrac{\eta_-}{\sqrt3}$	$-\dfrac{2\eta_-}{\sqrt3}$	$-\dfrac{\eta_-}{\sqrt3}$	$\dfrac{\eta_-}{\sqrt3}$
10	0	$-\eta_-$	$-\eta_-$	0	η_-	η_-
11	$-\dfrac{2\eta_+}{\sqrt3}$	$-\dfrac{\eta_+}{\sqrt3}$	$\dfrac{\eta_+}{\sqrt3}$	$\dfrac{2\eta_+}{\sqrt3}$	$\dfrac{\eta_+}{\sqrt3}$	$-\dfrac{\eta_+}{\sqrt3}$
12	0	η_+	η_+	0	$-\eta_+$	$-\eta_+$

Mode	$E_{\pi O}$	$E_{\pi G_1}$	$E_{\pi G_2}$	$E_{\pi G_3}$	$E_{\pi G_4}$	$E_{\pi G_5}$
1	$\dfrac{1}{\sqrt6}$	$\dfrac{1}{\sqrt6}$	$\dfrac{1}{\sqrt6}$	$\dfrac{1}{\sqrt6}$	$\dfrac{1}{\sqrt6}$	$\dfrac{1}{\sqrt6}$
2	$\dfrac{1}{\sqrt6}$	$-\dfrac{1}{\sqrt6}$	$\dfrac{1}{\sqrt6}$	$-\dfrac{1}{\sqrt6}$	$\dfrac{1}{\sqrt6}$	$-\dfrac{1}{\sqrt6}$
3	0	0	0	0	0	0
4	0	0	0	0	0	0
5	0	ξ_-	$-\xi_-$	0	ξ_-	$-\xi_-$
6	$\dfrac{2\xi_-}{\sqrt3}$	$-\dfrac{\xi_-}{\sqrt3}$	$-\dfrac{\xi_-}{\sqrt3}$	$\dfrac{2\xi_-}{\sqrt3}$	$-\dfrac{\xi_-}{\sqrt3}$	$-\dfrac{\xi_-}{\sqrt3}$
7	0	ξ_+	$-\xi_+$	0	ξ_+	$-\xi_+$
8	$\dfrac{2\xi_+}{\sqrt3}$	$-\dfrac{\xi_+}{\sqrt3}$	$-\dfrac{\xi_+}{\sqrt3}$	$\dfrac{2\xi_+}{\sqrt3}$	$-\dfrac{\xi_+}{\sqrt3}$	$-\dfrac{\xi_+}{\sqrt3}$
9	0	η_+	η_+	0	$-\eta_+$	$-\eta_+$
10	$\dfrac{2\eta_+}{\sqrt3}$	$\dfrac{\eta_+}{\sqrt3}$	$-\dfrac{\eta_+}{\sqrt3}$	$-\dfrac{2\eta_+}{\sqrt3}$	$-\dfrac{\eta_+}{\sqrt3}$	$\dfrac{\eta_+}{\sqrt3}$
11	0	η_-	η_-	0	$-\eta_-$	$-\eta_-$
12	$\dfrac{2\eta_-}{\sqrt3}$	$\dfrac{\eta_-}{\sqrt3}$	$-\dfrac{\eta_-}{\sqrt3}$	$-\dfrac{2\eta_-}{\sqrt3}$	$-\dfrac{\eta_-}{\sqrt3}$	$\dfrac{\eta_-}{\sqrt3}$

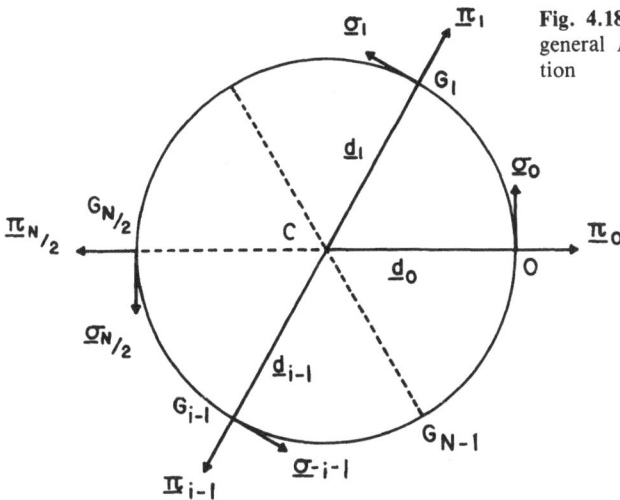

Fig. 4.18. Polarization vectors in a general N-beam symmetric diffraction

consist of standing and traveling waves. Let us reconsider the wavefield in a general N-beam case, and rewrite the component of the Bloch wavefunction:

$$E_{G_m}(j) = [\hat{\sigma}_m E_{\sigma G_m}(j) + \hat{\pi}_m E_{\pi G_m}(j)] \exp(-2\pi i g_m \cdot r)$$
$$\times \exp(-2\pi i K_O(j) \cdot r) . \tag{4.165}$$

Since all the g's lie in a plane, the vector r can be chosen, for simplicity, to be in this plane. Defining the vectors d as shown in Fig. 4.18, such that

$$g_1 = -d_O + d_2, \quad g_m = -d_O + d_m, \quad |d_m| = |d| , \tag{4.166}$$

and choosing $r = d_O/d^2$, we obtain

$$\exp[-2\pi i(g_m \cdot r)] = \exp\left[-2\pi i \left(\frac{nm}{N}\right)\right], \tag{4.167}$$

for mode j, with $j = 2n+1$ for the σ component and $j = 2n$ for the π component (i.e., the σ modes are associated with odd numbers j and the π modes with even j). By substituting (4.167) into (4.165), the wavefield amplitude can be written as

$$E_{G_m}(j) = \hat{\sigma}_m \left[A_m \cos\left(\frac{2\pi nm}{N}\right) + B_m \sin\left(\frac{2\pi nm}{N}\right) \right]$$
$$+ \hat{\pi}_m \left[A'_m \cos\left(\frac{2\pi nm}{N}\right) + B'_m \sin\left(\frac{2\pi nm}{N}\right) \right], \tag{4.168}$$

where $n = (j-1)/2$ for odd j and $n = j/2$ for even j. From the geometric relation between σ and π, it can be seen from Fig. 4.18 that

$$\hat{\sigma}_m = -\hat{\sigma}_{m'}, \quad \hat{\pi}_m = \hat{\pi}_{m'}, \tag{4.169}$$

if $m' = N - m$. Similarly, the wavefield amplitudes retain this character:

$$E_{G_m}(j = 2n - 1) = E_{\sigma G_m} = -E_{\sigma G_{N-m}},$$

$$E_{G_m}(j = 2n) = E_{\pi G_m} = E_{\pi G_{N-m}}. \tag{4.170}$$

The coefficients A_m, B_m, A'_m, and B'_m can be further defined by the constraint imposed by (4.169) so that

$$E_{G_m}(j = 2n - 1) = \hat{\sigma}_m A_m \cos\left(\frac{2\pi nm}{N}\right) + \hat{\pi}_m B'_m \sin\left(\frac{2\pi nm}{N}\right),$$

$$E_{G_m}(j = 2n) = \hat{\sigma}_m B_m \sin\left(\frac{2\pi nm}{N}\right) + \hat{\pi}_m A'_m \cos\left(\frac{2\pi nm}{N}\right), \tag{4.171}$$

where

$$\sum_{m=0}^{N-1} (A_m^2 + B'^2_m) = \sum_{m=0}^{N-1} (B_m^2 + A'^2_m) = 1. \tag{4.172}$$

When $n = N$,

$$E_{G_m}(j = 2N - 1) = \hat{\sigma}_m A_m, \quad E_{G_m}(j = 2N) = \hat{\pi}_m A'_m, \tag{4.173}$$

for all m ($m = 0, 1, 2, 3, \ldots, N-1$). The amplitudes $A_m = A'_m = 1/\sqrt{N}$. When $n = N/2$, if N is even,

$$E_{G_m}(j = N - 1) = (-1)^m \hat{\pi}_m B'_m, \quad E_{G_m}(j = N) = (-1)^m \hat{\sigma}_m B_m, \tag{4.174}$$

for all m. Again $B'_m = B_m = 1/\sqrt{N}$.

These two situations, i.e., $n = N$ and $n = N/2$, can be found in Tables 4.2, 3. The corresponding modes are non-degenerate. Their eigenvalues 2ε can be obtained by substituting (4.171) into (4.36). For $n \neq N$ and $n \neq N/2$, the ratios A/B', A'/B and the eigenvalues can be determined from (4.36) with the wavefield amplitudes taken from (4.171). For each eigenvalue 2ε, there are two values of A, B' and A', B, which take the forms [4.15]

$$A_m = -B'_m = \sqrt{\frac{2}{N}} \frac{1}{\sqrt{1 + Z_\pm^2}}, \quad A'_m = B_m = -\sqrt{\frac{2}{N}} \frac{Z_\pm}{\sqrt{1 + Z_\pm^2}}, \tag{4.175}$$

where $Z_\pm = \xi \pm \sqrt{1 + \xi^2}$. The term ξ can be determined from the dispersion equation involving 2ε. The eigenvalue is therefore doubly degenerate. This approach leads to essentially the same results as obtained in Tables 4.2, 3.

4.5.2 Number of Permitted Modes

For a given mode of propagation j, there are $2N$ wavefields, $E_{\sigma O}(j)$, $E_{\pi O}(j)$, $E_{\sigma G_1}(j)$, $E_{\pi G_1}(j), \ldots$, in an N-beam diffraction. For a given reflection G, the wavefields involved are $E_{\sigma G}(1)$, $E_{\pi G}(1)$, $E_{\sigma G}(2)$, $E_{\pi G}(2), \ldots, E_{\sigma G}(j)$, $E_{\pi G}(j), \ldots$. The number of modes is therefore equal to the number of existing wavefields of any one of these reflections.

It has been well established, in two-beam cases, that there are two modes for transmission (Laue), if only one polarization, say σ polarization, is considered. However, there is only one mode allowed in a symmetric two-beam Bragg reflection for a thick crystal [4.16, 17]. This can be demonstrated by considering the Poynting vectors at the excited tie points on the dispersion surface. In Fig. 4.9, the tie points for a symmetric Laue reflection are T_1' and T_2'. The corresponding Poynting vectors S_1' and S_2' approximately perpendicular to the dispersion surface (Sect. 4.4) at T_1' and T_2' point towards the interior of the crystal. For a Bragg reflection, the Poynting vectors are S_1 and S_2 at the corresponding tie points T_1 and T_2. Each tie point defines a mode of wave propagation. Only the mode associated with T_2 has an energy flow towards the crystal. A positive absorption coefficient is associated with this mode. A more rigorous argument, according to *Kato* et al. [4.18], can be given, namely, that this mode has a negative Riemann sheet. Due to this positive absorption coefficient, the wavefield amplitude is always attenuated within the crystal. The requirement of the conservation of total energy is hence fulfilled. The other mode with the Poynting vector S_1 has a negative absorption coefficient. For a negative absorption coefficient, the total energy is increased as the beam penetrates through the crystal. This has no physical meaning when a very thick crystal is involved. Therefore, the mode at T_1 should not be included in the dynamical calculation. Only one mode is permitted.

For N-beam cases, a general rule for determining the number of permitted modes can be derived from the dispersion equation at the exact N-beam diffraction position [4.19].

For simplicity, let us consider a general three-beam diffraction, O, G_1 and G_2, whose dispersion equation is given in (4.41). From Fig. 4.9, it can be seen that the dispersion sheets lying below the Laue point, with respect to the crystal surface, at the exact diffracting position, always have Poynting vectors towards the interior of the crystal. Therefore, the number of permitted modes equals the number of positive δ in (4.86) at the exact three-beam diffracting position.

Actually, the number of positive δ can be easily determined using Descartes' rule of signs or the properties of the eigenvalue equation. Since the location of tie points only depends on the real part δ_r (if the approximation of a small imaginary part is employed), the dispersion equation (4.41) for a σ-polarized wave can be expressed in terms of δ_r in a matrix form:

$$\begin{vmatrix} \dfrac{\chi^r_{0-0}}{2\gamma_0} + \delta_r & \dfrac{C_1\chi^r_{0-G_1}}{2\gamma_0} & \dfrac{C_2\chi^r_{0-G_2}}{2\gamma_0} \\[3mm] \dfrac{C_1\chi^r_{G_1-0}}{2\gamma_{G_1}} & \dfrac{\chi^r_{0-0}}{2\gamma_{G_1}} + \delta_r & \dfrac{C_3\chi^r_{G_1-G_2}}{2\gamma_{G_1}} \\[3mm] \dfrac{C_2\chi^r_{G_2-0}}{2\gamma_{G_2}} & \dfrac{C_3\chi^r_{G_2-G_1}}{2\gamma_{G_2}} & \dfrac{\chi^r_{0-0}}{2\gamma_{G_2}} + \delta_r \end{vmatrix} = 0 . \tag{4.176}$$

As the susceptibility for the incident reflection is always greater than for the diffracted one, i.e., $|\chi^r_{0-0}| > |\chi^r_{0-G_1}|$, $|\chi^r_{0-0}| > |\chi^r_{0-G_2}|$ and $|\chi^r_{0-0}| > |\chi^r_{G_1-G_2}|$, the signs of the eigenvalues δ_r are independent of the off-diagonal elements of the secular determinant of (4.176). By ignoring $\chi^r_{0-G_1}$, $\chi^r_{0-G_2}$, $\chi^r_{G_1-0}$, $\chi^r_{G_2-0}$, $\chi^r_{G_1-G_2}$ and $\chi^r_{G_2-G_1}$, (4.176) becomes

$$(\delta_r - a_{0,r})(\delta_r - a_{1,r})(\delta_r - a_{2,r}) = 0 , \quad \text{where} \tag{4.177}$$

$$a_{0,r} = -\frac{\chi^r_{0-0}}{2\gamma_0} , \quad a_{1,r} = -\frac{\chi^r_{0-0}}{2\gamma_{G_1}} , \quad a_{2,r} = -\frac{\chi^r_{0-0}}{2\gamma_{G_2}} . \tag{4.178}$$

For an N-beam case,

$$\prod_{j=0}^{N-1} (\delta_r - a_{j,r}) = 0 , \quad \text{where} \tag{4.179}$$

$$a_{j,r} = -\frac{\chi^r_{0-0}}{2\gamma_j} \tag{4.180}$$

for all reflections $j = 0,1,2,\dots, N-1$. χ^r_{0-0} is a negative quantity.

It is clear that the sign of δ_r depends on whether the direction cosine γ_j is positive or negative. Since the corresponding direction cosines are negative for Bragg reflections, the δ_r are always negative. The number of permitted modes is then equal to $N - N_R$, where N_R is the number of Bragg reflections involved. If the two polarizations are considered, the number of permitted modes N_p is then

$$N_p = 2(N - N_R) . \tag{4.181}$$

For an N-beam Borrmann diffration, there are $2N$ permitted modes because no Bragg reflections are involved.

Although this relation is quite general, it is not applicable to those cases which involve extremely asymmetric reflections, for which the angle between the incident (or diffracted) beam and the crystal surface is less than two degrees [4.20, 21]. For such cases, higher-order terms of ΔK in (4.39) need to be considered. This leads to more permitted modes since the equation of dispersion has the form of a high-order polynomial. Extra modes may also be introduced by some related physical phenomena, such as the specular reflection of x-rays from the crystal surface where the glancing angle of the incident beam is less than one or two degrees (Sect. 5.4.1). Nevertheless, without extremely asymmetric reflections for a thick crystal in an N-beam case, the relation (4.181) holds as a general rule for determining the number of permitted modes of wave propagation. For a thin crystal where the product of the linear absorption coefficient μ and the crystal thickness t is less than one, the $2N$ possible modes should be included in the dynamical calculation.

4.6 Absorption

It is well known that the absorption of x-rays in crystals results mainly from the inelastic scattering of x-rays by atoms, including photoelectric absorption, the Compton effect and thermal diffuse scattering − the ineleastic scattering from phonons. However, in most cases of dynamical diffraction, photoelectric absorption dominates over the other two effects by two order of magnitude. As discussed in Sect. 4.1, the fundamental equation has been derived with the condition that only the dipole term be included in the susceptibility. The contributions from quadrupole and higher-order multipoles are negligibly small. In this section, the expression for photoelectric absorption due to the dipole term is derived from the fundamental equation.

We rewrite the fundamental equation (4.36) for the component wavefield $E_{\sigma G}$:

$$(\chi_{O-O} - 2\varepsilon_G)E_{\sigma G} = - \sum_{\sigma M \neq \sigma G}' \chi_{G-M}(\hat{\sigma}_M \cdot \hat{\sigma}_M)E_{\sigma M}, \tag{4.182}$$

where σM denotes all σ's and π's except for σG. Multiplying by $E_{\sigma G}^*$, the complex conjugate of $E_{\sigma G}$, on both sides of (4.182) and summing over all σ_G involved, we obtain

$$\sum_{\sigma G}(\chi_{O-O} - 2\varepsilon_G)|E_{\sigma G}|^2 = - \sum_{\sigma G}\sum_{\sigma M \neq \sigma G}' \chi_{G-M}(\hat{\sigma}_G \cdot \hat{\sigma}_M)(E_{\sigma G}^* E_{\sigma M}). \tag{4.183}$$

Since the imaginary part of $2\varepsilon_G$ is proportional to the absorption coefficient, we shall take the imaginary part of each term in (4.183) and rewrite the equation for the imaginary parts only:

$$\sum_{\sigma G}(\chi_{O-O}^i - 2\varepsilon_G^i)|E_{\sigma G}|^2 = - \sum_{\sigma G}\sum_{\sigma M \neq \sigma G}' (\hat{\sigma}_G \cdot \hat{\sigma}_M)\chi_{G-M}^i(E_{\sigma G}^* E_{\sigma M}). \tag{4.184}$$

From (4.44) and (4.87),

$$2\varepsilon_G^i = -2\delta_i \gamma_G. \tag{4.185}$$

The imaginary part of the eigenvalue can then be written as

$$-2\delta_i \frac{\sum\limits_{\sigma G} \gamma_G |E_{\sigma G}|^2}{\chi^i_{0-O} \sum\limits_{\sigma G} |E_{\sigma G}|^2} = 1 + \frac{\sum\limits_{\sigma G} \sum\limits_{\sigma M \neq \sigma G}{}' (\hat{\sigma}_G \cdot \hat{\sigma}_M) \chi^i_{G-M}(E^*_{\sigma G} E_{\sigma M})}{\chi^i_{0-O} \sum\limits_{\sigma G} |E_{\sigma G}|^2}.$$

(4.186)

According to the expression (4.45),

$$\mu = -4\pi k \delta_i = \mu_0 \frac{\sum\limits_{\sigma G} |E_{\sigma G}|^2}{\sum\limits_{\sigma G} \gamma_G |E_{\sigma G}|^2}$$

$$\times \left[1 + \frac{\sum\limits_{\sigma G} \sum\limits_{\sigma M \neq \sigma G}{}' (\hat{\sigma}_G \cdot \hat{\sigma}_M) \left(\frac{\chi^i_{G-M}}{\chi^i_{0-O}} \right) E^*_{\sigma G} E_{\sigma M}}{\sum\limits_{\sigma G} |E_{\sigma G}|^2} \right],$$

(4.187)

where $\mu_0 = -2\pi k \chi^i_{0-O}$. The absorption coefficient along the direction of the transport velocity v_e is then

$$\mu_e = \frac{\mu_0 c}{|v_e|} \left[1 + \frac{\sum\limits_{\sigma G} \sum\limits_{\sigma M \neq \sigma G}{}' (\hat{\sigma}_G \cdot \hat{\sigma}_M) \left(\frac{\chi^i_{G-M}}{\chi^i_{0-O}} \right) E^*_{\sigma G} E_{\sigma M}}{\sum\limits_{\sigma G} |E_{\sigma G}|^2} \right],$$

(4.188)

where v_e is defined in (4.114).

It should be noted that (4.188) is valid for non-degenerate cases. For degenerate cases, the wavefields of the degenerate mode should be represented by a linear combination of wavefields satisfying the fundamental equation. Equation (4.188) is again valid for these final forms of the wavefields.

The absorption coefficient defined in (4.187) is the linear absorption coefficient of a given mode j in the direction normal to the entrance crystal surface. This coefficient can be considered to be composed of parts referring to the normal absorption μ_0 and abnormal absorption correction $\Delta\mu_j$, i.e.,

$$\mu(j) = \mu_0 + \Delta\mu_j, \quad \text{where}$$

(4.189)

$$\Delta\mu_j = 2\pi k y^i_1(j) = 2\pi k [2\varepsilon_i(j) - \chi^i_{0-O}].$$

(4.190)

Since the trace of the matrix Φ in (4.41) is zero, then

$$\sum\limits_j [2\varepsilon_i(j) - \chi^i_{0-O}] = 0.$$

(4.191)

This means

$$\sum\limits_j \Delta\mu_j = 0,$$

(4.192)

and the normal absorption coefficient is proportional to the averaged sum of the eigenvalues:

$$\mu_0 = 2\pi k \frac{\sum\limits_{i=1}^{N_{\mathrm{p}}} 2\varepsilon_i(j)}{N_{\mathrm{p}}}. \tag{4.193}$$

Of particular interest is the minimum absorption coefficient, defined as

$$\mu_{\min} = -2\pi k \, \mathrm{Min}[2\varepsilon_i(1), 2\varepsilon_i(2), \ldots, 2\varepsilon_i(N_{\mathrm{p}})]. \tag{4.194}$$

In the case of the two-beam transmission reflection G,

$$\mu_{\min} = \frac{\mu_0}{\cos\theta_G}(1-\Omega), \quad \text{where} \tag{4.195}$$

$$\Omega = \frac{\chi^i_{G-O}}{\chi^i_{O-O}} \simeq e^{-M}. \tag{4.196}$$

The quantity Ω is the Debye-Waller factor [4.22, 23].

In a symmetric two-beam Bragg reflection, the presence of total reflection divides the angular setting of the crystal into three ranges. Range II is the total reflection region in which the reflected wave has a maximum intensity. Ranges I and III are the regions beyond and behind the total reflection region [4.24]. The minimum absorption is therefore different in different regions. This is a well-known fact in two-beam dynamical diffraction [4.25]. We shall only present the results. A derivation can be made by looking for the eigenvalues of (4.36) and considering the solution valid for the imaginary parts of the eigenvalues in these three ranges. Further details can be found in [4.6, 24, 26].

The linear absorption coefficient μ in range III is smaller than the ordinary value μ_0. μ approaches μ_0 asymptotically as the angle of incidence θ decreases from θ_G towards lower angles. The minimum absorption is

$$\mu_{\min}^{\mathrm{III}} = \frac{\mu_0}{\gamma_0}\sqrt{1-\Omega\cos^2\Psi_G}, \tag{4.197}$$

where Ψ_G is the phase angle of the structure factor F_G. In range I, μ is a monotonically decreasing function of θ. μ is greater than μ_0 and approaches μ_0 as θ increases. The absorption in the total reflection region is always much greater than μ_0, due to the primary extinction caused by the total reflection from the perfect crystal lattice. The maximum value $\mu_{\max}$, instead of $\mu_{\min}$, is

$$\mu_{\max}^{\mathrm{II}} = \frac{2\pi C|\chi_{G-O}|}{\lambda\sin\theta_G}, \tag{4.198}$$

at the center of the range of total reflection. $\mu_{\max}^{\mathrm{II}}$ is usually an order of magnitude larger than μ_0 for a strong reflection.

4.7 Boundary Conditions

We have discussed in Sect. 4.3.2 the boundary condition for matching the wavevectors at the crystal boundary so that they conserve momentum. The directions of the accommodations (eigenvalues) are thus defined. For the eigenvectors the fundamental equation, as described before, gives only the amplitude ratios between the component waves. The ratios can be expressed in terms of the cofactors (Sect. 4.3.1).

To determine the actual amplitudes, the wavefields inside and outside the crystal should be matched at the crystal boundary. As usual in classical electromagnetic theory, the boundary conditions consist of (i) the continuity of the tangential components of the electric and magnetic fields and (ii) the continuity of the normal components of the electric displacements at the boundary. These lead to the equality of the electric fields inside and outside the crystal at the boundary, i.e.,

$$E(\text{inside}) = E'(\text{outside}) , \tag{4.199}$$

as long as the polarizability is small − in x-ray cases, it is of the order of 10^{-5}.

Let us consider a general N-beam case consisting of N_T transmission reflections, $O, L_1, \ldots, L_{N_T-1}$, and N_R Bragg reflections, $M_1, \ldots, M_{N_R}$. The incident beam is O. The boundary conditions are, according to (4.199),

$$\sum_j \alpha_j E_{\sigma O}(j) \exp[-2\pi i(K_O(j) \cdot r_e)] = E'_{\sigma O} \exp(-2\pi i k_O \cdot r_e) ,$$

$$\sum_j \alpha_j E_{\sigma L}(j) \exp(-2\pi i K_L(j) \cdot r_e) = 0 \quad \text{for} \quad L = L_1, \ldots, L_{N_T-1}, \tag{4.200}$$

$$\sum_j \alpha_j E_{\sigma M}(j) \exp(-2\pi i K_M(j) \cdot r_o) = 0 \quad \text{for} \quad M = M_1, \ldots, M_{N_R} ,$$

where r_e and r_o are the vectors denoting the entrance and exit surfaces of the crystal. Similarly, (4.200) holds for the π-component waves, πO, πL and πM. The constants α_j are the matching coefficients. $E'_{\sigma O}$ is the wavefield amplitude of the incident wave. Because of the boundary conditions for the wavevectors, the relations between k_O, K_O and K_L are

$$k_O \cdot r_e = K_O(j) \cdot r_e = K_L(j) \cdot r_e , \tag{4.201}$$

for all modes j. If the crystal is a plane-parallel plate of thickness T,

$$K_M(j) \cdot r_o = k \delta(j) T . \tag{4.202}$$

Substituting (4.201, 202) into (4.200), we obtain

$$\sum_j \alpha_j E_{\sigma O}(j) = E'_{\sigma O} ,$$

$$\sum_j \alpha_j E_{\sigma L}(j) = 0 \,, \tag{4.203}$$

$$\sum_j \alpha_j E_{\sigma M}(j) \exp[-2\pi i k \delta(j) T] = 0 \,.$$

By introducing the parameters $\lambda_\sigma(j)$ and $\lambda_\pi(j)$, where

$$\lambda_\sigma(j) = \alpha_j E_{\sigma O}(j)/\mathrm{Cof}(\sigma O, \sigma O)_j \,, \tag{4.204}$$

$$\lambda_\pi(j) = \alpha_j E_{\pi O}(j)/\mathrm{Cof}(\pi O, \pi O)_j \,,$$

(4.203) can be expressed in terms of the cofactors:

$$\sum_j \lambda_\sigma(j) \,\mathrm{Cof}(\sigma G, \sigma O)_j V_G(j) = E'_{\sigma O} \delta_{\sigma O, \sigma G} \,, \tag{4.205}$$

$$\sum_j \lambda_\pi(j) \,\mathrm{Cof}(\sigma G, \pi O)_j V_G(j) = E'_{\pi O} \delta_{\pi O, \sigma G} \,,$$

where σG can be any one of the σ's and π's, and

$$V_G(j) = \begin{cases} 1 & \text{for} \quad G = O \text{ and } L \text{ (Laue)} \,, \\ \exp[-2\pi i k \delta(j) T] & \text{for} \quad G = M \text{ (Bragg)} \,. \end{cases} \tag{4.206}$$

Equation (4.205) is a set of $4N$ linear equations for $4N$ unknown λ's. The purpose of introducing the two λ's is to take care of the correlation between the σ and π components. Physically speaking, this correlation results from the excitations of σ wavefields and π wavefields by the incident $E'_{\sigma O}$ and $E'_{\pi O}$, respectively. In other words, $E'_{\sigma O}$ and $E'_{\pi O}$ can excite all σ and π waves. The excited σG wave of a given mode j is therefore proportional to the sum of $\lambda_\sigma \mathrm{Cof}(\sigma G, \sigma O)$ and $\lambda_\pi \mathrm{Cof}(\sigma G, \sigma O)$. The first term represents the excitation of the σG wave by $E'_{\sigma O}$ and the second term, the excitation of the σG wave by $E'_{\pi O}$. Thus, the wavefield of the G reflection can be written in the form

$$\begin{aligned} E_G(r) &= \sum_{j=1}^{2N} \{\hat{\sigma}_G \mathrm{Cof}(\sigma G, \sigma O)_j [\lambda_\sigma(j) + \lambda_\pi(j)] \\ &\quad + \hat{\pi}_G \mathrm{Cof}(\pi G, \pi O)_j [\lambda_\sigma(j) + \lambda_\pi(j)]\} \exp[-2\pi i K_G(j) \cdot r] \\ &= \sum_{j=1}^{2N} E_G(j) \exp[-2\pi K_G(j) \cdot r] \end{aligned} \tag{4.207}$$

for all G's inside the crystal,

$$E_G^t(r) = \exp(-2\pi i k_G \cdot r) \sum_{j=1}^{2N} E_G(j) \exp[-2\pi i k \delta(j) T] \tag{4.208}$$

for $G(=O$ and $L)$ waves emerging from the crystal through the exit surface and

$$E_G^{r}(r) = \exp(-2\pi i k_G \cdot r) \sum_{j=1}^{2N} E_G(j) \tag{4.209}$$

for $G(=M)$ waves reflected from the entrance surface into empty space. The coefficients of transmission T_L and reflection R_M are then determined as follows:

$$T_L = \frac{\langle S_t \cdot \hat{K}_L \rangle}{\langle S_o \cdot \hat{k}_O \rangle_{T=0}} = \frac{\gamma_L}{\gamma_O} \frac{\left| \sum_j E_L(j) \right|^2}{|E'_{\sigma O}|^2 + |E'_{\pi O}|^2},$$

$$T_O = \frac{\left| \sum_j E_O(j) \right|^2}{|E'_{\sigma O}|^2 + |E'_{\pi O}|^2}, \tag{4.210}$$

$$R_M = \frac{\langle S_r \cdot \hat{k}_M \rangle}{\langle S_o \cdot \hat{k}_O \rangle_{T=0}} = \frac{\gamma_M}{\gamma_O} \frac{\left| \sum_j E_M(j) \right|^2}{|E'_{\sigma O}|^2 + |E'_{\pi O}|^2},$$

where S_r and S_t are the Poynting vectors, defined in (4.93), in the empty spaces above the entrance crystal surface and below the exit surface, respectively. S_o is the Poynting vector of the incident wave. The corresponding reflected intensities are

$$I_G = \frac{\left| \sum_j E_G(j) \right|^2}{|E'_{\sigma O}|^2 + |E'_{\pi O}|^2}. \tag{4.211}$$

If the crystal is a semi-infinitely thick plate, the third boundary condition in (4.200) is excluded. This signifies as discussed in Sect. 4.5.2, the fact that the total number N_p of permitted modes reduces from $2N$ to $2(N-N_R)$. In this case, an additional boundary condition for the non-zero wavefield amplitudes, $E'_{\sigma M}$ and $E'_{\pi M}$, of the waves M reflected at the entrance surface of the crystal, should be considered:

$$\sum_j^{N_p} \alpha_j E_{\sigma M}(j) = E'_{\sigma M}, \qquad \sum_j^{N_p} \alpha_j E_{\pi M}(j) = E'_{\pi M}, \tag{4.212}$$

for all M's. By taking into account the excitation correlation between σ and π, $E'_{\sigma M}$ and $E'_{\pi M}$ can be decomposed into four components, i.e., $E_{\sigma M}(\sigma)$, $E_{\sigma M}(\pi)$, $E_{\pi M}(\sigma)$ and $E_{\pi M}(\pi)$. The argument inside the parentheses indicates the source of excitation. For example, $E_{\sigma M}(\pi)$ means the excitation of the σM wave by the incident wave $E'_{\pi O}$. The boundary conditions now become

$$\sum_{j=1}^{N_p} \lambda_\sigma(j) \, \mathrm{Cof}(\sigma G, \sigma O)_j = E'_{\sigma O} \delta_{\sigma O, \sigma G},$$

$$\sum_{j=1}^{N_p} \lambda_\pi(j) \, \mathrm{Cof}(\sigma G, \pi O)_j = E'_{\pi O} \delta_{\pi O, \sigma G}, \tag{4.213}$$

for $G = O$ and L, and

$$\sum_{j=1}^{N_p} \lambda_\sigma(j) \, \mathrm{Cof}(\sigma M, \sigma O)_j = E_{\sigma M}(\sigma) \,,$$

$$\sum_{j=1}^{N_p} \lambda_\sigma(j) \, \mathrm{Cof}(\pi M, \pi O)_j = E_{\pi M}(\sigma) \,,$$

$$\sum_{j=1}^{N_p} \lambda_\pi(j) \, \mathrm{Cof}(\sigma M, \sigma O)_j = E_{\sigma M}(\pi) \,,$$

$$\sum_{j=1}^{N_p} \lambda_\pi(j) \, \mathrm{Cof}(\pi M, \pi O)_j = E_{\pi M}(\pi) \,,$$

(4.214)

for $G = M$. There are $4(N - N_R)$ linear equations in (4.213) and $4N_R$ equations in (4.214). These $4N$ equations are soluble for $4N$ unknowns consisting of $4(N - N_R)$ λ's and $4N_R$ E_M's.

The expressions for the wavefields of the transmitted waves remain the same as in (4.208) except that the sum should be taken over the N_p permitted modes. The reflected wavefields can be written as

$$E_M^r(r) = \{\hat{\sigma}_M[E_{\sigma M}(\sigma) + E_{\sigma M}(\pi)] + \hat{\pi}_G[E_{\pi M}(\sigma) + E_{\pi M}(\pi)]\}$$

$$\times \exp(-2\pi i k_G \cdot r) \,.$$

(4.215)

The corresponding reflected intensity is therefore

$$I_M^r = \frac{|E_{\sigma M}(\sigma) + E_{\sigma M}(\pi)|^2 + |E_{\pi M}(\sigma) + E_{\pi M}(\pi)|^2}{|E'_{\sigma O}|^2 + |E'_{\pi O}|^2} \,.$$

(4.216)

Let us now consider N-beam Borrmann diffraction for which all the diffracted beams are transmitted through the crystal. This case is actually a special case of the general diffraction situation discussed previously because no reflections of the Bragg type are involved. The boundary conditions are the same as (4.205), except that V_G is always unity. The wavefields have exactly the same expression as in (4.207) and (4.208).

A more direct derivation for the wavefields in N-beam Borrmann diffraction in non-absorbing crystals can be carried out starting from the fundamental equation of the wavefield, (4.36). We rewrite the equations for the wavefield of mode j_p and the complex-conjugate field of mode j_q:

$$[\chi_{0-O} - 2\varepsilon_G(j_p)]E_{\sigma G}(j_p) = - \sum_{\sigma L \neq \sigma G}' \chi_{G-L}(\hat{\sigma}_G \cdot \hat{\sigma}_L)E_{\sigma L}(j_p) \,,$$

(4.217)

$$[\chi^*_{0-O} - 2\varepsilon_G^*(j_q)]E^*_{\sigma G}(j_q) = - \sum_{\sigma L \neq \sigma G}' \chi^*_{G-L}(\hat{\sigma}_G \cdot \hat{\sigma}_L)E^*_{\sigma L}(j_q) \,.$$

Multiplying the first equation by $E_{\sigma G}^*(j_q)$ and the second equation by $E_{\sigma G}(j_p)$ and subtracting one from the other, we obtain

$$(2\,\varepsilon_G(j_p) - 2\,\varepsilon_G^*(j_q)]\,E_{\sigma G}(j_p)\,E_{\sigma G}^*(j_q) = 0 , \tag{4.218}$$

where we assume $\chi_{G-L} = \chi_{G-L}^*$ as in the case of non-absorbing crystals. Referring to (4.87b), (4.218) becomes in the non-degenerate case,

$$\sum_{\sigma G} \gamma_G E_{\sigma G}(j_p) E_{\sigma G}^*(j_q) = 0 , \tag{4.219}$$

for $j_p \neq j_q$. In the N-beam transmission case, γ_G is constant for all the G reflections. This leads to the orthogonality of the eigenvectors, i.e.,

$$\sum_{\sigma G} E_{\sigma G}(j_p) E_{\sigma G}^*(j_q) = 0 , \tag{4.220}$$

for $j_p \neq j_q$.

Returning to the boundary conditions for transmission, i.e., the first two equations of (4.203), we rewrite them in the form

$$\sum_{j_p=1}^{2N} \alpha_{j_p} E_{\sigma G}(j_p) = E'_{\sigma O} \delta_{\sigma O, \sigma G} . \tag{4.221}$$

Multiplying (4.221) by $E_{\sigma G}^*(j_p)$ and summing over all σG, we obtain

$$\sum_{\sigma G} \alpha_{j_p} E_{\sigma G}(j_p) E_{\sigma G}^*(j_p) = E'_{\sigma O} E_{\sigma O}^*(j_p) . \tag{4.222}$$

By introducing the parameters λ as given in (4.204) and using (4.42, 96), (4.222) provides the solutions for λ:

$$\lambda_\sigma(j_p) = \frac{E'_{\sigma O}}{\sum\limits_{\sigma G} \mathrm{Cof}(\sigma G, \sigma G)_{j_p}} ,$$

$$\lambda_\pi(j_p) = \frac{E'_{\pi O}}{\sum\limits_{\sigma G} \mathrm{Cof}(\sigma G, \sigma G)_{j_p}} . \tag{4.223}$$

The wavefields are then obtained according to (4.207, 208).

It is worthwhile stressing that the approach for N-beam transmission cannot be applied to cases involving Bragg reflections. Since the direction cosine γ_G in (4.219) is positive for a transmission and negative for a Bragg reflection, the orthogonality condition (4.220) cannot be obtained.

More rigorously speaking, the difference between Laue and Bragg cases can be analyzed through the eigenvalue equation (4.36). For simplicity, let us redefine the components of the eigenvectors as $A_{\sigma G}$,

$$A_{\sigma G} = \sqrt{\gamma_G} E_{\sigma G} , \tag{4.224}$$

such that the eigenvalue equation (4.36) takes the form

$$\underset{\sim}{S} A = \lambda A \, , \tag{4.225}$$

where the matrix λ has diagonal elements δ and zeros otherwise. The matrix $\underset{\sim}{S}$ is the scattering matrix with its elements defined as

$$S_{\sigma G, \sigma M} = -\frac{1}{2} \frac{1}{\sqrt{\gamma_G \gamma_M}} \chi_{G-M} (\hat{\sigma}_G \cdot \hat{\sigma}_M) \tag{4.226}$$

for $G \neq M$, and

$$S_{\sigma G, \sigma G} = -\frac{1}{2} \frac{1}{\gamma_G} (\chi_{O-O} - \alpha_G) \, . \tag{4.227}$$

The subscripts σ_G and σ_M indicate the contributions from both σ and π components of the corresponding wavefields. α_G has been defined in (4.87a).

For Laue cases, because the direction cosines γ are positive, the matrix $\underset{\sim}{S}$ is Hermitian. This leads to the relation (4.220). For a Bragg reflection involved in a Bragg-type multiple diffraction, γ is negative. The corresponding $\underset{\sim}{S}$ is non-Hermitian and the wavefield amplitude $A_{\sigma G}$ is equal to $i\sqrt{|\gamma_G|} E_{\sigma G}$. This fact describes the total reflection in Bragg cases. This special characteristic provides favorable conditions for Bragg-type multiple diffraction to exhibit the information about reflection phases via the asymmetry of its intensity profile. We shall come to this point in Chap. 7.

4.8 Excitation of Mode and Excitation of Beam

The amount of energy associated with a given mode or with a given reflection is of physical importance when we consider the distributions of the incident energy to the existing modes and to the diffraction directions. This amount of energy can be defined in terms of the wavefield amplitudes determined in Sect. 4.7:

$$\mathrm{Ex}_G(j) = \frac{|E_{\sigma G}^*(j) E_{\sigma G}(j) + E_{\sigma G}^*(j) E_{\pi G}(j)|}{|E_{\sigma O}'|^2 + |E_{\pi O}'|^2} \, , \tag{4.228a}$$

$$\mathrm{Ex}(G) = \sum_j \mathrm{Ex}_G(j) \, , \tag{4.228b}$$

$$\mathrm{Ex}(j) = \sum_G \mathrm{Ex}_G(j) \, . \tag{4.228c}$$

$\mathrm{Ex}_G(j)$ is the excitation of mode j for reflection G. The quantity $\mathrm{Ex}(j)$ is called the excitation of mode j, and $\mathrm{Ex}(G)$ the excitation of beam. Considering the definition of Poynting vectors (4.93), we can write the Poynting vector S in terms of the excitation of beam, i.e.,

$$\frac{S}{|S_O|} = \sum_G \hat{k}_G \text{Ex}(G) \exp(-\mu T)$$

$$= \sum_G \frac{S_G}{|S_O|} \exp(-\mu T), \quad \text{where} \tag{4.229}$$

$$|S_O| = \frac{c}{8\pi} (|E'_{\sigma O}|^2 + |E'_{\pi O}|^2). \tag{4.230}$$

Hence the excitation of beam G is actually the modulus of the Poynting vectors of that beam. In an N-beam transmission case,

$$\sum_j \text{Ex}(j) = \sum_G \text{Ex}(G) = 1. \tag{4.231}$$

However, in the case involving Bragg reflections, only the relation

$$\sum_G \left(\frac{\gamma_G}{\gamma_O} \right) \text{Ex}(G) = 1 \tag{4.232}$$

holds. Equation (4.232) is consistent with the fact that the sum of transmittance and reflectance (4.210) is unity, i.e.,

$$\sum_{L=0}^{L_{N_T}-1} T_L + \sum_{M=1}^{M_{N_R}} R_M = 1. \tag{4.233}$$

This relation also indicates that the total energy is conserved. The excitation of mode can be, on the other hand, considered as a weighting factor or probability density, which is a measure of the participation of the mode in the diffraction process. Since each mode has its own value μ_j of absorption, the effective, or expectation, value of the overall absorption is equal to

$$\mu_{\text{eff}} = \langle \mu \rangle = \frac{\sum_j \mu_j \text{Ex}(j)}{\sum_j \text{Ex}(j)}. \tag{4.234}$$

4.9 Intensity of Wavefield (Standing-Wave) in Crystal

Modes of wave propagation have been discussed in Sect. 4.5 with respect to the direction of the wavefields and the relative amplitudes among the wavefields. However, the actual absolute wavefield amplitude cannot be obtained until the boundary conditions are taken into account. After the equations for the boundary conditions have been solved, the excitation of mode can then be calculated quantitatively. Unfortunately excitation is a physically immeasurable quantity. We shall therefore introduce a quantity, related to the excitation of mode, which is experimentally measurable. This is the intensity of the total wavefield associated with a given mode j in the crystal, defined as

$$I_F(j) = \frac{|E(r,t)_j|^2}{|E'_O(r,t)|^2}, \tag{4.235}$$

where $E(r,t)_j$ is the vector sum of all the wavefields, associated with the mode j, which usually form standing waves within the crystal:

$$E(r,t)_j = \sum_{\sigma G} E_{\sigma G}(j) \exp[i\omega t - 2\pi i K_G(j) \cdot r], \tag{4.236}$$

and $E'_O(r,t)$ is the incident wavefield, such that

$$E'_O(r,t) = (\hat{\sigma}_O E'_{\sigma O} + \hat{\pi}_O E'_{\pi O}) \exp[i\omega t - 2\pi i k_O \cdot r]. \tag{4.237}$$

The term $|E'_O(r,t)|^2$ serves as the normalization constant which is equal to $|E'_{\sigma O}|^2 + |E'_{\pi O}|^2$. By simple manipulation, (4.235) can be written

$$I_F(j) = \frac{\exp[-4\pi k \delta_i(j)\hat{n}_e \cdot r]}{|E'_{\sigma O}|^2 + |E'_{\pi O}|^2}$$

$$\times \left\{ \sum_{\sigma G} |E_{\sigma G}(j)|^2 + \sum_{\substack{\sigma G \;\; \sigma G' \\ G \neq G'}} E_{\sigma G}(j) E^*_{\sigma G'}(j) \exp[-2\pi i(g_G - g_{G'}) \cdot r] \right\}, \tag{4.238}$$

where the sums are taken over all σ's and π's, including σO and πO. If we consider separately the σ and π components, (4.238) can be expressed in terms of the wavefield-amplitude ratios:

$$I_F(j) = I_F^\sigma(j) + I_F^\pi(j) + I_F^{\sigma\pi}(j), \quad \text{where} \tag{4.239}$$

$$I_F^\sigma(j) = \frac{|E_{\sigma O}(j)|^2}{|E'_O|^2} \left\{ 1 + \sum_G \left| \frac{E_{\sigma G}(j)}{E_{\sigma O}(j)} \right|^2 + 2\,\text{Re}\left[\sum_{G \neq O} \left(\frac{E^*_{\sigma G}(j)}{E^*_{\sigma O}(j)} (\hat{\sigma}_O \cdot \hat{\sigma}_G) \right. \right. \right.$$

$$\left. \left. + \frac{E^*_{\pi G}(j)}{E^*_{\sigma O}(j)} (\hat{\sigma}_O \cdot \hat{\pi}_G) \right) \exp(2\pi i g_G \cdot r) \right] + \frac{|E_{\sigma O}(j)|^2}{|E'_O|^2}$$

$$\times \sum_{G \neq O} \sum_{G' \neq O} \frac{E_{\sigma G}(j)}{E_{\sigma O}(j)} \frac{E^*_{\sigma G'}(j)}{E^*_{\sigma O}(j)} \exp[-2\pi i(g_G - g_{G'}) \cdot r]$$

$$\times \exp[-4\pi k \delta_i(j)\hat{n}_e \cdot r] \bigg\}, \tag{4.240a}$$

$$I_F^{\sigma\pi}(j) = 2\,\mathrm{Re}\left\{\frac{|E_{\sigma O}(j)|^2}{|E_O'|^2}\sum_G\sum_{G'}\frac{E_{\sigma G}(j)E_{\pi G'}^*(j)}{E_{\sigma O}(j)E_{\pi O}^*(j)}\exp[-2\pi i(\mathbf{g}_G-\mathbf{g}_{G'})\cdot\mathbf{r}]\right\}$$

$$\times\exp[-4\pi k\delta_i(j)\hat{n}_e\cdot\mathbf{r}]\,,\tag{4.240b}$$

and $I_F^\pi(j)$ has a similar form to $I_F^\sigma(j)$, with the σ's and π's interchanged.

Equations (4.238–240) are functions of the wavefield-amplitude ratio, which has a quite different form in the reflection case than for transmission because of the different characteristics associated with the matrix Φ in the two cases. The function I_F therefore depends heavily on whether the G reflections involved are transmissions or reflections. In view of this difference, we shall review two-beam Bragg reflection and Laue transmission.

In a two-beam, O and G, case, (4.238) is simplified for a singly polarized wave to

$$I_F(j) = \frac{|E_{\sigma O}|^2}{|E_{\sigma O}'|^2}[1+|Y_j|^2+2\,\mathrm{Re}\{Y_j\exp(\mathbf{g}_G\cdot\mathbf{r})\}]$$

$$\times\exp[-4\pi k\delta_i(j)\hat{n}_e\cdot\mathbf{r}]\,,\quad\text{where}\tag{4.241}$$

$$Y_j = \frac{E_{\sigma G}(j)}{E_{\sigma O}(j)}\,.\tag{4.242}$$

For simplicity, let us first consider a singly polarized two-beam symmetric Laue transmission at the exact two-beam diffraction position. The wavefield-amplitude ratios for the two modes, say modes 1 and 4, associated with the σ polarization (Sect. 4.5.1), are, according to [Ref. 4.26, Eq. (30.13)],

$$Y_1 = |X|\exp(v)\,,\qquad Y_4 = -|X|\exp(-v)\,,\tag{4.243}$$

where X is defined as

$$X = (-1)^\tau\frac{|\gamma_O|}{|\gamma_G|}\exp(i\Psi_G)\,,\tag{4.244}$$

with $\tau = 0$. The parameter v depends on the parameter p from [4.26] such that

$$p = \sinh v\,,\tag{4.245}$$

where p is directly proportional to the angle of incidence. By substituting (4.243) into (4.241), the intensity I_F of mode 1 then takes the following form:

$$I_F(1) = \tfrac{1}{2}[1+|X|\exp(v)+2|X|\exp(v)\cos(\mathbf{g}_G\cdot\mathbf{r})]$$

$$\times\exp[-4\pi k\delta_i(j)\hat{n}_e\cdot\mathbf{r}]\,.\tag{4.246}$$

If we restrict the position vector $\mathbf{r}$ to the atomic planes, i.e.,

$$g_G \cdot r = 2\pi m, \tag{4.247}$$

with m an integer, the intensity of the wavefield at the atomic planes is

$$I_F(1) = \tfrac{1}{2}(1 + |X|)^2$$
$$\propto (|E_{\sigma O}(1)| + |E_{\sigma G}(1)|)^2. \tag{4.248}$$

Similarly, for mode 4,

$$I_F(4) = \tfrac{1}{2}(1 - |X|)^2$$
$$\propto (|E_{\sigma O}(4)| - |E_{\sigma G}(4)|)^2. \tag{4.249}$$

$I_F(1)$ has its maximum value, 1, at the atomic planes while $I_F(4)$ is zero, the minimum intensity. This is consistent with (4.121).

For a two-beam symmetric Bragg reflection, the wavefield-amplitude ratio has a different value in different angular ranges. Using [Ref. 4.26, Sect. 30], we obtain the following wavefield-amplitude ratios:

In range (I), $p = -\cosh v < -1$, $Y_1 = -X \exp(-v)$.
In range (II), $-1 < p = \cos v < 1$, $Y_1 = X \exp(-iv)$.
In range (III), $p = \cosh v > 1$, $Y_1 = X \exp(-v)$.

Note that $E_{\sigma O}(1) = E'_{\sigma O}$. The factor X in this case is given in (4.244) with $\tau = 1$. The corresponding intensities for $\Psi_G = \pi$ at the atomic planes G are

$$I_F(1) = \tfrac{1}{2}[1 + |X|\exp(-v)]^2$$
$$\propto (|E_{\sigma O}(1)| + |E_{\sigma G}(1)|)^2 \tag{4.250a}$$

for range I;

$$I_F(1) = \tfrac{1}{2}[1 - |X|\exp(-v)]^2$$
$$\propto (|E_{\sigma O}(1)| - |E_{\sigma G}(1)|)^2 \tag{4.250b}$$

for range III;

$$I_F(1) = \tfrac{1}{2}(1 + |X|^2 - 2|X|\cos v) \tag{4.250c}$$

for range II. Note that there is only one mode allowed for the two-beam Bragg reflection of singly polarized incident radiation in an absorbing crystal. The intensity of the wavefield in the Bragg case is quite different from that in the Laue case. In the Bragg case, $E_{\sigma O}(1)$ and $E_{\sigma G}(1)$ are out of phase in range III. For the crystal setting with the angle of incidence smaller than the Bragg angle θ_B (range III), I_F is a minimum at $p = +1$ because $E_{\sigma O}(1) = E_{\sigma G}(1)$. When the crystal is set in the total reflection region, range II, the intensity I_F increases as the angle of incidence increases. In the range I, the intensity is maximum at

$p = -1$ since $E_{\sigma O}(1)$ and $E_{\sigma G}(1)$ are in phase and $|E_{\sigma O}(1)| = |E_{\sigma G}(1)|$. On the contrary, in the Laue case, whether the intensity I_F is proportional to $(|E_{\sigma O}(1)| + |E_{\pi O}(1)|)^2$ or $(|E_{\sigma O}(1)| - |E_{\sigma G}(1)|)^2$ does not depend on the crystal setting.

This special feature of the wavefield intensity in Bragg reflections greatly influences the diffracted intensity in multiple beam cases. The asymmetry, i.e., minimum at $p = +1$ and maximum at $p = -1$, remains in the diffracted intensity of Bragg-type multiple diffraction, especially in three-beam cases. In analogy to the fact that I_F depends on the crystallographic angle Ψ_G of the G reflection in the two-beam case, the diffracted intensity of the three-beam Bragg reflection is influenced by the relative phases between the primary, the secondary and the coupling reflections. We shall return to this point in Chap. 7.

4.10 Consideration of the Spherical-Wave Nature of the Incident X-Rays

We have discussed dynamical diffraction in multi-beam cases, under the assumption that the incident x-ray is a plane monochromatic wave. As a matter of fact, the incident wave possesses a spherical coherent wavefront and has a large ray divergence. In a typical x-ray diffraction experiment, the incident beam usually has an angular divergence of a few minutes of arc, which is one order of magnitude higher than the intrinsic width of a reflection. With this relatively large divergence, the incident x-rays may be regarded as a spherical wave.

To consider this sphericity of the incident wave, *Kato* [4.27] developed the spherical-wave dynamical theory, which has been successfully used to account for the experimental investigation of Pendellösung fringes obtained from a two-beam transmission in a wedge-shaped crystal [4.28]. Although this theory is designed for the two-beam case, the same consideration should also be extended to the multi-beam case because of the same spherical-wave nature of the incident x-rays involved in all diffraction studies.

Following *Kato*'s treatment, the expression for spherical waves of a wavefield can be obtained by considering the spontaneous emission of electromagnetic radiation in empty space. The electromagnetic field far away from the source is governed by

$$E(r) = \frac{i}{k} \operatorname{curl} H(r) , \qquad (4.251)$$

where H depends solely on the current density. With the first-order approximation in r^{-1}, E and H have the form

$$H(r) = \mathrm{i}k(J \times \hat{r})\,\varPsi, \tag{4.252a}$$

$$E(r) = \mathrm{i}k[(J \times \hat{r}) \times \hat{r}]\,\varPsi, \tag{4.252b}$$

where $\varPsi$ is a scalar spherical wave, i.e.,

$$\varPsi = \frac{1}{4\pi r}\exp(\mathrm{i}kr). \tag{4.253}$$

The current density J, induced by the spontaneous emission of the photon, is a constant vector. The term $\hat{r}$ is a unit vector from the source to the observation point. The Fourier transform of $\varPsi$ propagating in a positive z direction takes the form

$$\varPsi = \frac{\mathrm{i}}{8\pi^2}\int\limits_{-\infty}^{\infty}\!\!\int \frac{1}{k_z}\exp(\mathrm{i}k \cdot r)\,dk_x dk_y. \tag{4.254}$$

Since the incident beam, in most cases, is collimated in a small solid angle, the polarization factors $J \times \hat{r}$ and $(J \times \hat{r}) \times \hat{r}$ are assumed to be constant. Furthermore, when the polarization vectors defined in Sect. 4.2 are employed as reference vectors, each decomposed wave in a reference direction can be treated as if it were a scalar wave. Under these two assumptions, the wavefield amplitude of each component wave can be written as

$$E_{\sigma G}(r) = \frac{\mathrm{i}}{8\pi}\int\limits_{-\infty}^{\infty}\!\!\int \frac{1}{k_z}e_{\sigma G}dk_x dk_y, \quad \text{where} \tag{4.255}$$

$$e_{\sigma G} = \sum_{j=1}^{N_\mathrm{p}} E_{\sigma G}(j)\exp[\mathrm{i}(K_G - k) \cdot r], \tag{4.256}$$

for all σ's, π's and G's. Equation (4.255) indicates that the spherical wave is a superposition of plane waves. The quantity $E_{\sigma G}(j)$ can be determined from the dispersion relation (4.36) and the boundary conditions (4.213, 214). For simplicity, let us define k_x along the incident direction and k_y perpendicular to k_x in the plane of incidence of one of the reflections, say reflection G. The vector k_y is then orthogonal to k_x and k_z. In two-beam cases, the wavefield amplitudes E_O and E_G are independent of k_y. The integration in (4.255) can be carried out by means of Kelvin's stationary-phase method [4.29]. The integration with respect to k_x, with the aid of a two-dimensional oblique coordinate system (with two coordinates along the K_O and K_G) leads to a Bessel function. In multi-beam cases, $E_O, E_G, E_H, \ldots$ depend on both k_x and k_y. An analytical expression cannot be obtained because of the difficulty in the integration and the lack of a general analytical expression for $E_{\sigma O}, E_{\sigma G}, \ldots$ to work with. Numerical calculation may therefore be adopted and a new three-dimensional coordinate system should be used. In addition, the interaction between the wavefields of the modes of propagation which have a common Poynting vector should be included as a constraint in the integration.

5. Approximations, Numerical Computing, and Other Approaches

This chapter is an extension of Chap. 4. In Chap. 4, we have given the general formalism for the dynamical theory of x-ray diffraction. However, analytical expressions for diffracted intensity are usually desirable to have in order to reveal the connection between diffracted intensities and various physical parameters for special circumstances. In this chapter, analytical expressions for intensity are derived, using the two-beam approximation. Calculation procedures developed on the basis of Chap. 4 are also presented here to provide full numerical information about the diffraction mechanism and to complement the two-beam approximation. A section on the quantum mechanical approach then follows, with a comparison between the classical and quantum approaches. Other phenomena, which can be described in terms of the two-beam approximation, are discussed at the end of this chapter as examples of treating radiation-solid interactions as multi-beam cases. In the following, we shall begin with the dispersion equation (4.41) for the discussion on the two-beam approximation.

5.1 Two-Beam Approximation for Three-Beam Diffraction

5.1.1. General Considerations

Analytical solution of the dispersion equation (4.41) can be obtained under special circumstances. One example is the case involving symmetric reflection. Symmetric N-beam transmissions have been discussed in Sect. 4.5.1. However, the dispersion equation for a symmetric N-beam Bragg reflection cannot be solved analytically, because the direction cosines γ are not equal, namely, the quantities 2ε are different. When the presence of a secondary reflection in a general three-beam case provides a very small perturbation to the two-beam diffraction of the incident and the primary reflection, the analytical solution can be found from the two-beam solution with modifications to include the small perturbation. We shall concentrate on these three-beam cases, since higher-order diffraction can be treated as perturbed three-beam diffraction.

We rewrite the fundamental equation similar to (4.36), for a general three-beam reflection, O, G and H:

$$2\xi_O E_{\sigma O} + P_\sigma \chi_{O-G} E_{\sigma G} + d_2 \chi_{O-G} E_{\sigma H} = 0,$$

$$2\xi_O E_{\pi O} + d_0 \chi_{O-G} E_{\pi G} + d_1 \chi_{O-H} E_{\sigma H} + d_3 \chi_{O-H} E_{\pi H} = 0,$$

$$P_\sigma \chi_{G-O} E_{\sigma O} + 2\xi_G E_{\sigma G} + d_2 \chi_{G-H} E_{\sigma H} = 0,$$

$$P_\pi \chi_{G-O} E_{\pi O} + 2\xi_G E_{\pi G} + d_1' \chi_{G-H} E_{\sigma H} + d_3' \chi_{G-H} E_{\pi H} = 0,$$ (5.1)

$$d_2 \chi_{H-O} E_{\sigma O} + d_1 \chi_{H-O} E_{\pi O} + d_2 \chi_{H-G} E_{\sigma G} + d_1' \chi_{H-G} E_{\pi G} + 2\xi_H E_{\sigma H} = 0,$$

$$d_3 \chi_{H-O} E_{\pi O} + d_3' \chi_{H-G} E_{\pi G} + 2\xi_H E_{\pi H} = 0,$$

where

$$P_\sigma = \hat{\sigma}_O \cdot \hat{\sigma}_G = 1,$$

$$P_\pi = \hat{\pi}_O \cdot \hat{\pi}_G = \cos 2\theta_G,$$

$$d_1 = \hat{\pi}_O \cdot \hat{\sigma}_H = -\sin\psi \sin(\theta_G - \alpha),$$

$$d_2 = \hat{\sigma}_O \cdot \hat{\sigma}_H = \hat{\sigma}_G \cdot \hat{\sigma}_H = \cos\psi,$$ (5.2)

$$d_3 = \hat{\pi}_O \cdot \hat{\pi}_H = \cos(\theta_G - \alpha),$$

$$d_1' = \hat{\pi}_G \cdot \hat{\sigma}_H = \sin\psi \sin(\theta_G + \alpha),$$

$$d_3' = \hat{\pi}_G \cdot \hat{\pi}_H = \cos(\theta_G + \alpha),$$

$$\hat{\sigma}_O \cdot \hat{\pi}_O = \hat{\sigma}_O \cdot \hat{\pi}_H = \hat{\pi}_O \cdot \hat{\sigma}_G = \hat{\sigma}_G \cdot \hat{\pi}_H = 0.$$

The angles θ_G and ψ are, respectively, the Bragg angle of the G reflection and the angle between the wavevector of the H reflection and the plane LOG containing the reciprocal lattice points O and G and the center of the Ewald sphere L. α is the angle between the wavevector LO of the O reflection and the plane LMH containing the wavevector of the H reflection. The plane LMH is

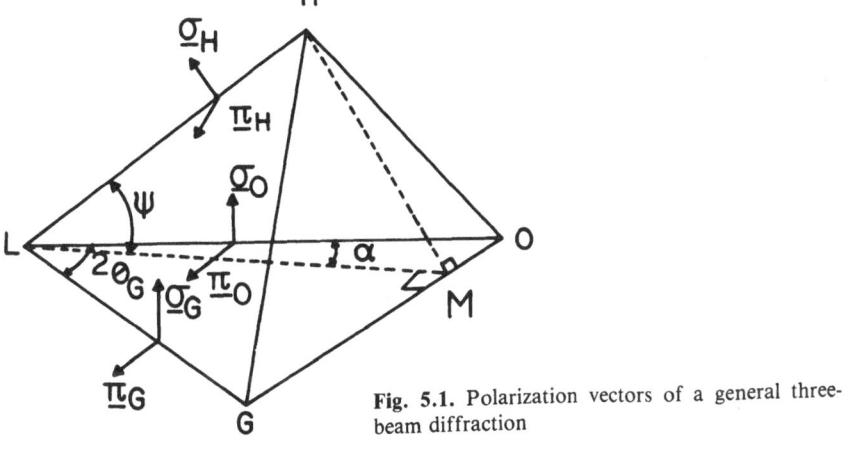

Fig. 5.1. Polarization vectors of a general three-beam diffraction

perpendicular to the LOG plane (Fig. 5.1). The polarization vectors σ_O, σ_G and π_O, π_G and π_H are perpendicular and parallel to the LOG plane, respectively. The vector σ_H is perpendicular to π_H and LH. The quantities 2ξ are defined as

$$2\xi_O = \chi_{O-O} - 2\varepsilon_O,$$
$$2\xi_G = \chi_{O-O} - 2\varepsilon_G, \tag{5.3}$$
$$2\xi_H = \chi_{O-O} - 2\varepsilon_H,$$

where the 2ε's came from (4.39). Eliminating $E_{\sigma H}$ and $E_{\pi H}$ in (5.1), we obtain the following four equations:

$$E_{\sigma O}\left(2\xi_O - \frac{d_2^2 \chi_{O-H}\chi_{H-O}}{2\xi_H}\right) - E_{\pi O}\left(\frac{d_1 d_2 \chi_{O-H}\chi_{H-O}}{2\xi_H}\right)$$
$$+ E_{\sigma G}\left(P_\sigma \chi_{O-G} - \frac{d_2^2 \chi_{O-H}\chi_{H-G}}{2\xi_H}\right) - E_{\pi G}\left(\frac{d_1' d_2 \chi_{O-H}\chi_{H-G}}{2\xi_H}\right) = 0,$$

$$E_{\sigma O}\left(P_\sigma \chi_{G-O} - \frac{d_2^2 \chi_{H-O}\chi_{G-H}}{2\xi_H}\right) - E_{\pi O}\left(\frac{d_1 d_2 \chi_{H-O}\chi_{G-H}}{2\xi_H}\right)$$
$$+ E_{\sigma G}\left(2\xi_G - \frac{d_2^2 \chi_{G-H}\chi_{H-G}}{2\xi_H}\right) - E_{\pi G}\left(\frac{d_1' d_2 \chi_{G-H}\chi_{H-G}}{2\xi_H}\right) = 0,$$

$$-E_{\sigma O}\left(\frac{d_1 d_2 \chi_{O-H}\chi_{H-O}}{2\xi_H}\right) + E_{\pi O}\left(2\xi_O - \frac{(d_1^2 + d_3^2)\chi_{O-H}\chi_{H-O}}{2\xi_H}\right) \tag{5.4}$$
$$-E_{\sigma G}\left(\frac{d_1 d_2 \chi_{O-H}\chi_{H-G}}{2\xi_H}\right) + E_{\pi G}\left(P_\pi \chi_{O-G} - \frac{(d_1 d_1' + d_3 d_3')\chi_{O-H}\chi_{H-G}}{2\xi_H}\right)$$
$$= 0,$$

$$-E_{\sigma O}\left(\frac{d_1' d_2 \chi_{H-O}\chi_{G-H}}{2\xi_H}\right) + E_{\pi O}\left(P_\pi \chi_{G-O} - \frac{(d_1 d_1' + d_3 d_3')\chi_{H-O}\chi_{G-H}}{2\xi_H}\right)$$
$$-E_{\sigma G}\left(\frac{d_1' d_2 \chi_{G-H}\chi_{H-G}}{2\xi_H}\right) + E_{\pi G}\left(2\xi_G - \frac{(d_1'^2 + d_3'^2)\chi_{G-H}\chi_{H-G}}{2\xi_H}\right) = 0.$$

In the first-order approximation, (5.4) can be written as the two sets of equations

$$E_{\sigma O}\left(2\xi_O - \frac{d_2^2 \chi_{O-H}\chi_{H-O}}{2\xi_H}\right) + E_{\sigma G}\left(P_\sigma \chi_{O-G} - \frac{d_2^2 \chi_{O-H}\chi_{H-G}}{2\xi_H}\right) = 0,$$
$$\tag{5.5a}$$
$$E_{\sigma O}\left(P_\sigma \chi_{G-O} - \frac{d_2^2 \chi_{H-O}\chi_{G-H}}{2\xi_H}\right) + E_{\sigma G}\left(2\xi_G - \frac{d_2^2 \chi_{G-H}\chi_{H-G}}{2\xi_H}\right) = 0$$

and

$$E_{\pi O}\left(2\xi_O - \frac{(d_1^2+d_3^2)\chi_{O-H}\chi_{H-O}}{2\xi_H}\right) + E_{\pi G}\left(P_\pi\chi_{O-G} - \frac{(d_1 d_1'+d_3 d_3')}{2\xi_H}\right.$$

$$\left. \cdot \chi_{O-H}\chi_{H-G}\right) = 0,$$

(5.5b)

$$E_{\pi O}\left(P_\pi\chi_{G-O} - \frac{(d_1 d_1'+d_3 d_3')\chi_{H-O}\chi_{G-H}}{2\xi_H}\right) + E_{\pi G}\left(2\xi_G - \frac{(d_1'^2+d_3'^2)}{2\xi_H}\right.$$

$$\left. \cdot \chi_{G-H}\chi_{H-G}\right) = 0.$$

Let us consider the situation where the crystal is set so that the Bragg condition for the G reflection is almost satisfied, while the reflection condition is not fulfilled for the H reflection. In this case, ξ_H is much greater than ξ_O and ξ_G, since the crystal setting is far away from the exact three-beam diffraction position, but always very close to the exact two-beam (G-reflection) position. The wavevectors, K_O and K_G inside the crystal can then be expressed approximately in terms of two-beam geometry. In Fig. 5.2, O and G are the reciprocal lattice points of the O and G reflections. Points A, T and La are the entrance point, the excitation point, and the Laue point for an incident wavevector AO, deviating by $\Delta\theta$ from the Bragg angle θ_G for the G reflection. LaA is the wavefront of the incident wave, LaM is perpendicular to AG, $LaG = MG = k(=1/\lambda)$, $LaA = k\Delta\theta$, $AT = k\delta\hat{n}_e$, where $\hat{n}_e$ is the unit vector normal to the crystal surface. The wavevectors K_O and K_G inside the crystal are

$$K_G = TG = AG - k\delta\hat{n}_e, \quad K_O = TO = AO - k\delta\hat{n}_e,$$

(5.6)

where

$$AG = AM + MG \simeq k - k\Delta\theta\sin\theta, \quad AO = k.$$

(5.7)

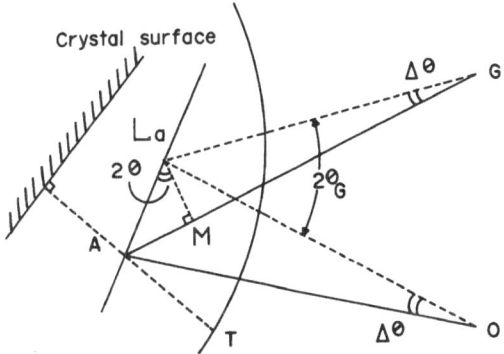

Fig. 5.2. Geometrical relation between $\Delta\theta$ and the wavevectors

Referring to (4.39, 5.3), we obtain

$$2\xi_O = \chi_{O-O} + 2\delta\gamma_O,$$
$$2\xi_G = \chi_{O-O} + 2(\Delta\theta)\sin 2\theta_B + 2\delta\gamma_G. \tag{5.8}$$

Without losing generality, let us first consider the equation (5.5) for π-component wavefields, in which rather complicated polarization factors are involved. Define

$$a_{\pi O} = \frac{d_1^2 + d_3^2}{\xi_{H\pi}},$$

$$a'_{\pi O} = \frac{d_1'^2 + d_3'^2}{\xi_{H\pi}}, \tag{5.9}$$

$$\alpha_\pi = \frac{d_1 d_1' + d_3 d_3'}{\xi_{H\pi}},$$

and the generalized susceptibilities $\chi^\pi_{GO,HO}/4\pi$ and $\chi^\pi_{OG,OH}/4\pi$:

$$\chi^\pi_{OG,OH} = P_\pi \chi_{O-G} - \frac{\alpha_\pi}{2} \chi_{O-H}\chi_{H-G},$$

$$\chi^\pi_{GO,HO} = P_\pi \chi_{G-O} - \frac{\alpha_\pi}{2} \chi_{H-O}\chi_{G-H}. \tag{5.10}$$

The quantities $\xi_{O\pi}$, $\xi_{G\pi}$ and $\xi_{H\pi}$ are used to replace ξ_O, ξ_G and ξ_H for the π-polarized wavefields. The dispersion equation obtained from (5.5b) takes the form

$$4\xi_{O\pi}\xi_{G\pi} - \xi_{O\pi}a'_{\pi O}\chi_{G-H} - \xi_{G\pi}a_{\pi O}\chi_{O-H}\chi_{H-O} + \frac{a_{\pi O}a'_{\pi O}}{4}\chi_{O-H}\chi_{H-O}$$

$$\cdot \chi_{G-H}\chi_{H-G} - \chi^\pi_{GO,HO}\chi^\pi_{OG,OH} = 0. \tag{5.11}$$

By eliminating the accommodation, δ, in (5.8), the term $\xi_{G\pi}$ can be expressed in terms of $\xi_{O\pi}$ as follows:

$$\xi_{G\pi} = \beta + \frac{\gamma_G}{\gamma_O}\xi_{O\pi}, \tag{5.12}$$

where

$$\beta = \frac{\chi_{O-O}}{2}\left(1 - \frac{\gamma_G}{\gamma_O}\right) + (\Delta\theta)\sin 2\theta_G. \tag{5.13}$$

Using (5.12), the dispersion relation (5.11) becomes

$$\xi_{O\pi}^2 + \xi_{O\pi}\left(\frac{\gamma_O}{\gamma_G}\right)\left(\beta - \frac{1}{4}a'_{\pi O}\chi_{G-H}\chi_{H-G} - \frac{\gamma_G}{4\gamma_O}a_{\pi O}\chi_{O-H}\chi_{H-O}\right) - \frac{\gamma_O}{4\gamma_G}$$

$$\cdot \left(\beta a_{\pi O}\chi_{O-H}\chi_{H-O} + \frac{a_{\pi O}a'_{\pi O}}{4}\chi_{O-H}\chi_{H-O}\chi_{G-H}\chi_{H-G} + \chi_{GO,HO}^{\pi}\chi_{OG,OH}^{\pi}\right)$$

$$= 0 . \tag{5.14}$$

The solution for $\xi_{O\pi}$ is

$$\left.\begin{array}{c}\xi_{O\pi}(1)\\\xi_{O\pi}(2)\end{array}\right\} = \frac{1}{2}(z \mp \sqrt{z^2 + q}) + \frac{1}{4}a_{\pi O}\chi_{O-H}\chi_{H-O} , \tag{5.15}$$

where

$$z = -\frac{\gamma_O}{\gamma_G}\left(\beta - \frac{1}{4}a'_{\pi O}\chi_{G-H}\chi_{H-G} + \frac{\gamma_G}{4\gamma_O}a_{\pi O}\chi_{O-H}\chi_{H-O}\right), \tag{5.16a}$$

$$q = \frac{\gamma_G}{\gamma_O}\chi_{GO,HO}^{\pi}\chi_{OG,OH}^{\pi} . \tag{5.16b}$$

The two eigenvalues, $\xi_{O\pi}(1)$ and $\xi_{O\pi}(2)$, define the modes of wave propagation. The corresponding eigenvectors, referring to (5.5b), are

$$\frac{E_{\pi G}(1)}{E_{\pi O}(1)} = -\frac{(\gamma_O/\gamma_G)\chi_{GO,HO}^{\pi}}{z + \sqrt{z^2 + q^2}} = B_1 ,$$

$$\frac{E_{\pi G}(2)}{E_{\pi O}(2)} = -\frac{(\gamma_O/\gamma_G)\chi_{GO,HO}^{\pi}}{z - \sqrt{z^2 + q^2}} = B_2 . \tag{5.17}$$

Similarly, the σ components of the wavefield-amplitude ratios can be obtained by replacing $\chi_{GO,HO}^{\pi}$ and $\chi_{OG,OH}^{\pi}$ by

$$\chi_{GO,HO}^{\sigma} = P_{\sigma}\chi_{G-O} - \frac{\alpha_{\sigma}}{2}\chi_{H-O}\chi_{G-H} , \tag{5.18}$$

$$\chi_{OG,OH}^{\sigma} = P_{\sigma}\chi_{O-G} - \frac{\alpha_{\sigma}}{2}\chi_{O-H}\chi_{H-G} ,$$

where

$$\alpha_{\sigma} = \frac{d_0^2}{\xi_{H\sigma}} , \quad P_{\sigma} = 1 . \tag{5.19}$$

The corresponding eigenvalues have the same expressions as in (5.15) except that the coefficients a are replaced by

$$a_{\sigma O} = a'_{\sigma O} = \frac{d_2^2}{\xi_{H\sigma}} . \tag{5.20}$$

In order to derive an expression for the reflection intensities, the boundary conditions should be considered. However, these conditions depend on whether the reflection is a transmission (Laue) or reflection (Bragg) case. For simplicity, let us consider diffractions for a plane-parallel crystal plate. Both transmission and Bragg reflection cases are discussed in the following sections.

5.1.2 Three-Beam Transmission Case (Borrmann Diffraction)

The boundary conditions at the entrance crystal surface, i.e., at $t = 0$, for the two-beam approximation are

$$E_{\pi O}(1) + E_{\pi O}(2) = E'_O, \qquad E_{\pi G}(1) + E_{\pi G}(2) = 0, \qquad (5.21)$$

where E'_O is the wavefield amplitude of the incident wave in empty space. Substituting the wavefield-amplitude ratios, B_1 and B_2, of (5.17) into the boundary conditions, we obtain

$$E_O(1) = \frac{B_2}{B_2 - B_1} E'_O, \qquad E_O(2) = -\frac{B_1}{B_2 - B_1} E'_O, \qquad (5.22)$$

and

$$E_G(1) = \frac{B_1 B_2}{B_2 - B_1} E'_O, \qquad E_G(2) = -\frac{B_1 B_2}{B_2 - B_1} E'_O. \qquad (5.23)$$

At the exit surface of the crystal plate, i.e., $t = T_0$, where T_0 is the crystal thickness, the transmitted and reflected wavefields inside the crystal are

$$E_O(T_0) = \exp(-2\pi i k_0 \cdot r)[E_O(1) \exp(-i\phi_1 T_0) + E_O(2) \exp(-i\phi_2 T_0)],$$

$$E_G(T_0) = \exp[-2\pi i(k_0 + g_G) \cdot r][B_1 E_O(1) \exp(-i\phi_1 T_0)$$
$$+ B_2 E_O(2) \exp(-i\phi_2 T_0)], \qquad (5.24)$$

and the incident wavefield in empty space is

$$E^e_O = E'_O \exp(-2\pi i k_0 \cdot r), \qquad (5.25)$$

where

$$\phi_1 = -2\pi k \delta(1), \qquad \phi_2 = -2\pi k \delta(2). \qquad (5.26)$$

The accommodation δ can be obtained from (5.8, 15):

$$\left. \begin{matrix} \delta(1) \\ \delta(2) \end{matrix} \right\} = \frac{1}{2\gamma_O}\left(z \mp \sqrt{z^2 + q} + \frac{1}{2} a_{\pi O} \chi_{O-H} \chi_{H-O} - \chi_{O-O}\right). \qquad (5.27)$$

The intensity ratios between the transmitted beam, the reflected beam and the incident beam are therefore equal to

$$\frac{I_O}{I_O^e} = \frac{|E_O(T_0)|^2}{|E_O^e|^2} = \left| \frac{B_1 B_2}{B_2 - B_1} (p_1 - p_2) \right|^2 ,$$

$$\frac{I_G}{I_O^e} = \frac{|E_G(T_0)|^2}{|E_O^e|^2} = \left| \frac{B_2 p_1 - B_1 p_2}{B_2 - B_1} \right|^2 ,$$

(5.28)

where

$$p_1 = \exp(-i\phi_1 T_0) , \quad p_2 = \exp(-i\phi_2 T_0) .$$

(5.29)

Let us define

$$z^2 + q = m + in , \quad \text{and}$$

(5.30)

$$\sqrt{z^2 + q} = v + iw ,$$

(5.31)

where

$$v = \frac{1}{\sqrt{2}} \sqrt{\sqrt{m^2 + n^2} + m} , \quad w = \frac{1}{\sqrt{2}} \sqrt{\sqrt{m^2 + n^2} - m} ,$$

$$m = z_r^2 - z_i^2 + q_r , \quad n = 2 z_r z_i + q_i .$$

(5.32)

The terms z_r, z_i and q_r, q_i are the real and imaginary parts of z and q, respectively. With the aid of the following relations obtained from (5.17, 29 – 32):

$$B_1 B_2 = -\frac{(\chi_{GO,HO})^2}{q} ,$$

$$B_2 - B_1 = -\frac{2(v + iw)}{(\gamma_O/\gamma_G) q} ,$$

(5.33)

$$\cosh\left(\frac{\phi^+}{2}\right) + \sinh\left(\frac{\phi^+}{2}\right) = \exp(-\mu T) ,$$

$$\sin^2\left(\frac{\phi^-}{2}\right) + \sinh^2\left(\frac{\phi^+}{2}\right) = \sin^2(av) + \sinh^2(aw) ,$$

the intensity ratio I_G/I_O^e can be derived as

$$\frac{I_G}{I_O^e} = \left(\frac{\gamma_O}{\gamma_G}\right)^2 |\chi_{GO,HO}|^2 \exp(-\mu T) \left(\frac{\sin^2(av) + \sinh^2(aw)}{v^2 + w^2}\right) ,$$

(5.34)

where

$$\phi^+ = (\phi_1 + \phi_2) T_0 , \quad \phi^- = (\phi_1 - \phi_2) T_0 ,$$

$$T = \frac{1}{2}\left(\frac{1}{\gamma_O} + \frac{1}{\gamma_G}\right) T_0 ,$$

$$\mu = \mu_0 + \sigma_0, \quad \mu_0 = -\frac{2\pi k}{\gamma_O} \chi^i_{O-O}, \tag{5.35}$$

$$\sigma_0 = \frac{\pi k}{\gamma^2_O} \frac{1}{\left(\dfrac{1}{\gamma_O} + \dfrac{1}{\gamma_G}\right)} \left(\frac{\gamma_O}{\gamma_G} \operatorname{Im}\{a'_O \chi_{G-H} \chi_{H-G}\} - \operatorname{Im}\{a_O \chi_{O-H} \chi_{H-O}\}\right),$$

$$a = \frac{\pi k}{\gamma_O} T_0.$$

Equation (5.34) is the familiar expression for the intensity of a two-beam transmission case (for example, see [Ref. 5.1, Eq. (3.133)]), except that the susceptibility of two-beam reflection is now replaced by the generalized susceptibility $\chi_{GO,HO}/4\pi$ of the three-beam diffraction. The ordinary linear absorption coefficient μ_0 of the two-beam case is also modified by the extra term σ_0, which often causes anomalous transmission in multi-beam Borrmann diffraction.

5.1.3 Diffracted Intensity of Three-Beam Bragg Reflection

Three-beam Bragg diffraction involves at least one Bragg reflection, say reflection G. The corresponding direction cosine γ_G is therefore negative. The boundary conditions at the entrance surface of the crystal in the two-beam approximation are

$$E_O(1) + E_O(2) = E'_O, \quad E_G(1) + E_G(2) = E'_G. \tag{5.36}$$

The subscripts indicating the polarization states are left out. The quantity E'_G is the reflected wavefield amplitude in vacuum. In addition to (5.36), we have the following condition at the exit surface [see also the third equation of (4.200)]:

$$E_G(1)p_1 + E_G(2)p_2 = 0. \tag{5.37}$$

By considering (5.17, 29, 36, 37), the wavefield amplitudes of the direct waves are obtained as

$$E_O(1) = \frac{p_2 B_2 E'_O}{p_2 B_2 - p_1 B_1}, \quad E_O(2) = -\frac{p_1 B_1 E'_O}{p_2 B_2 - p_1 B_1}. \tag{5.38}$$

The reflected intensity at the entrance surface of the crystal can therefore be derived from (5.36) by using (5.38):

$$\frac{I_G}{I^e_O} = \left|\frac{E'_G}{E'_O}\right|^2 = \left|\frac{B_1 B_2 (P_1 - p_2)}{p_2 B_2 - p_1 B_1}\right|^2. \tag{5.39}$$

Similarly, the transmitted intensity at the exit surface takes the form

$$\frac{I_O(T_0)}{I_O^e} = \left| \frac{p_1 p_2 (B_2 - B_1)}{p_2 B_2 - p_1 B_1} \right|^2 . \tag{5.40}$$

A more explicit expression can be written for (5.39), by following the similar manipulation given in Sect. 5.1.2:

$$\frac{I_G}{I_O^e} = \frac{1}{R} \left(\frac{\gamma_O}{\gamma_G} \right)^2 |\chi_{GO,HO}|^2 [\sin^2(av) + \sinh^2(aw)] , \tag{5.41}$$

where

$$R = |q + z^2| + R_1 \sinh^2(aw) - R_2 \sin^2(av) + \tfrac{1}{2} \sqrt{|R_1^2 - |q^2||}$$
$$\times \sinh|2aw| + \tfrac{1}{2} \sqrt{|R_2^2 - |q^2||} |\sin|2av| , \tag{5.42}$$

with

$$R_1 = |q + z^2| + |z|^2, \quad R_2 = |q + z^2| - |z|^2 . \tag{5.43}$$

This expression has been given by *Zachariasen* [5.1] for two-beam Bragg reflections.

5.1.4 Integrated Reflection for Non-Absorbing Crystals

The expressions for diffracted intensities given in (5.34, 41) are so complicated that intensities can only be calculated numerically. However, in some special cases, simple forms can be obtained. We shall therefore confine ourselves to a few particular cases. In addition, in order to compare the theory with experiments, the integrated intensities are also dealt with.

The integrated reflection R_G for a given reflection G over an incident angle θ is defined in the usual way as the integrated power ratio P_G/P_O^e between the reflected and the incident waves, i.e.,

$$R_G^\theta = \int \frac{P_G}{P_O^e} d\theta . \tag{5.44}$$

Since the diffraction power is a product of the reflected intensity and the cross section Λ of the reflected beam, the power ratio takes the form

$$\frac{P_G}{P_O^e} = \frac{I_G}{I_O^e} \frac{\Lambda_G}{\Lambda_O} = \frac{I_G}{I_O^e} \frac{\gamma_G}{\gamma_O} = \frac{I_G}{|b| I_O^e} . \tag{5.45}$$

The ratio of the cross sections for a plane-parallel crystal plate is equal to the ratio of the direction cosines, which is denoted by the parameter $1/b$.

The intensity expressions (5.34, 41) are exactly the same as those in two-beam diffractions [5.1]. Discussions of the behaviors of the intensities in special two-beam cases have been given in many books. We shall not go through in detail the derivation of the simplified formulas for (5.34, 41) but rather present the final expressions for the reflected intensity and integrated reflection for various special situations.

a) Power ratios of non-absorbing Laue cases. The assumption that there is no absorption means that the electric susceptibility is real. The conditions $\chi^*_{GO,HO} = \chi_{OG,OH}$, $\chi_{OG} = \chi^*_{GO}$, and $|\chi_{GO}\chi_{OG}| = |\chi_{OG}|^2$ are fulfilled. In a Laue case, b is positive. The parameter q is therefore a positive real quantity. The imaginary parts of $\sqrt{q+z^2}$ and of z are zero, i.e., $w = 0$, $z = z_r$. Similarly, the absorption coefficient μ, defined in (5.35), is also null.

Let us define the quantities M, ϱ, A, as

$$M = \sqrt{|b|}\,|\chi_{GO,HO}|, \tag{5.46a}$$

$$A = aM = \frac{\pi k T_0 |\chi_{GO,HO}|}{\sqrt{|\gamma_O \gamma_G|}}, \tag{5.46b}$$

$$\varrho = \frac{z_r}{M}$$

$$= \frac{-b}{\sqrt{|b|}\,|\chi_{GO,HO}|}\left((\Delta\theta)\sin 2\theta_G + \chi_{O-O}\frac{(1-b)}{2} - \frac{a'_O\chi_{G-H}\chi_{H-G}}{4}\right.$$

$$\left. + \frac{(a_O\chi_{O-H}\chi_{H-O})}{4b}\right). \tag{5.46c}$$

In terms of these quantities,

$$v = \sqrt{q+z^2} = M\sqrt{1+\varrho^2}, \tag{5.47}$$

and the power ratio (5.45) can be written in the form

$$\frac{P_G}{P^e_O} = \frac{\sin^2(A\sqrt{1+\varrho^2})}{1+\varrho^2}. \tag{5.48}$$

The power ratio is therefore a function of the angular deviation from the exact Bragg angle θ_G of the reflection G.

b) Power ratios of non-absorbing Bragg cases. The parameter b is negative in the Bragg case. Hence, $q = -|b|\,|\chi_{GO,HO}|^2$, which is also negative. This implies that $\sqrt{q+z^2}$ is real for $q+z^2 > 0$ and imaginary for $q+z^2 < 0$.

For $q+z^2 > 0$, the imaginary part w of $\sqrt{q+z^2}$ is null and the real part v is $A\sqrt{\varrho^2-1}$. The power ratio of (5.45) becomes

$$\frac{P_G}{P_O^e} = \frac{\sin^2(A\sqrt{\varrho^2-1})}{\varrho^2-1+\sinh^2(A\sqrt{\varrho^2-1})} = \frac{1}{\varrho^2+(\varrho^2-1)\cot^2(A\sqrt{\varrho^2-1})} . \quad (5.49)$$

For $q+z^2 < 0$, $q+z^2 = iA\sqrt{1-\varrho^2}$ and $v = 0$, and the power ratio is

$$\frac{P_G}{P_O^e} = \frac{\sinh^2(A\sqrt{1-\varrho^2})}{1-\varrho^2+\sinh^2(A\sqrt{1-\varrho^2})} = \frac{1}{\varrho^2+(1-\varrho^2)\coth^2(A\sqrt{1-\varrho^2})} .$$

$$(5.50)$$

It should be noted that (5.48 – 50) are exactly the same as the expressions for the power ratios in the two-beam Laue and Bragg cases [5.1], except that the parameters A and ϱ are modified by the generalized Fourier components $\chi_{GO,HO}$ and the components involving the secondary and coupling reflections, i.e., $\chi_{O-H}\chi_{H-O}$ and $\chi_{G-H}\chi_{H-G}$. This modification has the following physical significance.

The diffraction patterns, represented by (5.48 – 50), are functions of ϱ. They are symmetric about $\varrho = 0$. The angular positions corresponding to $\varrho = 0$ are the centers of the diffraction patterns. These positions deviate from θ_G according to (5.46c) by

$$\Delta\theta_{3\text{-beam}} = -\frac{\chi_{O-O}}{2}(1-b) + \frac{1}{4}a_O'\chi_{G-H}\chi_{H-G} - \frac{1}{4b}a_O\chi_{O-H}\chi_{H-O}$$

$$= \Delta\theta_{2\text{-beam}} + \theta_3 , \quad (5.51)$$

where $\Delta\theta_{2\text{-beam}} = -\chi_{O-O}(1-b)/2$ is the shift of the center of the diffraction pattern away from the Bragg angle for two-beam Bragg cases. This quantity is zero for two-beam symmetric Laue diffraction because $b = 1$. The center of the two-beam Bragg diffraction pattern is further shifted by θ_3 due to the presence of the additional reflection H.

The parameter A defined in (5.46b) is usually used in two-beam diffraction to classify whether the crystal is thick or thin [5.1]. We shall adopt this concept to distinguish 'thick' crystals from the thin ones in three-beam cases. Following *Zachariasen* [5.1], 'thick' crystals and 'thin' crystals have values of A greater and smaller than unity, respectively. For $A \simeq 1$, the crystal is regarded as having an intermediate thickness. 'Thick' crystals may therefore involve either a large crystal thickness or strong reflection while 'thin' crystals are those with small T or subjected to weak reflection. For the intermediately thick crystal, the thickness may be thin for a strong reflection or thick for a weak reflection. We shall, in the following, discuss these three cases separately.

c) Laue diffraction in thick crystals. As a sine function always provides oscillating values, (5.48) does not approach any finite value for increasing A. When A is very large, $\sin^2(A\sqrt{1+\varrho^2})$ may be replaced by its average value, i.e., 1/2, provided that the variation in A is greater than $\pi/2$. Under this assumption, the average power ratio is equal to

$$\left\langle \frac{P_G}{P_O^e} \right\rangle = \frac{1}{2(1+\varrho^2)} \, . \tag{5.52}$$

The maximum of (5.52) is 1/2 at $\varrho = 0$. The width at half maximum of the diffraction pattern is then

$$\Delta W = \Delta \theta(\varrho = 1) - \Delta \theta(\varrho = -1) \, . \tag{5.53}$$

Using (5.46c), we obtain

$$\Delta W = \frac{1}{|b|\sin 2\theta_G} |\chi_{GO, HO}| \, . \tag{5.54}$$

The integrated reflection over θ is

$$R_G^\theta = \frac{d\theta}{d\varrho} \int \left\langle \frac{P_G}{P_O^e} \right\rangle d\varrho = \frac{\pi |\chi_{GO, HO}|}{2\sqrt{|b|}\sin 2\theta_G} \, , \qquad \text{since} \tag{5.55}$$

$$\int_{-\infty}^{\infty} \frac{d\varrho}{1+\varrho^2} = \frac{\pi}{2} \, , \qquad \text{and} \tag{5.56}$$

$$\frac{d\theta}{d\varrho} = \frac{|\chi_{GO, HO}|}{\sqrt{|b|}\sin 2\theta_G} \, . \tag{5.57}$$

If the incident beam is unpolarized, both σ and π components of the wavefields should be considered. The corresponding integrated reflection takes the form

$$R_G^\theta = \tfrac{1}{2}(R_{G\sigma}^\theta + R_{G\pi}^\theta) \, , \tag{5.58}$$

where $R_{G\sigma}^\theta$ and $R_{G\pi}^\theta$ are the integrated reflections of the σ and π waves.

d) Bragg reflection in thick crystals. In Bragg reflection, the values of ϱ are such that $|\varrho| < 1$ or $|\varrho| > 1$. For $|\varrho| < 1$, (5.50) should be employed for the power ratio. The power ratio approaches its limit as A increases:

$$\frac{P_G}{P_O^e} = \lim_{A \to \infty} \frac{\sinh^2(A\sqrt{1-\varrho^2})}{1-\varrho^2+\sinh^2(A\sqrt{1-\varrho^2})} = 1 \, . \tag{5.59}$$

Since the conservation of energy holds for all circumstances in diffraction, i.e.,

$$\frac{P_O}{P_O^e} + \frac{P_G}{P_O^e} = 1 \, , \tag{5.60}$$

the diffracted power ratio of the transmitted beam, $P_O/P_O^e = 0$. This indicates that total reflection takes place for $|\varrho| < 1$ and $A \gg 1$.

In the range $|\varrho| > 1$, the power ratio has no definite limit as A increases because of the sine function. Following the same argument as in the Laue case, the power ratio can be represented by its average value:

$$\frac{P_G}{P_O^e} = 1 - \sqrt{1 - \varrho^{-2}} \, . \tag{5.61}$$

The integrated reflection can then be written as

$$R_G^\theta = \frac{\pi \, |\chi_{GO,HO}|}{\sqrt{|b|} \sin 2\theta_G} \, , \qquad \text{since} \tag{5.62}$$

$$\int_1^\infty (1 - \sqrt{1 - \varrho^{-2}}) \, d\varrho = \frac{\pi}{2} - 1 \, . \tag{5.63}$$

e) Diffraction in thin crystals. The power ratios for both Laue and Bragg cases approach the following expression for very small A:

$$\frac{P_G}{P_O^e} \simeq \frac{\sin^2(A\varrho)}{\varrho^2} \, . \tag{5.64}$$

The corresponding integrated reflection is

$$R_G^\theta = \frac{\pi A \, |\chi_{GO,HO}|}{\sqrt{|b|} \sin 2\theta_G} \, , \qquad \text{where} \tag{5.65}$$

$$\int_0^\infty \frac{\sin^2(A\varrho)}{\varrho^2} = \frac{\pi}{2} A \, . \tag{5.66}$$

Substituting A by (5.46b), the integrated reflection can be written as

$$R_G^\theta = \frac{Q_G T_0}{\gamma_G} \, , \tag{5.67}$$

where the reflectivity

$$Q_G = \frac{\pi^2}{\lambda} \frac{|\chi_{GO,HO}|^2}{\sin 2\theta_G} \tag{5.68}$$

is proportional to $|\chi_{GO,HO}|^2$. Since the reflectivity does not depend on $|\chi_{GO,HO}|$ but on $|\chi_{GO,HO}|^2$, the diffraction in thin crystals is kinematical [see (3.15) for the reflectivity in the kinematical approach] rather than dynamical.

f) Laue diffraction in crystals of intermediate thickness. The approximate solutions of P_G/P_O^e for $A \gg 1$ and $A \ll 1$ are not valid for $A \simeq 1$. No analytical expression can be formed to describe the power ratio. Numerical calculation can be made as in two-beam cases [5.1, 2]. These calculations depend very much on whether A approaches $\pi/2$. The integrated reflection can, however, be obtained, according to *Waller* [5.3], by replacing

$$R_G^\varrho = \int_{-\infty}^{\infty} \frac{\sin^2(A\sqrt{1+\varrho^2})}{1+\varrho^2} d\varrho \quad \text{by} \tag{5.69}$$

$$R_G^\varrho = \int_0^{\pi/2} \frac{\sin(2A\sin\zeta_0)}{\sin\zeta_0} d\zeta_0. \tag{5.70}$$

Since the zeroth-order Bessel function, $J_0(\xi_1)$, may be defined, with $\xi_1 = 2A$, as

$$J_0(\xi_1) = \frac{2}{\pi} \int_0^{\pi/2} \cos(\xi_1 \sin\zeta_0) d\zeta_0, \quad \text{and} \tag{5.71}$$

$$\frac{dR_G^\varrho}{d\xi_1} = \frac{\pi}{2} J_0(\xi_1), \tag{5.72}$$

the integrated reflection can be written as

$$R_G^\varrho = \int_0^{2A} \frac{dR_G^\varrho}{d\xi_1} d\xi_1 = \frac{\pi}{2} \int_0^{2A} J_0(\xi_1) d\xi_1 = \pi W_0, \quad \text{where} \tag{5.73}$$

$$W_0 = \sum_{n=0}^{n=\infty} J_{2n+1}(2A) = \begin{cases} A, & \text{for } A \ll 1, \\ \frac{1}{2}, & \text{for } A \gg 1. \end{cases} \tag{5.74}$$

Equation (5.73) is in agreement with the power ratios for diffraction in thin and thick crystals.

g) Bragg reflection in crystals of intermediate thickness. With the aid of the integral given by *Darwin* [5.4],

$$\int_{-\infty}^{\infty} \frac{d\varrho}{\varrho^2 + (1-\varrho^2)\coth^2(A\sqrt{1-\varrho^2})} = \pi \tanh A, \tag{5.75}$$

the integrated reflection can be obtained:

$$R_G^\theta = \frac{|\chi_{GO,HO}|}{\sqrt{|b|}\sin 2\theta_G} \pi \tanh A. \tag{5.76}$$

For $A \gg 1$, $\tanh A \simeq 1$ and for $A \ll 1$, $\tanh A = A$, and (5.76) approaches (5.62) and (5.65), respectively.

5.1.5 Diffraction in Absorbing Crystals

In this section, we shall discuss the approximate solutions of (5.34, 41) for diffraction powers and integrated reflections in absorbing crystals. Since absorption is governed by the imaginary part of χ (4.187), we shall first consider the general expression for χ. The complex χ_G can be written in terms of the real part χ_G^r and the imaginary part χ_G^i:

$$\chi_G = \chi_G^r + i\chi_G^i. \tag{5.77a}$$

Both χ_G^r and χ_G^i are still complex, such that

$$\chi_G^r = \chi_{G,r}' + i\chi_{G,r}'', \quad \text{and} \tag{5.77b}$$

$$\chi_G^i = \chi_{G,i}' + i\chi_{G,i}''. \tag{5.77c}$$

Although the three quantities $|\chi_G|^2$, $|\chi_{\bar{G}}|^2$ and $|\chi_G\chi_{\bar{G}}|$ are identical in non-absorbing crystals, they are different in the presence of absorption. The terms χ_G and $\chi_{\bar{G}}$ are the abbreviations of $\chi_{GO,HO}$ and $\chi_{OG,OH}$. However, these three quantities can be expressed in terms of $|\chi_{GO,HO}^r|$ as follows:

$$|\chi_G|^2 = |\chi_G^r|^2 + |\chi_G^i|^2 + 2(\chi_{G,r}''\chi_{G,i}' - \chi_{G,r}'\chi_{G,i}'') = |\chi_G^r|^2(1 + \varkappa^2 + 2s),$$

$$|\chi_{\bar{G}}|^2 = |\chi_G^r|^2(1 + \varkappa^2 - 2s), \tag{5.78}$$

$$|\chi_G\chi_{\bar{G}}| = |\chi_G^r|^2(1 - \varkappa^2 - 2ip_0), \quad \text{where}$$

$$\varkappa = \frac{|\chi_G^i|}{|\chi_G^r|},$$

$$s = \frac{\chi_{G,r}''\chi_{G,i}' - \chi_{G,r}'\chi_{G,i}''}{|\chi_G^r|^2}, \tag{5.79}$$

$$p_0 = \frac{\chi_{G,r}'\chi_{G,i}' + \chi_{G,r}''\chi_{G,i}''}{|\chi_G^r|^2}.$$

For a centrosymmetric structure, which possesses a center of inversion, $s = 0$ and $\varkappa = p_0$, and $|\chi_G\chi_{\bar{G}}| = |\chi_G|^2$. For generality, we shall maintain the form of (5.77, 78) for the χ's in the following discussion.

The imaginary part $|\chi_G^i|$ is usually very small in comparison with the real part $|\chi_G^r|$ when the reflection is not very weak and when the incident wavelength is not very close to an absorption edge λ_E.

We shall first consider this situation for a three-beam diffraction. Assume $|\chi_{GO,HO}^r| \gg |\chi_{GO,HO}^i|$, and redefine, for the sake of convenience, the quantities M, A, ϱ, m_0 and g_0 as

$$M = \sqrt{|b|}\,\sqrt{\text{Re}\{\chi_{GO,HO}\chi_{OG,OH}\}}, \tag{5.80a}$$

$$m_0 = \frac{\text{Im}\{\chi_{GO,HO}\chi_{OG,OH}\}}{\text{Re}\{\chi_{GO,HO}\chi_{OG,OH}\}}, \tag{5.80b}$$

$$A = \frac{\pi k T_O}{\sqrt{|\gamma_O \gamma_G|}} \sqrt{\text{Re}\{\chi_{GO,HO}\chi_{OG,OH}\}}, \quad \text{and} \tag{5.80c}$$

$$\varrho = \frac{z_r}{M}, \quad g_0 = \frac{z_i}{M}. \tag{5.80d}$$

The parameters v and w defined in (5.32) take the following forms for Laue cases:

$$\left.\begin{array}{c} v \\ w \end{array}\right\} = \frac{M}{\sqrt{2}} \sqrt{\Xi_1 \pm (\varrho^2 - g_0^2 + 1)}, \quad \text{with} \tag{5.81}$$

$$\Xi_1 = \sqrt{(\varrho^2 - g_0^2 + 1)^2 + (2\varrho g_0 + m_0)^2},$$

and for Bragg cases:

$$\left.\begin{array}{c} v \\ w \end{array}\right\} \frac{M}{\sqrt{2}} \sqrt{\Xi_2 \pm (\varrho^2 - g_0^2 - 1)}, \tag{5.82}$$

with

$$\Xi_2 = \sqrt{(\varrho^2 - g_0^2 - 1)^2 + (2\varrho g_0 - m_0)^2}.$$

Special Laue and Bragg cases are discussed separately as follows:

a) Laue case with $g_0 \ll m_0 < 1$ and $b = 1$. In this case, the parameters v and w of (5.81) are simplified as

$$v = M\sqrt{1 + \varrho^2}, \quad w = \frac{M m_0}{\sqrt{1 + \varrho^2}}, \tag{5.83}$$

and

$$v^2 + w^2 \simeq M^2 (1 + \varrho^2). \tag{5.84}$$

Substituting (5.83) into (5.34), we obtain

$$\frac{P_G}{P_O^e} = \frac{|\chi_{GO,HO}|^2}{\text{Re}\{\chi_{GO,HO}\chi_{OG,OH}\}} \frac{\exp(-\mu T)}{(1 + \varrho^2)} \left[\sin^2(A\sqrt{1 + \varrho^2}) \right.$$
$$\left. + \sinh^2\left(\frac{m_0 A}{\sqrt{1 + \varrho^2}}\right) \right]. \tag{5.85}$$

When A becomes very large, i.e., the crystal is considered to be very 'thick', the square of the sine function approaches its average value of 1/2. The reflection power ratio thus becomes

$$\frac{P_G}{P_O^e} \simeq \frac{|\chi_{GO,HO}|^2}{2\,|\mathrm{Re}\{\chi_{GO,HO}\chi_{OG,OH}\}|} \cdot \frac{\exp(-\mu T)}{(1+\varrho^2)} \left[1 + 2\sinh^2\left(\frac{m_0 A}{\sqrt{1+\varrho^2}}\right)\right].$$

(5.86)

If the imaginary part $\mathrm{Im}\{\chi_{GO,HO}\chi_{OG,OH}\}$ is very small in comparison with its real part, such that $|m_0|A < 0.4$, (5.86) can be further simplified to [5.1]

$$\frac{P_G}{P_O^e} = \frac{|\chi_{GO,HO}^r|^2}{2\,|\mathrm{Re}\{\chi_{GO,HO}\chi_{OG,OH}\}|} \cdot \frac{\exp(-\mu T)}{(1+\varrho^2)} \left[1 + \frac{2(m_0 A)^2}{1+\varrho^2}\right],$$

(5.87)

which leads to the expression for the integrated reflection,

$$R_G^\theta = \frac{\pi}{2\sin 2\theta_G} \frac{|\chi_{GO,HO}^r|}{\sqrt{\mathrm{Re}\{\chi_{GO,HO}\chi_{OG,OH}\}}} [1 + (m_0 A)^2] \exp(-\mu T).$$

(5.88)

b) Bragg reflection in thick crystals. The complicated expression for the intensity ratio, given in (5.41), can be greatly simplified by assuming the crystal thickness to be so large that $\sinh^2(aw)$ and $\sinh|2aw|$ have very large values. Using this assumption, (5.41) can be reduced to the simple form

$$\frac{I_G}{I_O^e} = \frac{b^2\,|\chi_{GO,HO}|^2}{|q+z^2|+|z|^2+\sqrt{(|q+z^2|+|z|^2)^2-|q|^2}}.$$

(5.89)

The same expression can also be derived by selecting the only permitted mode (Sect. 4.5.2) associated with a two-beam diffraction. The eigenvalue of this mode, referring to (5.15), is

$$\xi_0(1) = \tfrac{1}{2}(z - \sqrt{z^2 + q}) + \tfrac{1}{4}a_0\chi_{O-H}\chi_{H-O},$$

(5.90)

so that its corresponding wavefield-amplitude ratio $E_G(1)/E_O(1)$ remains small, i.e., $|b\,|\chi_{GO,HO}/(z + \sqrt{z^2 + q})$ compared with $|b\,|\chi_{GO,HO}/(z - \sqrt{z^2 + q})$. According to the following simple boundary conditions at the entrance surface of the crystal:

$$E_O(1) = E_O', \qquad E_G(1) = E_G',$$

(5.91)

the intensity ratio is directly proportional to the square of the wavefield-amplitude ratio, i.e.,

$$\frac{I_G}{I_O^e} = \left|\frac{E_G(1)}{E_O(1)}\right|^2 = \frac{b^2\,|\chi_{GO,HO}|^2}{|z + \sqrt{z^2 + q}|^2}.$$

(5.92)

By employing the parameters m, n and v, w, the following relation is obtained:

$$z_r \sqrt{\sqrt{m^2+n^2}+m} + z_i \sqrt{\sqrt{m^2+n^2}-m} = \frac{1}{\sqrt{2}}\sqrt{(|z^2+q|+|z|^2)^2-|q|^2}.$$

$$(5.93)$$

Considering the real and imaginary parts of the denominator of (5.92) and using (5.93), we obtain the same expression as (5.89) for the intensity ratio of the reflected beam.

According to *Miller* [5.5], (5.89) can be written in the alternative form

$$\frac{I_G}{I_O^e} = \frac{|b||\chi_{GO,HO}|^2}{|\chi_{GO,HO}\chi_{OG,OH}|} \cdot \frac{1}{|q|}[|q+z^2|+|z|^2$$

$$- \sqrt{(|q+z^2|+|z|^2)^2-|q|^2}].$$

$$(5.94)$$

Defining

$$Na = \frac{|\chi_{GO,HO}|^2}{|\chi_{GO,HO}\chi_{OG,OH}|},$$

$$(5.95)$$

and using (5.82), we obtain a much simpler form for the intensity ratio:

$$\frac{I_G}{I_O^e} = |b|Na(L-\sqrt{L^2-1}),$$

$$(5.96)$$

where

$$L = \frac{\varrho^2+g_0^2+\sqrt{(\varrho^2-g_0^2-1)^2+(2\varrho g_0-m_0)^2}}{\sqrt{1+m_0^2}}.$$

$$(5.97)$$

The integrated reflection then takes the form

$$R_G^\theta = \frac{Na\sqrt{\mathrm{Re}\{\chi_{GO,HO}\chi_{OG,OH}\}}}{\sqrt{|b|}\sin 2\theta_G}\int_{-\infty}^{\infty}(L-\sqrt{L^2-1})d\varrho.$$

$$(5.98)$$

If $m_0 \ll 1$ and $g_0 < 1$, i.e., $z_i < \sqrt{|b|}\sqrt{\mathrm{Re}\{\chi_{GO,HO}\chi_{OG,OH}\}}$, the three-beam case is considered to be a strong diffraction. The integral in (5.98) assumes the form [5.1]

$$\int_{-\infty}^{\infty}(L-\sqrt{L^2-1})d\varrho = \frac{8}{3}.$$

$$(5.99)$$

The integrated reflection is thus proportional to $\sqrt{\mathrm{Re}\{\chi_{GO,HO}\chi_{OG,OH}\}}$, i.e.,

$$R_G^\theta = \frac{8Na}{3\sqrt{|b|}\sin 2\theta_G}\frac{\sqrt{\mathrm{Re}\{\chi_{GO,HO}\chi_{OG,OH}\}}}{\sqrt{1+m_0^2}}.$$

$$(5.100)$$

If $g_0 \gg 1$ and $g_0 > m_0$, i.e., $z_i > \sqrt{|b|}\,\sqrt{\mathrm{Re}\{\chi_{GO,HO}\chi_{OG,OH}\}}$, which implies that the three-beam case is a weak three-beam diffraction, L is much greater than unity. The integral in (5.98) is then simplified to

$$\int\limits_{-\infty}^{\infty} (L - \sqrt{L^2-1})\,d\varrho = \frac{1+m_0^2}{4} \int\limits_{-\infty}^{\infty} \frac{1}{\varrho^2 + g_0^2}\,d\varrho = \frac{\pi(1+m_0^2)}{4g_0} . \qquad (5.101)$$

The final expression for the integrated reflection is thus proportional to $(\sqrt{\mathrm{Re}\{\chi_{GO,HO}\chi_{OG,OH}\}})^2$:

$$R_G^\theta = \frac{\pi Na\,(1+m_0^2)}{4\sqrt{|b|}\,|g_0|\sin 2\theta_G}\,(\sqrt{\mathrm{Re}\{\chi_{GO,HO}\chi_{OG,OH}\}})^2 . \qquad (5.102)$$

Equations (5.100, 102) imply that the diffraction is dynamical when strong reflections are involved in the three-beam case, while the diffraction becomes kinematical when weak reflections participate in the three-beam case, provided that $|\mathrm{Re}\{\chi_{GO,HO}\chi_{OG,OH}\}| \gg |\mathrm{Im}\{\chi_{GO,HO}\chi_{OG,OH}\}|$. These two expressions have been derived by *Juretschke* [5.6] to give an analytical account of the phase-determination experiments of *Chang* [5.7] (Sect. 7.1).

When the reflections involved in a three-beam case are all very weak and when the wavelength of the radiation used is very close to and slightly less than the absorption edge λ_E, the imaginary part of $\chi_{GO,HO}$ may sometimes be greater than or comparable to its real part. Under these conditions, the real part of $\chi_{GO,HO}\chi_{OG,OH}$ in (5.100, 102) should be replaced by its imaginary part:
i) for $|\varrho' - g_0'| < 1$ and $|m_0'| \ll 1$, or for $|\varrho' + g_0'| < 1$ and $|m_0'| \ll 1$,

$$R_G^\theta = \frac{8Na}{3\sqrt{|b|}\sin 2\theta_G}\,\frac{\sqrt{|\mathrm{Im}\{\chi_{GO,HO}\chi_{OG,OH}\}|}}{\sqrt{1+m_0'^2}} ; \qquad (5.103)$$

ii) for $|\varrho' - g_0'| \gg 1$ and $|\varrho' - g_0'| > |m_0'|$, or for $|\varrho' + g_0'| \gg 1$ and $|\varrho' + g_0'| > |m_0'|$,

$$R_G^\theta = \frac{\pi Na(1+m_0'^2)}{4\sqrt{|b|}\sin 2\theta_G}\,\frac{\sqrt{|\mathrm{Im}\{\chi_{GO,HO}\chi_{OG,OH}\}|}}{|\varrho' \mp g_0'|} , \qquad (5.104)$$

where

$$\varrho' = \frac{z_r}{M'} , \qquad g_0' = \frac{z_i}{M'} , \qquad m_0' = \frac{1}{m} , \qquad (5.105)$$

with

$$M' = \sqrt{|b|}\,\sqrt{|\mathrm{Im}\{\chi_{GO,HO}\chi_{OG,OH}\}|} . \qquad (5.106)$$

When the imaginary part of $\chi_{GO,HO}$ is comparable to its real part, the above analytical expressions for integrated reflection are not valid because the conditions mentioned above for m_0, g_0, $\varrho - g_0$ and $\varrho + g_0$ cannot be fulfilled. Under these circumstances, numerical calculation may be useful to find the integrated reflection.

It is also worth mentioning that diffraction with a wavelength near a critical absorption edge λ_E should, in principle, be treated rigorously by using the quantum mechanical theory of diffraction [5.8 – 17]. Nevertheless, by taking into account the anomalous dispersion effect in the structure factor near λ_E, the classical dynamical theory can still give satisfactory results [5.18].

5.2 Procedures for Numerical Computing

The analytical expression for diffracted intensities for three-beam cases near the exact three-beam point, as given in Sects. 5.1.1 – 5, are derived from the fundamental equation of the wavefield, using the two-beam approximation. It is, however, difficult to obtain an analytical form for the diffracted intensities in N-beam cases at the exact N-beam point, except for the symmetric N-beam Borrmann diffraction. An attempt has been described by *Ewald* and *Heno* [5.19]. No expression as analytical as those in the two-beam or approximate two-beam cases has been obtained.

Numerical calculation with the aid of a computer is almost a necessary step to providing more exact information, complementary to the analytical two-beam approximation, about the configuration of the dispersion surface, the wavefield amplitude, the absorption and the diffracted intensities. Since transmission (Laue) cases are different from Bragg cases in geometry, we should consider these two cases separately. According to previous discussion in Sect. 4.7, N-beam Laue diffraction is merely a special case of N-beam Bragg diffraction. We shall therefore consider first the N-beam Bragg case, and give details of the computing procedures for this case.

Multiple diffraction of x-ray, as described in Chap. 2, involved the Bragg angle θ_G for the primary reflection G and the azimuthal angle ϕ for the rotation of the crystal about the reciprocal lattice vector of the primary reflection. Supposing multiple diffraction occurs at $\theta = \theta_G$ and $\phi = \phi_O$, the deviation of an incident beam from θ_G and ϕ_O may be denoted as $\Delta\theta$ and $\Delta\phi$. The crystal rotation can actually be regarded as being equivalent to the rotation of the incident beam about the reciprocal lattice vector g of the primary reflection G, when the crystal is considered to be fixed. This rotation can be represented by the motion of the starting point A of the incident wavevector k_O around g. If the incident beam only satisfies the Bragg condition for the G reflection, the point A coincides with the Laue point L of this two-beam reflection. The locus of this Laue point, shown in Fig. 5.3, is a circle of radius $k \cos\theta_G$, centered at the mid-point W of the vector g. When an N-beam diffraction

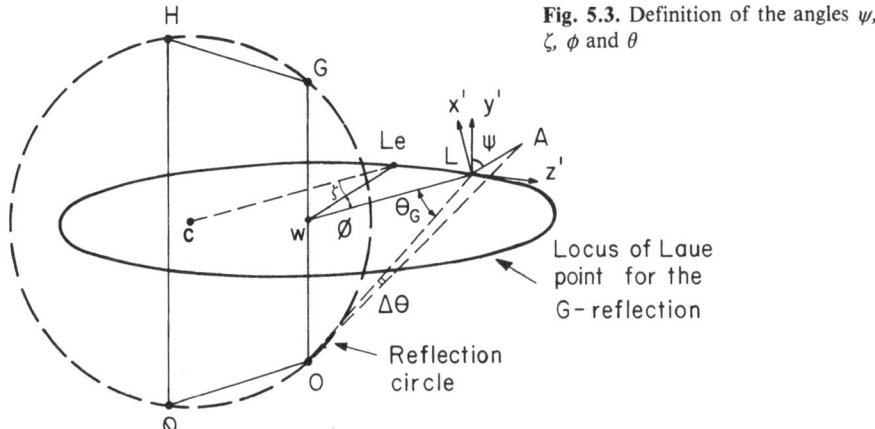

Fig. **5.3.** Definition of the angles ψ, ζ, ϕ and θ

occurs, point L moves to point Le, such that the distances between Le and the reciprocal lattice points involved are equal to k.

For simplicity, let us consider a four-beam case, in which G and H (Fig. 5.3) are symmetric and asymmetric Bragg reflections, respectively; Q is a Laue transmission. Point C is the center of the Ewald circle connecting the reciprocal lattice points, O, G, H and Q. The angle between LeC and LeW is ζ. Suppose an incident beam has its starting point at A, whose coordinates are x', y', z' with the origin at L. The angular deviation of the origin with respect to the N-beam position Le is ϕ. The distance AL is proportional to the deviation from the Bragg angle θ_G of the reflection G:

$$\overline{AL} = \frac{\Delta\theta}{\lambda}. \qquad (5.107)$$

The angle between AL and the y' axis is ψ. The coordinates of A, in terms of $\Delta\theta$ and ψ, are therefore expressed as

$$x' = -k(\Delta\theta)\sin\left(\frac{\Delta\theta}{2}\right),$$

$$y' = k(\Delta\theta)\cos\left(\frac{\Delta\theta}{2}\right)\cos\psi, \qquad (5.108)$$

$$z' = k(\Delta\theta)\cos\left(\frac{\Delta\theta}{2}\right)\sin\psi.$$

We define a new coordinate system with the origin at Le, the x axis along the direction of CLe, the y axis parallel to the vector OG and the z axis perpendicular to the x and y axes. For convenience, the coordinates of A can be expressed, after rotation and translation operations on the coordinate system, in terms of the new coordinates defined above, as

$$[X] = T_\xi(T_\phi T_\theta[X'] + [B_\phi]) , \tag{5.109}$$

where X and X' represent the new and the old coordinates, x, y, z and x', y', z', respectively. The matrices T_ζ, T_ϕ, T_θ and the vector $[B_\phi]$ are defined as

$$T_\zeta = \begin{pmatrix} \cos\zeta & 0 & -\sin\zeta \\ 0 & 1 & 0 \\ \sin\zeta & 0 & \cos\zeta \end{pmatrix} ,$$

$$T_\phi = \begin{pmatrix} \cos\phi & 0 & \sin\phi \\ 0 & 1 & 0 \\ -\sin\phi & 0 & \cos\phi \end{pmatrix} ,$$

$$T_\theta = \begin{pmatrix} \cos\theta_G & -\sin\theta_G & 0 \\ \sin\theta_G & \cos\theta_G & 0 \\ 0 & 0 & 1 \end{pmatrix} , \tag{5.110}$$

$$[B_\phi] = \begin{pmatrix} -k\phi & \cos\theta_G & \sin(\phi/2) \\ & 0 & \\ k\phi & \cos\theta_G & \cos(\phi/2) \end{pmatrix} .$$

The reason for choosing the y axis parallel to the vector OG is that the G reflection is a symmetric Bragg reflection. The atomic planes which are perpendicular to OG are parallel to the crystal surface. In order to fulfill the condition of continuity of the tangential components of the wavevectors at the crystal surface, the accommodation vector $k\delta$ must be normal to the crystal surface. The y axis is therefore chosen along the same direction as $k\delta$ to facilitate the calculation. Suppose the vector AT, shown in Fig. 5.4, represents $k\delta$. The coordinates of point A are (x, y, z). Point T is the tie point on

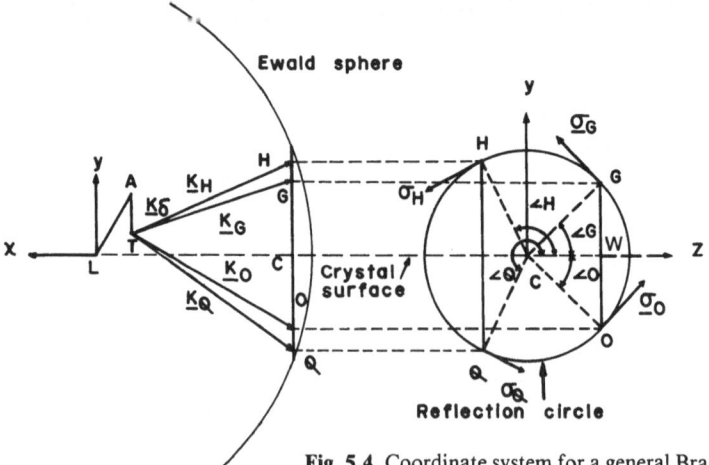

Fig. 5.4. Coordinate system for a general Bragg-type four-beam diffraction

the dispersion surface. Thus, the wavevectors within the crystal are the vectors from the point T to each reciprocal lattice point, i.e.,

$$K_O = TO,$$
$$K_G = TG,$$
$$K_H = TH,$$
$$K_Q = TQ.$$

(5.111)

The incident wavevector is $k_O^e = AO$. All these wavevectors can be expressed in terms of the angles $\sphericalangle O = \sphericalangle OCW$, $\sphericalangle G = \sphericalangle GCW$, $\sphericalangle H = \sphericalangle HCW$ and $\sphericalangle Q = \sphericalangle QCW$ by the following general form:

$$k_O^e = \hat{x}k_x + \hat{y}k_y + \hat{z}k_z,$$
$$K_M = \hat{x}K_{Mx} + \hat{y}K_{My} + \hat{z}K_{Mz},$$

(5.112)

for all $M = O, G, H, Q$, where

$$k_x = -\overline{LeC} - x,$$
$$k_y = \overline{OC}\sin(\sphericalangle O) - y,$$
$$k_z = \overline{OC}\cos(\sphericalangle O) - z,$$
$$K_{Mx} = k_x,$$
$$K_{My} = \overline{OC}\sin(\sphericalangle M) + k\delta,$$
$$K_{Mz} = \overline{OC}\cos(\sphericalangle M) - z.$$

(5.113)

The σ-polarization vectors are chosen according to the discussion in Sects. 4.2, 4.5.1, to lie in the plane containing the circle of reflection. The coordinates of the σ vectors are also functions of the angles $\sphericalangle M$, i.e.,

$$\hat{\sigma}_M = \frac{-\hat{y}K_{Mz} + \hat{z}K_{My}}{\sqrt{K_{My}^2 + K_{Mz}^2}},$$

(5.114)

for $M = O, G, H$ and Q. The corresponding π vectors are then defined by (4.35). The term $\chi_{O-O} - 2\varepsilon$ in (4.38), after geometric manipulation, can therefore be written as

$$\chi_{O-O} - 2\varepsilon_M = \chi_{O-O} - B'_M\delta + \alpha_M,$$

(5.115)

where

$$B'_M = \frac{2\overline{OC}\sin(\sphericalangle M)}{k},$$

$$\alpha_M = \frac{2\overline{OC}}{k^2}\{y[\sin(\not{\angle}M) - \sin(\not{\angle}O)] + z[\cos(\not{\angle}M) - \cos(\not{\angle}O)]\}. \quad (5.116)$$

The first-order approximation has been employed, and the second-order terms in x, y, z and $k\delta$, of their products were neglected because the order of magnitude of these quantities is 10^{-5} of the wavevectors. By substituting (5.115, 116) into (4.38), the dispersion equation takes the form of an eigenvalue equation. For simplicity, the dispersion equation is written for a singly polarized wavefield:

$$\begin{pmatrix} \Phi_{OO}-\delta & \Phi_{OG} & \Phi_{OH} & \Phi_{OQ} \\ \Phi_{GO} & \Phi_{GG}-\delta & \Phi_{GH} & \Phi_{GQ} \\ \Phi_{HO} & \Phi_{HG} & \Phi_{HH}-\delta & \Phi_{HQ} \\ \Phi_{QO} & \Phi_{QG} & \Phi_{QH} & \Phi_{QQ}-\delta \end{pmatrix} \begin{pmatrix} E_{\sigma O} \\ E_{\sigma G} \\ E_{\sigma H} \\ E_{\sigma Q} \end{pmatrix} = 0, \quad (5.117)$$

where

$$\Phi_{MM} = \frac{\chi_{O-O} + \alpha_M}{B'_M}, \quad (5.118a)$$

for all M's, and

$$\Phi_{MS} = \frac{\chi_{M-S}P}{B'_M}, \quad (5.118b)$$

for all $M \neq S$. In the actual calculation, an (8×8) matrix equation like (5.117) is used to include the two polarizations. The polarization factor P is then the product of the corresponding $\hat{\sigma}$'s and $\hat{\pi}$'s.

It should be noted that the first-order approximation is invalid when the angle $\not{\angle}M = 0°$ or $180°$. This condition for surface diffraction leads to a singularity in (5.118). In this case, a higher-order approximation needs to be used [5.20].

The eigenvalues and eigenvectors of (5.117) can be obtained, as usual, by transforming the matrix in (5.117) into a Hessenberg matrix [5.21]. The number of permitted modes N_p is four, according to (4.181), since there are two Bragg reflections, G and H. Matrix form can also be adopted for the boundary conditions (4.213, 214) for very thick crystals:

$$\begin{bmatrix} A & C \\ B & -D \end{bmatrix}[\lambda] = [E], \quad (5.119)$$

where

$$A = \begin{bmatrix} \text{Cof}(\sigma O, \sigma O)_1 & \text{Cof}(\sigma O, \sigma O)_2 & \text{Cof}(\sigma O, \sigma O)_3 & \text{Cof}(\sigma O, \sigma O)_4 \\ \text{Cof}(\pi O, \pi O)_1 & \text{Cof}(\pi O, \pi O)_2 & \text{Cof}(\pi O, \pi O)_3 & \text{Cof}(\pi O, \pi O)_4 \\ \text{Cof}(\sigma Q, \sigma O)_1 & \text{Cof}(\sigma Q, \sigma O)_2 & \text{Cof}(\sigma Q, \sigma O)_3 & \text{Cof}(\sigma Q, \sigma O)_4 \\ \text{Cof}(\pi Q, \pi O)_1 & \text{Cof}(\pi Q, \pi O)_2 & \text{Cof}(\pi Q, \pi O)_3 & \text{Cof}(\pi Q, \pi O)_4 \end{bmatrix},$$

$$B = \begin{bmatrix} \text{Cof}(\sigma G, \sigma O)_1 & \text{Cof}(\sigma G, \sigma O)_2 & \text{Cof}(\sigma G, \sigma O)_3 & \text{Cof}(\sigma G, \sigma O)_4 \\ \text{Cof}(\pi G, \pi O)_1 & \text{Cof}(\pi G, \pi O)_2 & \text{Cof}(\pi G, \pi O)_3 & \text{Cof}(\pi G, \pi O)_4 \\ \text{Cof}(\sigma H, \sigma O)_1 & \text{Cof}(\sigma H, \sigma O)_2 & \text{Cof}(\sigma H, \sigma O)_3 & \text{Cof}(\sigma H, \sigma O)_4 \\ \text{Cof}(\pi H, \pi O)_1 & \text{Cof}(\pi H, \pi O)_2 & \text{Cof}(\pi H, \pi O)_3 & \text{Cof}(\pi H, \pi O)_4 \end{bmatrix},$$

$$C = [0] \ (4 \times 4 \ \text{null matrix}), \tag{5.120}$$

$$D = [1] \ (4 \times 4 \ \text{unit matrix}),$$

$$[\lambda] = \begin{bmatrix} \lambda_\sigma(1) & \lambda_\pi(1) \\ \lambda_\sigma(2) & \lambda_\pi(2) \\ \lambda_\sigma(3) & \lambda_\pi(3) \\ \lambda_\sigma(4) & \lambda_\pi(4) \\ E_{\sigma G}(\sigma) & E_{\sigma G}(\pi) \\ E_{\pi G}(\sigma) & E_{\pi G}(\pi) \\ E_{\sigma H}(\sigma) & E_{\sigma H}(\pi) \\ E_{\pi H}(\sigma) & E_{\pi H}(\pi) \end{bmatrix}, \quad [E] = \begin{bmatrix} E'_{\sigma O} & 0 \\ 0 & E'_{\pi O} \\ 0 & 0 \\ 0 & 0 \\ 0 & 0 \\ 0 & 0 \\ 0 & 0 \\ 0 & 0 \end{bmatrix}.$$

The dimensions of matrices A, B, C, D, λ and E in a general N-beam diffraction are $(N_T \times N_T)$, $(N_R \times N_T)$, $(N_T \times N_R)$, $(N_R \times N_R)$, $(N \times 2)$, $(N \times 2)$, where N_T and N_R are the numbers of transmitted and reflected beams, respectively. For simplicity, the components of the incident wavefields $E'_{\sigma O}$ and $E'_{\pi O}$ are set equal to unity in the calculation.

The linear complex matrix equation (5.119) can be easily solved with the aid of Gaussian reduction [5.22]. The reflected intensities can therefore be obtained according to (4.216).

For multiple Bragg diffractions in thin crystals, the matrices A and E involved in (5.119) are the same as in (5.120), while the elements of the matrix B must be multiplied by $V_G(j)$ (abbreviated as V_j) which is defined in (4.206). Matrices C, D and λ take the following different forms:

$$C = \begin{bmatrix} \text{Cof}(\sigma O, \sigma O)_5 & \text{Cof}(\sigma O, \sigma O)_6 & \text{Cof}(\sigma O, \sigma O)_7 & \text{Cof}(\sigma O, \sigma O)_8 \\ \text{Cof}(\pi O, \pi O)_5 & \text{Cof}(\pi O, \pi O)_6 & \text{Cof}(\pi O, \pi O)_7 & \text{Cof}(\pi O, \pi O)_8 \\ \text{Cof}(\sigma Q, \sigma O)_5 & \text{Cof}(\sigma Q, \sigma O)_6 & \text{Cof}(\sigma Q, \sigma O)_7 & \text{Cof}(\sigma Q, \sigma O)_8 \\ \text{Cof}(\pi Q, \pi O)_5 & \text{Cof}(\pi Q, \pi O)_6 & \text{Cof}(\pi Q, \pi O)_7 & \text{Cof}(\pi Q, \pi O)_8 \end{bmatrix},$$

$$D = - \begin{bmatrix} V_5\text{Cof}(\sigma G, \sigma O)_5 & V_6\text{Cof}(\sigma G, \sigma O)_6 & V_7\text{Cof}(\sigma G, \sigma O)_7 & V_8\text{Cof}(\sigma G, \sigma O)_8 \\ V_5\text{Cof}(\pi G, \pi O)_5 & V_6\text{Cof}(\pi G, \pi O)_6 & V_7\text{Cof}(\pi G, \pi O)_7 & V_8\text{Cof}(\pi G, \pi O)_8 \\ V_5\text{Cof}(\sigma H, \sigma O)_5 & V_6\text{Cof}(\sigma H, \sigma O)_6 & V_7\text{Cof}(\sigma H, \sigma O)_7 & V_8\text{Cof}(\sigma H, \sigma O)_8 \\ V_5\text{Cof}(\pi H, \pi O)_5 & V_6\text{Cof}(\pi H, \pi O)_6 & V_7\text{Cof}(\pi H, \pi O)_7 & V_8\text{Cof}(\pi H, \pi O)_8 \end{bmatrix},$$

$$\tag{5.121}$$

$$\lambda = \begin{bmatrix} \lambda_\sigma(1) & \lambda_\pi(1) \\ \lambda_\sigma(2) & \lambda_\pi(2) \\ \cdot & \cdot \\ \cdot & \cdot \\ \cdot & \cdot \\ \lambda_\sigma(8) & \lambda_\pi(8) \end{bmatrix},$$

where the number of permitted modes is 8.

For N-beam Borrmann diffractions, a coordinate system for wavevectors in reciprocal space is chosen so that the x axis is along the accommodation vector $k\delta$, and the reciprocal lattice point O lies on the z axis. This is shown in Fig. 5.5. The σ- and π-polarization vectors are defined in the same way as those in Bragg diffraction. The term $\chi_{O-O} - 2\varepsilon$ has the same form as (5.115), except that

$$B'_M = \frac{2}{k}\sqrt{k^2 - \overline{OC}^2}, \tag{5.122}$$

and $\measuredangle O = 0°$. Since B'_M is a constant for all the M reflections, the dispersion equation is readily in the form of an eigenvalue equation. The matrix form for the boundary conditions is also simplified as

$$[A][\lambda] = [E], \tag{5.123}$$

where the matrices A and λ are similar to the A and λ in (5.119), except that the dimensions are different. The dimensions of A and λ are $(2N \times 2N)$ and $(2N \times 2)$, respectively. The intensities can be calculated according to (4.211, 216).

This calculation procedure gives the value of the diffracted intensity for a given entrance point A, i.e., for a given $\Delta\theta$ and $\Delta\phi$. For experimental comparison, one can either calculate the intensity distribution versus $(\Delta\theta, \Delta\phi)$ or

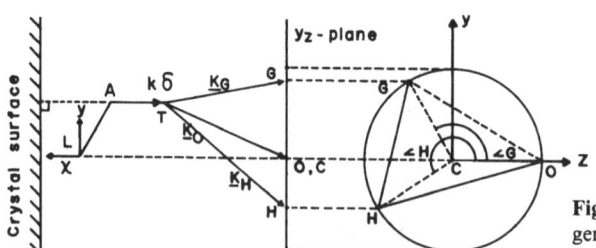

Fig. 5.5. Coordinate system for a general transmission-type three-beam diffraction

calculate the integrated intensity over $\Delta\theta$ and plot it versus $\Delta\phi$. The integration over $\Delta\phi$ is necessary for a large incident beam divergence, say, greater than 10 min of arc. For a small beam divergence, the integration over $\Delta\phi$ may not be needed, since the calculation always exhibits qualitatively the features of the diffracted intensity. This fact is discussed in Chap. 6. The calculated intensity distribution can be verified by the divergent-beam photographs (Sect. 2.2.2) which reveal the diffracted intensity in the vicinity of the N-beam diffracting position. The integrated intensity can be compared directly with the multiple diffraction patterns obtained with the collimated-beam method (Sect. 2.2.1).

This calculation procedure can be programmed for any N-beam case, as long as the capacity of a computer permits. This is because the geometric arrangement for the reciprocal lattice points, i.e., the chosen coordinate system, provides a common expression for calculating the angles $\not\downarrow O$, $\not\downarrow G$, ... However, care must be taken when the indices of the reflections involved are introduced in the input of the calculation. The sequence in which the reflections are introduced should be consistent with the actual geometrical relation between the reciprocal lattice points. In the four-beam case just discussed, the sequence is O, G, H and Q, i.e., along the reflection circle (Fig. 5.4). If the reflections are introduced in erroneous order, for example, O, H, G, Q, the calculation leads to wrong values for $\not\downarrow H$ and $\not\downarrow G$.

The Bragg-type multiple diffraction discussed is restricted to cases involving symmetric Bragg reflections as the primary reflections. The calculation procedure is, however, applicable to cases in which the primary reflection is an inclined Bragg reflection, with a minor modification to the transformation of the coordinates.

5.3 Quantum Mechanical Approach

A quantum mechanical approach for x-ray diffraction in crystals was first given by *Kohler* [5.23] to remedy von Laue's macroscopic dynamical theory. A rather complete development of this quantum mechanical theory was reported by *Molière* [5.8] in 1939. On the one hand, the quantum mechanical theory of x-ray diffraction provides a more strictly microscopic view of the interaction between the incident electromagnetic waves and the electron distribution around the atomic cores. On the other hand, this approach is directly related to the origin of the absorption of x-rays in crystals, due to the photoelectric effect, Thomson scattering and Compton scattering. However, Molière's treatment is somewhat difficult to adopt for calculation. In 1964, *Ohtsuki* [5.10] adopted the *Yoshioka* dynamical theory [5.24] of inelastic scattering of electrons and derived a quantum mechanical theory to describe the elementary processes of inelastic scattering of x-rays by phonons in crystals. Independently *Afanas'ev* and *Kagan* [5.15] developed a similar theory to account for the interaction of γ-rays and neutrons with crystalline materials. Quantum field theory has also been employed to give the same formulation for x-ray diffraction by *Ashkin* and *Kuriyama* [5.11], *Ohtsuki* and *Yanagawa* [5.12], and *Kuriyama* [5.13]. In this section we shall give a brief derivation of this approach, following *Molière*'s and *Ohtsuki*'s treatments.

Let us return to Maxwell's equation (4.1). The electric and magnetic fields of external electromagnetic waves, like x-rays, can be written in terms of the vector and scalar potentials A and ϕ as

$$E = -\frac{1}{c}\frac{\partial A}{\partial t} - \operatorname{grad}\phi, \quad H = \operatorname{curl}A. \tag{5.124}$$

Since the radiation field is transverse, the Coulomb gauge is adopted:

$$\operatorname{div}A = 0. \tag{5.125}$$

By substituting (5.124, 125) into (4.1), Maxwell's equations take the form

$$\nabla^2 A - \frac{1}{c^2}\frac{\partial^2 A}{\partial t^2} = -\frac{4\pi}{c}J. \tag{5.126}$$

In an empty space, there is no interaction between the electromagnetic waves and the crystal. The current density J is therefore null. Under this condition, a typical plane-wave solution to

$$\nabla^2 A - \frac{1}{c^2}\frac{\partial^2 A}{\partial t^2} = 0, \quad \text{is} \tag{5.127}$$

$$A(r, t) = A(r)\exp(+i\omega t) + \text{c.c.} = A_0\exp(+i\omega t - 2\pi i k \cdot r) + \text{c.c.}, \tag{5.128}$$

where c.c. indicates the complex conjugate of the preceding term. Equation (5.125) implies that

$$A_0 \cdot k = 0. \tag{5.129}$$

Now let us turn to the interaction of the electromagnetic waves with the crystal. By ignoring the motion of the nuclei, the Hamiltonian can be expressed as follows

$$H = \frac{1}{2m}\sum_j\left[p_j - \frac{e}{c}A(r_j)\right]^2 - \sum_j e\phi(r_j) + \sum_j V(r_j), \tag{5.130}$$

where p is the canonical momentum operator and V is the potential energy binding the electrons. The corresponding time-dependent Schrödinger equation is then

$$i\hbar \frac{\partial \Psi}{\partial t} = \sum_j \left\{ -\frac{\hbar^2}{2m} \nabla_j^2 + \frac{ie\hbar}{mc} A(r_j) \cdot \nabla_j + \frac{ie\hbar}{2mc} [\operatorname{div} A(r_j)] \right.$$

$$\left. + \frac{e^2}{2mc^2} A^2(r_j) + e\phi(r_j) + V(r_j) \right\} \Psi, \tag{5.131}$$

where e and m are the charge and the mass of the electron. The quantity Ψ is the wavefunction of the system. The electromagnetic fields, A and ϕ, are so weak in x-ray cases that they can be considered as small perturbations. By neglecting $\operatorname{div} A$ and ϕ, (5.131) can be written as

$$i\hbar \frac{\partial \Psi}{\partial t} = (H_0 + H') \Psi, \quad \text{where} \tag{5.132}$$

$$H_0 = -\frac{\hbar^2}{2m} \sum_j \nabla_j^2 + \sum_j V(r_j), \quad \text{and} \tag{5.133}$$

$$H' = \frac{ie\hbar}{mc} \sum_j A(r_j) \cdot \nabla_j. \tag{5.134}$$

According to the time-dependent perturbation theory [5.25], the wavefunction Ψ can be represented by the superposition of the stationary eigenfunctions $u_l(r) = u_l(r_1, r_2, \ldots, r_n)$ of the unperturbed Hamiltonian H_0 with time-dependent coefficients $a_l(t)$, i.e.,

$$\Psi(r, t) = \sum_l a_l(t) u_l(r) \exp\left(\frac{iE_l^0 t}{\hbar}\right), \quad \text{where} \tag{5.135}$$

$$a_l(t) = a_l^{(0)} + a_l^{(1)}(t), \quad \text{with} \tag{5.136}$$

$$a_l^{(0)} = \langle l | n \rangle = \delta_{ln}, \tag{5.137}$$

$$a_l^{(1)}(t) = \frac{1}{i\hbar} \int_{-\infty}^t \langle l | H'(t') | n \rangle \exp(i\omega_{ln} t') dt'.$$

The 'ket' and 'bra' have their usual meanings in quantum mechanics, δ_{ln} is the Kronecker delta, and n and l are the initial and final states. The term

$$\omega_{ln} = \frac{E_l^0 - E_n^0}{\hbar}, \tag{5.138}$$

is the Bohr angular frequency where E^0 denotes the enery. Substituting (5.134) into (5.137), we obtain

$$a_l^{(1)}(t) = -\frac{\langle l | H_1 | n \rangle \exp[i(\omega_{ln} - \omega) t]}{\hbar(\omega_{ln} - \omega - i\delta)} - \frac{\langle l | H_2 | n \rangle \exp[i(\omega_{ln} + \omega) t]}{\hbar(\omega_{ln} + \omega - i\delta)}, \tag{5.139}$$

where

$$\langle l | H_1 | n \rangle = \frac{ie\hbar}{mc} \sum_j \langle l | [\exp(i\mathbf{k} \cdot \mathbf{r}_j)] A_0(\mathbf{r}_j) \cdot \mathrm{grad}_j | n \rangle,$$

$$\langle l | H_2 | n \rangle = \frac{ie\hbar}{mc} \sum_j \langle l | [\exp(i\mathbf{k} \cdot \mathbf{r}_j)] A_0(\mathbf{r}_j) \cdot \mathrm{grad}_j | n \rangle \tag{5.140}$$

and δ is an infinitesimal positive quantity.

As is well known, the general expression for the current of the perturbed system takes the form

$$J(\mathbf{r}) = \frac{e}{2m} \sum_j [p_j \delta(\mathbf{r} - \mathbf{r}_j) + \delta(\mathbf{r} - \mathbf{r}_j) p_j] - \frac{e^2}{mc} \sum_j A(\mathbf{r}_j, t) \delta(\mathbf{r} - \mathbf{r}_j), \tag{5.141}$$

where the summation is over all the electrons. The average value of J is then

$$\langle J \rangle = \sum_l [j_{nl}(\mathbf{r}) a_l^*(t) + j_{nl}^*(\mathbf{r}) a_l(t)] - \frac{e}{mc} \varrho(\mathbf{r}) A(\mathbf{r}, t), \tag{5.142}$$

where

$$j_{nl}(\mathbf{r}) = j_{ln}^*(\mathbf{r}) = \frac{e\hbar}{2mi} \sum_j \int (u_n \nabla_j u_l^* - u_n^* \nabla_j u_l) \delta(\mathbf{r} - \mathbf{r}_j) d\mathbf{r}_j, \tag{5.143}$$

$$\varrho(\mathbf{r}) = eN(\mathbf{r}) = \sum_j e \int \Psi^* \Psi \delta(\mathbf{r} - \mathbf{r}_j) d\mathbf{r}_j. \tag{5.144}$$

By expressing $a_l^{(1)}(t)$ of (5.137) in terms of j_{nl} according to (5.139, 140), the average current density can be written as

$$\langle J \rangle = -\frac{e\varrho(\mathbf{r})}{mc} A(\mathbf{r}, t) + \frac{1}{\hbar c} \sum_l \left\{ \frac{j_{nl}(\mathbf{r}) \int [A(\mathbf{r}', t), j_{nl}^*(\mathbf{r}')] d\mathbf{r}'}{\omega_{ln} - \omega - i\delta} \right.$$

$$\left. + \frac{j_{nl}^*(\mathbf{r}) \int [A(\mathbf{r}', t), j_{nl}(\mathbf{r}')] d\mathbf{r}'}{\omega_{ln} + \omega - i\delta} \right\}, \tag{5.145}$$

where $[A, j_{nl}]$ indicates the dyadic product (the second-rank tensor). Combining (5.124, 145) with Maxwell's equations (4.1) and setting $\phi = 0$, we obtain

$$E = -\frac{i\omega}{c} A,$$

$$\mathrm{curl} \, H = \frac{i\omega}{c} \int \tilde{\varepsilon}(\mathbf{r}, \mathbf{r}', \omega) E(\mathbf{r}') d\mathbf{r}', \tag{5.146}$$

where the dielectric tensor is defined as

$$\tilde{\varepsilon}(\mathbf{r}, \mathbf{r}', \omega) \simeq \delta(\mathbf{r} - \mathbf{r}') \left(1 - \frac{4\pi e\varrho(\mathbf{r}')}{m\omega^2} \right) + \frac{4\pi}{\hbar\omega^2} \sum_l A_l(\mathbf{r}, \mathbf{r}', \omega), \tag{5.147}$$

with

$$A_l(r, r', \omega) = \frac{[j_{nl}(r), j_{nl}^*(r')]}{\omega_{ln} - \omega - i\delta} + \frac{[j_{nl}^*(r), j_{nl}(r')]}{\omega_{ln} + \omega - i\delta}. \tag{5.148}$$

The wavefields E and H, following Laue's treatment (Sect. 4.1), have the form of the Bloch wave in (4.23). Since the vector potential is of short range, we can concentrate on the interaction of the electromagnetic waves within the crystal unit cells. The electron density $\varrho(r)$, which is the Fourier transform of the structure factor F or of the electric susceptibility $\chi/4\pi$, has been defined in (4.21).

By substituting (4.19 – 21) into (5.147), and following the same derivation as for (4.32) in Sect. 4.1, the fundamental equation of the wavefield can be written in the form

$$\frac{K_G^2 - k^2}{k^2} E_G + \sum_H \chi_{G-H} E_H + \sum_H C_{GH} E_H = 0, \tag{5.149}$$

where χ_{G-H} has been defined in (4.19) of Sect. 4.1 and $k = \omega/c$. The quantity C_{GH} is a tensor defined as

$$C_{GH}(\omega) = \frac{4\pi}{\hbar\omega^2} \sum_l \iint \exp(2\pi i K_G \cdot r)$$

$$\times A_l(r, r', \omega) \exp(-2\pi i K_H \cdot r') dr\, dr'. \tag{1.150}$$

Introducing the polarization factors in the same manner as in the classical case discussed in Sect. 4.2, we obtain the fundamental equation for a general three-beam (O, G, H) diffraction,

$$\frac{K_G^2 - k^2}{k^2} E_G + \sum_H \Phi_{G-H} E_H + \sum_H C_{GH} E_H = 0, \tag{5.151}$$

where

$$\Phi_{G-H} = (\hat{\sigma}_G \cdot \hat{\sigma}_H) \chi_{G-H},$$
$$C_{GH} = \hat{\sigma}_G \cdot C_{GH} \cdot \hat{\sigma}_H. \tag{5.152}$$

σ_G and σ_H represent all possible σ's and π's and E_G and E_H are the corresponding $E_{\sigma G}$ and $E_{\sigma H}$.

When the angular frequency ω is far away from the frequency ω_{ln}, C_{GH} is so small in comparison with χ_{0-0} and χ_{G-H} that the last term of (5.151) can be neglected. Equation (5.151) is then reduced to its classical form, (4.32). When the frequency ω is close to ω_{ln}, for example, at a critical absorption edge, C_{GH} becomes as important as χ. In this case, the quantum mechanical approach should be employed. For two-beam cases, absorption under the influence of photoelectric transitions has been studied both theoretically [5.9] and experimentally [5.26]. However, the quantum effects on the absorption in multi-beam cases have only been considered theoretically for symmetric transmission cases [5.27]. Experimental investigation of this effect in an N-beam regime has not been reported so far. Further development and study of this subject are necessary to reveal more fundamental insights into multi-beam quantum effects in x-ray diffraction.

It should also be noted that the classical formalism can still be used in N-beam cases when the anomalous scattering due to photoelectric absorption of electrons is taken into account as a correction in the form factors f. This classical treatment has been adopted in two-beam dynamical calculations. Theoretical results which compare satisfactorily with experimental ones have been reported [5.18, 26]. Numerical calculation of absorption coefficients, diffracted intensities and the related items discussed in Sects. 4.8, 9 in N-beam cases near an absorption edge can be carried out similarly.

5.4 N-Beam Diffraction in Other Types of Interaction

5.4.1 Two-Beam Diffraction with Specular Reflection

The effect of specular reflection of an incident x-ray from a crystal surface is usually neglected in the conventional theory of x-ray diffraction. The influence of the specular reflection cannot, however, be ignored when the incident glancing angle is smaller than the critical angle of the total reflection from the crystal surface. The occurrence of this specular reflection is often associated with asymmetric two-beam Bragg or two-beam Laue cases, in which the incident beam is nearly parallel to the crystal surface. In these cases, the two-beam diffraction involves three reflected beams and we shall therefore consider it as a multiple diffraction.

Let us consider first an asymmetric two-beam, O and G, Bragg reflection, which is shown schematically in Fig. 5.6, using a plane-parallel crystal plate of thickness T_0. The incident, the diffracted and the specularly reflected waves are denoted as E_O^e, E_G^e and E_S^e, respectively. The angle between the crystal surface and the diffracting atomic planes is α. θ_G and θ are the Bragg angle of

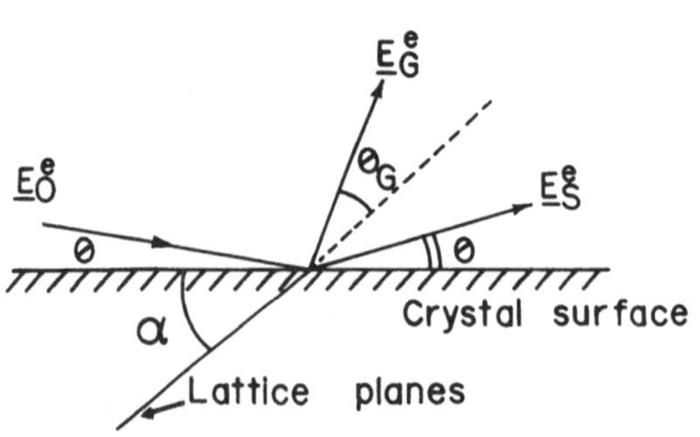

Fig. 5.6. Two-beam Bragg diffraction with a specularly reflected beam and a small angle of incidence (after *Kishino* and *Kohra* [5.28])

the reflection G and the glancing angle of the incident beam, respectively. The glancing angle θ is assumed to be comparable with the critical angle θ_c for the x-ray used.

The wavefields inside and outside the crystal plate are

$$E^i = \sum_j E_O(j) \exp(-2\pi i K_O(j) \cdot r) + E_G(j) \exp(-2\pi i K_G(j) \cdot r), \quad (5.153)$$

$$E^0 = E_O^e \exp(-2\pi i k_O \cdot r) + E_G^e \exp(-2\pi i k_G \cdot r) + E_S^e \exp(-2\pi i k_S \cdot r),$$

respectively, where $K_G(j) = K_O(j) + g$. By considering the continuity of the wavevectors inside and outside the crystal at the crystal boundary, the following relations are obtained:

$$K_O(j) \cdot \hat{t} = k_O \cdot \hat{t} = k_S \cdot \hat{t} = k \cos\theta,$$

$$K_G(j) \cdot \hat{t} = k_G \cdot \hat{t}, \quad (5.154)$$

$$k_O \cdot \hat{n} = -k_S \cdot \hat{n} = \Gamma_O = -\Gamma_S,$$

where $\hat{t}$ and $\hat{n}$ are the unit vectors tangential and normal to the crystal surface. The Γ's are the tangential components of the wavevectors K. The real part of Γ_O is equal to $k\gamma_O$, where γ_O is the direction cosine of the direct beam O relative to the crystal surface normal $\hat{n}$.

The fundamental equation of the wavefield, (4.36), should be considered for this two-beam diffraction case. For simplicity, only the σ-polarized wavefields are taken into account. The dispersion equation (4.41) gives

$$(K_O'^2 - K_O^2)(K_G'^2 - K_G^2) = k^4 \chi_{O-G} \chi_{G-O}. \quad (5.155)$$

The approximation $K_O = K_O' + \Delta K$ and $K_G = K_G' + \Delta K$ in the derivation (4.52) is not valid in the present case, because ΔK is no longer a small quantity compared with K_O' and K_G'. This can be seen clearly from the dispersion surface shown in Fig. 5.7. By considering the accommodation $k\delta$ as defined in (4.86) and the normal components of the vectors K_O and K_G, the dispersion equation (5.155) can be written as [5.28]

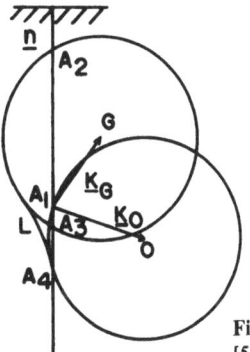

Fig. 5.7. The dispersion surface for Fig. 5.6 (after *Kishino* and *Kohra* [5.28])

$$(K_{O\perp,r}^2 - \xi_O^2)(K_{G\perp,r}^2 - \xi_G^2) = k^4 \chi_{O-G}\chi_{G-O}, \tag{5.156}$$

where

$$
\begin{aligned}
K_{O\perp,r}^2 &= K_O'^2 - k^2 \cos^2\theta, \\
K_{G\perp,r}^2 &= K_O'^2 - k^2 \cos^2(\theta + 2\alpha), \\
\xi_O^2(j) &= [K_{O\perp,r} - k\delta(j)]^2, \\
\xi_G^2(j) &= [g\cos\alpha - K_{O\perp,r}(j)]^2.
\end{aligned}
\tag{5.157}
$$

The subscript r indicates the real part of the vector. The wavevector modulus K' is defined in (4.51). For a given incident wave, there are four tie points excited, i.e., A_1, A_2, A_3 and A_4 in Fig. 5.7. The line $A_2 A_4$ is perpendicular to the crystal surface. It should be noted that the tie points in Fig. 5.7 are different from those in the conventional treatment (Sect. 4.3), because the dispersion surface in extremely asymmetric reflection gradually approaches the reflection circles of the Ewald sphere. In the conventional treatment these reflection circles are approximated by two straight lines. It is for this reason that the dispersion relation (4.52) has to be replaced by (5.156). The Poynting vectors at these tie points are directed normal to the dispersion surface. As discussed in Sect. 4.5.2, the tie point whose Poynting vector points towards the upper empty space need not be considered if the crystal is relatively thick, i.e., $\mu t \gtrsim 10$.

First let us assume the crystal to be infinitely thick, and that only the tie points A_1 and A_2 are excited. The boundary conditions for the wavefield amplitude, according to (4.199), lead to

$$E_O^e + E_S^e = E_O(1) + E_O(2), \qquad E_G^e = E_G(1) + E_G(2). \tag{5.158}$$

The continuity of the gradients of the wavefields normal to the surface at the boundary (boundary condition for magnetic induction) provides the two equations

$$
\begin{aligned}
(E_O^e - E_S^e)\Gamma_O &= E_O(1)\xi_O(1) + E_O(2)\xi_O(2), \\
E_G^e \Gamma_G &= E_G(1)\xi_G(1) + E_G(2)\xi_G(2).
\end{aligned}
\tag{5.159}
$$

The four unknown terms E_S^e, E_G^e, $E_O(1)$, $E_O(2)$ can be determined from (5.159). Following the same procedure given in Sect. 5.1.2, we have

$$
\begin{aligned}
E_S^e &= \frac{2(1+a)\Gamma_O E_O^e}{a(\Gamma_O + \xi_O(1)) + \Gamma_O + \xi_O(2)}, \\
E_G^e &= \frac{2\Gamma_O(h_1 a + h_2) E_O^e}{a(\Gamma_O + \xi_O(1)) + \Gamma_O + \xi_O(2)}, \\
E_O(1) &= a E_O(2),
\end{aligned}
\tag{5.160}
$$

$$E_O(2) = \frac{2\Gamma_O E_O^e}{a(\Gamma_O + \xi_O(1)) + \Gamma_O + \xi_O(2)},$$

where

$$a = -\frac{h_2[\Gamma_G - \xi_G(2)]}{h_1[\Gamma_G - \xi_G(1)]}. \tag{5.161}$$

The parameters h can be obtained from the two-beam fundamental equations (4.36, 115) by using the relations given in (5.157):

$$h_j = \frac{E_G(j)}{E_O(j)} = \sqrt{\frac{\xi_G(j)[\xi_O^2(j) - K_{O\perp,r}^2]}{\xi_O(j)[\xi_G^2(j) - K_{G\perp,r}^2]}}. \tag{5.162}$$

Since $|K_O(2) - k_O|$ is very large compared with $|K_O(1) - k_O|$, tie point A_2 is less excited than A_1. According to (4.116), $E_O(2)$ is almost null. Moreover, it can be seen from the geometry shown in Fig. 5.7 that $\xi_G(2) \simeq -\Gamma_G$, and $\xi_G(1) - \xi_G(2) \simeq 2\xi_G(1)$. With this approximation, we obtain

$$E_S^e = \frac{\Gamma_O - \xi_O(1)}{\Gamma_O + \xi_O(1)} E_O^e, \qquad E_G^e = \frac{2\xi_G(1)}{\Gamma_G + \xi_G(1)} E_G(1). \tag{5.163}$$

Thus, the intensities of the specularly reflected beam I_S and the diffracted beam I_G are obtained:

$$I_S \simeq |E_S^e|^2, \qquad I_G \simeq |E_G^e|^2. \tag{5.164}$$

The same consideration should also be given to cases involving a reflected beam nearly parallel to the crystal surface [5.29]. Figure 5.8 shows schematically this type of asymmetric reflection. The incident glancing angle θ is equal to $(\theta_G + \alpha)$. The angle $(\theta - \alpha)$ between the reflected wave E_G^e and the crystal surface is assumed to be very small. E_O^T is the transmitted wave in the incident

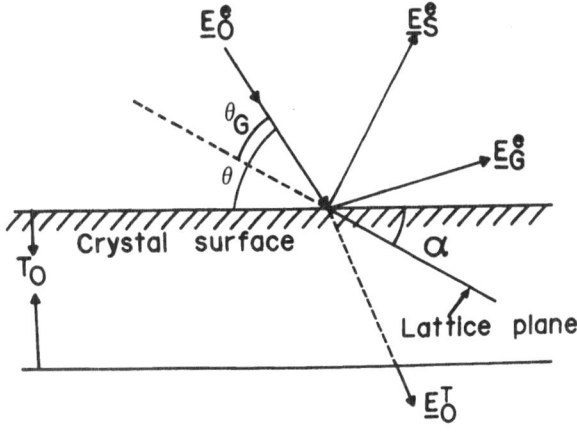

Fig. 5.8. Two-beam Bragg reflection with a specularly reflected beam and a large angle of incidence (after *Kishino* [5.29])

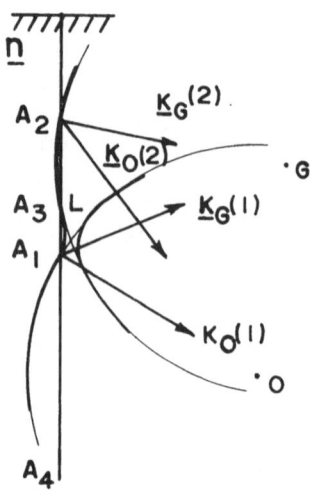

Fig. 5.9. The dispersion surface for Fig. 5.8 (after *Kishino* [5.29])

direction. The crystal thickness is T_0. The dispersion equation and the wave-fields inside and outside the crystal for the entrance surface of the crystal (upper surface) are exactly the same as given in (5.153, 157), except that the angle $(\theta + 2\alpha)$ has to be replaced by $(\theta - 2\alpha)$ in (5.157). The dispersion surface, similar to Fig. 5.7, is shown in Fig. 5.9.

If the crystal is relatively thin, the four tie points should all be considered. The tie points A_3 and A_4 which have their Poynting vectors directed towards the upper crystal surface can be regarded as the excitation points of the waves reflected from the lower crystal surface. Their contribution, especially from the tie point A_4, to the wavefield amplitude at the upper surface is relatively small compared with that from A_1 and A_2. It is justifiable to keep the two conditions (5.158, 159) for modes A_1 and A_2 and to add, referring to (4.207 – 209), the following additional conditions related to the crystal thickness effect and to the excitation of the tie point A_3:

$$E_O^{\mathrm{T}} \exp(2\pi i \Gamma_O T_0) = \sum_{j=1}^{3} E_O(j) \exp[2\pi i \xi_O(j) T_0] ,$$

$$0 = \sum_{j=1}^{3} E_G(j) \exp[2\pi i \xi_G(j) T_0] . \tag{5.165}$$

These are the boundary conditions at the lower crystal surface.

Combining (5.158, 159, 165), we obtain the transmitted reflected intensity I_G^{T} and the direct transmitted intensity I_O^{T} at the exit surface as:

$$I_O^{\mathrm{T}} \propto |E_O^{\mathrm{T}}|^2 , \qquad I_G^{\mathrm{T}} \propto |E_G^{\mathrm{T}}|^2 , \tag{5.166}$$

where

$$E_O^{\mathrm{T}} = \sum_{j=1}^{2} \left(1 - \frac{h_j}{h_3} \right) E_O(j) \exp[2\pi \xi_O^i(j) T_0] ,$$

$$E_G^T = \left(1 - \left|\frac{\Gamma_G - \xi_G(1)}{\Gamma_G - \xi_G(2)}\right|\right) h_1 E_O(1) . \tag{5.167}$$

The quantity ξ^i stands for the imaginary part of ξ.

If the two-beam diffraction is of the asymmetric Laue (transmission) type, the waves involved in the space above the upper crystal surface are more complicated than for asymmetric Bragg reflections [5.30]. Figure 5.10 shows the schematic representation of an asymmetric Laue diffraction. The quantities denoted as E_O^e, E_S^e, E_{GS}^e are the incident, specularly reflected and specularly diffracted waves above the upper surface. The waves E_O^T and E_G^T are the transmitted waves below the lower surface. The additional wave E_{GS}^e is due to the presence of the wavefields $E_G(3)$ and $E_G(4)$ inside the crystal, which are directed towards the upper surface. The corresponding Poynting vectors of these two fields are also indicated in Fig. 5.11. The resultant wavefield above the upper surface takes the form given in (5.153) for E_O, with E_{GS}^e and k_{GS} replacing E_G^e and k_G. The following relation concerning the boundary condition for the wavevectors at the upper surface has to be employed to replace the last equation of (5.154):

$$\boldsymbol{K}_G \cdot \hat{n} = \Gamma_G = -\boldsymbol{k}_{GS} \cdot \hat{n} . \tag{5.168}$$

In addition to (5.158, 159), the boundary conditions for the wavefields at the exit surface should be included:

$$E_O^T \exp(2\pi i \Gamma_O T_0) = \sum_{j=1}^{4} E_O(j) \exp[2\pi i \xi_O(j) T_0] ,$$

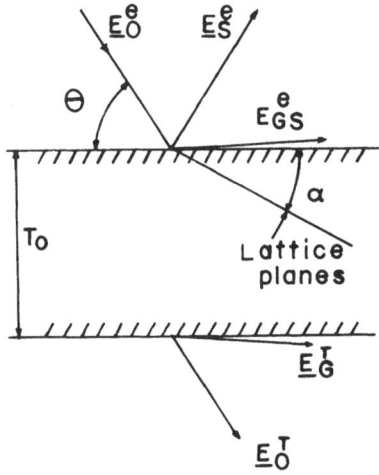

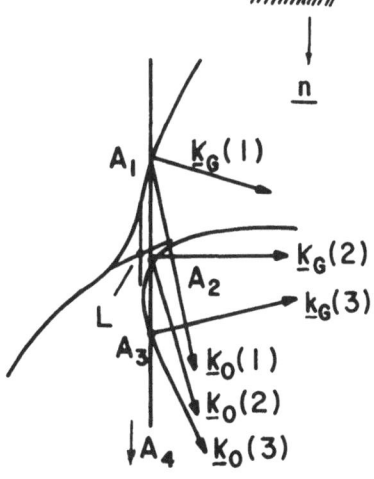

Fig. 5.10. Two-beam transmission diffraction with specularly diffracted and reflected beams (after *Kishino* et al. [5.30])

Fig. 5.11. The dispersion surface for Fig. 5.10 (after *Kishino* et al. [5.30])

$$E_G^T \exp(2\pi i \Gamma_G T_0) = \sum_{j=1}^{4} E_G(j) \exp[2\pi i \xi_G(j) T_0] ,$$

$$\Gamma_O E_O^T \exp(2\pi i \Gamma_O T_0) = \sum_{j=1}^{4} \xi_O(j) \exp[2\pi i \xi_O(j) T_0] , \qquad (5.169)$$

$$\Gamma_G E_G^T \exp(2\pi i \Gamma_G T_0) = \sum_{j=1}^{4} \xi_G(j) \exp[2\pi i \xi_G(j) T_0] ,$$

where the last two equations are related to the gradients of the wavefields normal to the exit surface.

By assuming that the tie point A_4 is not effective due to the large value of $|K_G(4) - K'|$ shwon in Fig. 5.11, the following expressions for the transmitted intensities can be derived:

$$I_O^T \propto \left| \sum_{j=1}^{2} \frac{\xi_O(j) - \xi_O(3)}{\Gamma_O - \xi_O(3)} E_O(j) \exp[2\pi \xi_O^i(j) T_0] \right|^2 ,$$

$$I_G^T \propto \left| \left| \frac{\Gamma_G}{\Gamma_O} \right|^2 \sum_{j=1}^{2} \frac{\xi_G(j) - \xi_G(3)}{\Gamma_G - \xi_G(3)} E_G(j) \exp[2\pi \xi_G^i(j) T_0] \right|^2 , \qquad (5.170)$$

where $|\Gamma_G/\Gamma_O|^2$ is the ratio of the density of x-ray waves per unit area between the two diffractions O and G. This ratio is equivalent to (γ_G/γ_O), where the γ's are the direction cosines.

Studies on specular reflection, which are similar to those discussed above, have also been reported in [5.20, 31].

It is worthwhile mentioning that the presence of the specular reflection, accompanied by an asymmetric two-beam diffraction, does not lead the fundamental equation of wavefields to the form of a three-beam diffraction. Rather, the excitation of the dispersion surface is affected via the geometrical relation of the crystal surface to the diffracted beams. The characteristics of two-beam diffraction remain in the dispersion relation as well as in the boundary conditions. This type of diffraction is therefore physically a 'pseudo' three-beam case.

5.4.2 Interaction of X-Rays with Phonons

It has been pointed out [5.32] that the dispersion surface of a two-beam diffraction is affected and modified by the interaction of x-rays with phonons in a crystal. This modification appears not only in the wavevectors but also in the energies, or the frequencies. In other words, instead of the wavevectors k, K_O, K_G and the wavefields E_O, E_G in a pure two-beam x-ray diffraction case, the modified wavevectors k, $K_{G\pm q} = K_G \pm q$, and the wavefields $E_O(0)$, $E_G(0)$, $E_O^\pm(q)$, $E_G^\pm(q)$ should be dealt with. The plus and minus signs on the superscripts stand for the increase and decrease in the wavevectors K due to the presence of the momenta q of the phonons. The quantities $E_O(0)$ and $E_G(0)$ are the unperturbed wavefields. The diffraction therefore actually involves a multi-beam interaction.

The fundamental equation of the wavefields for the unperturbed two-beam x-ray diffraction can be obtained from (5.5) by setting the terms involving the H reflection equal to zero:

$$2\xi_O E_O + P_G \chi_{O-G} E_G = 0 ,$$

$$P_G \chi_{G-O} E_O + 2\xi_G E_G = 0 , \quad \text{whence}$$

(5.171)

$$\xi_O \xi_G = \tfrac{1}{4} P_G^2 \chi_{O-G} \chi_{G-O} ,$$

(5.172)

where only one polarization with the polarization factor P_G ($= \cos 2\theta_G$) is concerned. The χ's include the temperature factor $\exp(-M)$, where M is the Debye-Waller parameter.

In the case of the inelastic scattering of x-rays by a single phonon of wavevector q in a phonon creation process, the structure factor F_q' can be defined as [5.33]

$$F_q' = \sum_j f_{q,j} \exp[2\pi i q \cdot (r_j + b_j)] ,$$

(5.173)

where r_j and b_j are the position vector of an undisplaced atom in the crystal and the displacement due to the phonon vibration, respectively. Since

$$\exp(2\pi i q \cdot b_j) \simeq 1 + i 2\pi q \cdot b_j$$

(5.174)

in the first-order approximation, and

$$\sum_j f_{q,j} \exp(2\pi i q \cdot r_j) = 0$$

(5.175)

for $q \neq G$, the corresponding χ_{q-O}' can be written as

$$\chi_{q-O}' = i \chi_{q-O} A^+(q) P_q ,$$

(5.176)

where χ_{q-O} has the same definition as χ_{G-O} except that G is replaced by q, and

$$A^+(q) = 2\pi \sqrt{\frac{\hbar}{2mn\omega_q}}\, a_{q,j}^+ ,$$

$$P_q = e_{q,j} \cdot \hat{q} .$$

(5.177)

In deriving (5.176), we employed the relation [5.34]

$$b_j = \sqrt{\frac{\hbar}{2mn\omega_q}}\ a_{q,j}^+ e_{q,j}, \tag{5.178}$$

where $a_{q,j}^+$, $e_{q,j}$, n and ω_q are the phonon creation operator [5.25], the unit polarization vector, the number of atoms in the crystal and the angular frequency of the phonon, respectively. Similarly, the χ'_{O-q} for the phonon annihilation process takes the form

$$\chi'_{O-q} = i\chi_{O-q}A^-(-q)P_q, \quad \text{where} \tag{5.179}$$

$$A^-(-q) = 2\pi \sqrt{\frac{\hbar}{2mn\omega_q}}\ a_{q,j}. \tag{5.180}$$

The quantity $a_{q,j}$ is the phonon annihilation operator. By replacing χ_{G-O} and χ_{O-G} in (5.171) by χ'_{q-O} and χ'_{O-q}, the wavefield equation for the interaction of the incident x-rays with a single phonon of wavevector q in phonon creation and annihilation processes can be written

$$2\xi_O E_O + iP_q\chi_{O-q}A^\mp(-q)E_O^\pm(q) = 0,$$
$$iP_q\chi_{q-O}A^\pm(q)E_O + 2\xi_G E_O^\pm(q) = 0. \tag{5.181}$$

In the multiple diffraction approach, the reflection G of the x-rays is the primary reflection, and the reflections with the reciprocal lattice vectors $g \pm q$ can be treated as the secondary reflections. The corresponding coupling reflections have therefore reciprocal lattice vectors equal to $\pm q$.

The interaction of the incident and the diffracted x-ray waves with the phonons can be described by the fundamental equation of multiple diffraction, taking into account the phonon creation and annihilation processes [5.34]:

$$2\xi_O E_O + P_G\chi_{O-G}E_G + i\sum_q P_q\chi_{O-(g+q)}A^\mp(-g-q)E_G^\pm(q)$$
$$+ i\sum_q P_q\chi_{O-q}A^\mp(-q)E_O^\pm(q) = 0,$$
$$2\xi_G E_G + P_G\chi_{G-O}E_O + i\sum_q P_q\chi_{O-q}A^\mp(-q)E_G^\pm(q)$$
$$+ i\sum_q P_{g-q}\chi_{(g-q)-O}A^\mp(g-q)E_O^\pm(q) = 0,$$
$$2\xi_{G+q}E_G^+(q) + iP_{g+q}\chi_{(g+q)-O}A^+(g+q)E_O + iP_q\chi_{q-O}A^+(q)E_G$$
$$+ P_G\chi_{G-O}E_O^+(q) = 0,$$
$$2\xi_{O+q}E_O^+(q) + iP_{g-q}\chi_{O-(g-q)}A^+(-g+q)E_G + iP_q\chi_{q-O}A^+(q)E_O$$
$$+ P_G\chi_{O-G}E_G^+(q) = 0,$$
$$\tag{5.182}$$

where the second- and higher-order phonon terms involved in $\exp(2\pi i q \cdot b_j)$ of (5.174) are not included. The summations in the first two equations are taken over all the first-order phonon processes. The last two equations are only for a given q. There are many equations like these two equations for different q's. The terms ξ_O and ξ_G are defined in (5.3), while ξ_{G+q} takes the form

$$2\xi_{G+q} = \chi_{O-O} + 2\gamma_{G+q}\delta - \Delta_{G+q}/k , \tag{5.183}$$

where Δ_{G+q} is the distance of the point Q from the modified Ewald sphere shown in Fig. 5.12. The quantity Δ_{G+q} depends therefore on the difference between k_{G+q} and k_ω. γ_{G+q} is the direction cosine of the wavevector K_{G+q} with respect to the inward crystal normal. Equation (5.182) can also be derived from Maxwell's equations for a periodic medium with the electron density a periodic function of space and time.

In the following, we shall apply (5.182) to deal with the interaction between x-rays and phonons in two simple cases.

a) One-beam x-ray and phonon interaction. For simplicity, let us first consider the interaction of phonons with the incident x-ray beam. Since phonon energies are very small compared with x-ray energies, k_ω is set to equal to k. The wavefield equation (5.182) of the phonon-scattered waves takes the following simple form:

$$2\xi_O E_O + i \sum_q P_{G+q}\chi_{O-(g+q)}A^{\mp}(-g-q)E_G^{\pm}(q)$$

$$+ i \sum_q P_q\chi_{O-q}A^{\mp}(-q)E_O^{\pm}(q) = 0 ,$$

$$2\xi_{O+q}E_O^+(q) + iP_q\chi_{q-O}A^+(q)E_O + P_G\chi_{O-G}E_G^+(q) = 0 , \tag{5.184}$$

$$2\xi_{G+q}E_G^+(q) + iP_{G+q}\chi_{(g+q)-O}A^+(g+q)E_O + P_G\chi_{G-O}E_O^+(q) = 0 .$$

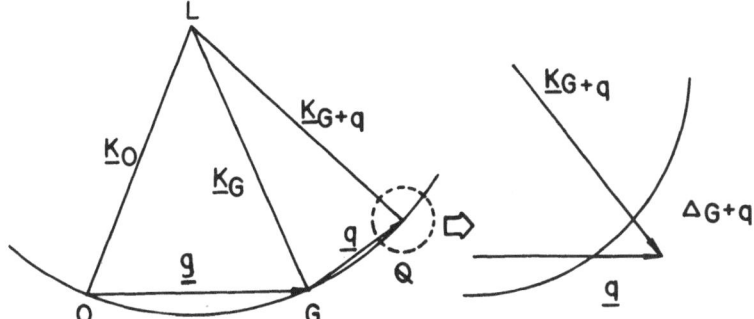

Fig. 5.12. Geometry of the phonon-photon interaction in a two-beam Bragg reflection (after O'Connor [5.34])

Since $E_O^+(q) \ll E_O$ and $E_G(q) = 0$ for the given q, (5.184) reduces in first-order approximation to the form of the two-beam interaction:

$$2\xi_O E_O + i \sum_q P_q \chi_{O-q} A^-(-q) E_O^+(q) = 0 \,,$$

$$2\xi_{O+q} E_O^+(q) + i P_q \chi_{q-O} A^+(q) E_O = 0 \,, \tag{5.185}$$

which is similar to (5.5) in Sect. 5.1.1. Eliminating E_O and $E_O(q)$, we obtain

$$\xi_O = -\frac{1}{4} \sum_q P_q^2 \chi_{O-q} \chi_{q-O} \left(\frac{A^+(q)A^-(-q)}{2\xi_{O+q}^+} + \frac{A^+(-q)A^-(q)}{2\xi_{O+q}^-} \right) , \tag{5.186}$$

where the distinction between ξ_{O+q}^+ and ξ_{O+q}^- is so small that we can set $\xi_{O+q}^+ = \xi_{O+q}^- = \xi_{O+q}$. Accordingly, the corresponding amplitude of a phonon-scattered wave can be written as

$$E_O^\pm(q) = -\frac{i P_q \chi_{q-O} A^\pm(q)}{2} \left(\frac{\xi_{O+q}^r - i \xi_{O+q}^i}{(\xi_{O+q}^r)^2 + (\xi_{O+q}^i)^2} \right) E_O , \tag{5.187}$$

which depends on ξ_{O+q} and ξ_O. Considering (5.8, 185, 186), and introducing $\xi_{O+q} = 0$ to satisfy the boundary condition that the phonon-scattered wave must vanish at one or other of the crystal surfaces, the wavefield amplitude can be found, following the procedure given in Sect. 4.7, for a plane-parallel crystal plate of a thickness T:

$$E_O^\pm(q) = \frac{P_q}{2} \chi_{q-O} A^\pm(q) E_O \left(\frac{\xi_{O+q}^i + i \xi_{O+q}^r}{|\xi_{O+q}|^2} \right) \exp(-2\pi i K_{O+q} \cdot r)$$

$$\times \left[\exp\left(\frac{2\pi i k \xi_{O+q} T}{\gamma_q} \right) - 1 \right]. \tag{5.188}$$

The intensity of the phonon-scattered wave at the exit surface can thus be obtained as

$$I_{O+q}^\pm = E_O^2 \chi_{O-q} \chi_{q-O} P_q^2 \left(\frac{\hbar}{2mn\omega_q} \right) [n_\omega(q) + 1 \pm 1] \exp(-4\pi k_{O+q,i} T/\gamma_q)$$

$$\times \frac{\sinh^2(2\pi k \xi^i T/2\gamma_q) + \sin^2(2\pi k \xi^r T/2\gamma_q)}{(\xi_{O+q}^r)^2 + (\xi_{O+q}^i)^2} , \tag{5.189}$$

which resembles (5.34) in Sect. 5.1.2. The quantity n_ω is the number of energy states. For small ξ^i, (5.189) can be further simplified, as is the case for diffractions in non-absorbing crystals discussed in Sect. 5.1.4.

b) *Two-beam x-ray and phonon interaction.* The fundamental equation
(5.182) should be considered for the phonon-photon interaction in a two-
beam x-ray diffraction. The dispersion surface of the two-beam x-ray diffraction, defined by (5.172), must be modified by the presence of the additional
wavefields $E_O^{\pm}(q)$ and $E_G^{\pm}(q)$. Eliminating $E_G^{\pm}(q)$ from the last two
equations in (5.182) with the aid of (5.171, 172), we obtain

$$E_O^+(q) = -\frac{i}{2\xi_{O+q}(1-\xi_O\xi_G/\xi_{O+q}\xi_{G+q})}[P_{g-q}\chi_{O-(g-q)}A^+(-g+q)$$

$$\times(1-\xi_G/\xi_{G+q})E_G+P_q\chi_{q-O}A^+(q)(1+\xi_O/\xi_{G+q})E_O]. \quad (5.190)$$

If the crystal setting is not at the exact interaction position and if ξ_{O+q} is
small, ξ_{G+q} is relatively large. Under this circumstance, (5.190) can be expressed as

$$E_O^+(q) = -\frac{i}{2\xi_{O+q}}[P_{g-q}\chi_{O-(g-q)}A^+(-g+q)E_G+P_q\chi_{q-O}A^+(q)E_O].$$
$$\quad (5.191)$$

The singularity in (5.190) when $\xi_{O+q}\xi_{G+q}$ is very close to $\xi_O\xi_G$ disappears accordingly.

The wavefield amplitudes E_O and E_G are always much greater than $E_O(q)$
and $E_G(q)$ in this phonon-photon interaction. Considering the presence of
phonons as a small perturbation to the system, we can set the unperturbed
$\xi_O = l\xi_G$ and the perturbed $\xi_O+\delta\xi_O$ with $\delta\xi_O \ll \xi_O$. The parameter l depends
on the angle of incidence. Using (5.191) for $E_O^+(q)$ and a similar expression
for $E_G^+(q)$ and combining them with (5.182), we obtain

$$\xi_O+\delta\xi_O = \pm\frac{\chi_{G-O}}{2\sqrt{l}} - \frac{\hbar}{16mnl}\sum_q\frac{n_\omega+\frac{1}{2}}{l\omega_q\xi_{O+q}\xi_{G+q}}\cdot[P_q^2\chi_{q-O}^2(l\xi_{G+q}$$

$$+\xi_{O+q})\mp 2P_qP_{g+q}\chi_{q-O}\chi_{(g+q)-O}\cdot\sqrt{l}(\xi_{G+q}+\xi_{O+q})$$

$$+P_{g+q}^2\chi_{(g+q)-O}^2(l\xi_{O+q}+\xi_{G+q})]. \quad (5.192)$$

A similar expression for $\xi_G+\delta\xi_G$ can also be obtained in the same way. This
expression (5.192) reduces to (5.186) when the crystal setting is far from the
Bragg diffraction position, i.e., when l is either very large or very small. When
$|q|$ approaches 0, $1/\omega$ tends to infinity. The last term in (5.192) then
dominates. This implies that the phonon scattering contributes as a diffuse
spot due to the long-wavelength phonons. The imaginary parts of (5.192) indicate that both the incident and elastically reflected waves suffer an extra attenuation due to phonon scattering. For the symmetric Bragg reflection, the
imaginary parts of ξ_O and ξ_G in the total reflection range are very large compared with the real parts, since $\xi_O^i \approx -\chi_{O-O}^r/2$. This effect which is due

entirely to the primary extinction (Sect. 4.6) overweighs the influence on the absorption of the phonon scattering. In the Laue case, ξ_O^i at the vertices of the dispersion hyperbola is equal to $-\chi_{O-O}^i/2$. The minimum absorption coefficient defined in (4.195) is related to $k(\chi_{O-O}^i - \chi_{G-O}^i)/2$. Since χ^i is a slowly varying function of the Bragg angle, the magnitudes of χ_{O-O}^i and χ_{G-O}^i differ only slightly from each other. The corresponding absorption is therefore very small. This leads to anomalously high transmission of the incident and scattered waves in crystals [5.35]. In this particular case, phonon scattering plays an important role in the transmission of the diffracted waves.

The intensity of the phonon scattering can also be calculated from (5.191) provided that E_O and E_G can be determined from the boundary conditions. Introducing a boundary wave with $\xi_{G+q} = 0$ as in the one-beam x-ray-phonon case, the wavefield amplitude at a depth t in the crystal is given by

$$
E_G^+(q) = -\frac{i}{2} \sum_j \frac{\exp[-2\pi i K_{G+q}(j) \cdot r]}{\xi_{G+q}(j)}
$$

$$
\times [P_{g-q}\chi_{O-(g-q)}A^+(-g+q)E_O(j) + P_q\chi_{q-O}A^+(q)E_G(j)]
$$

$$
\times \left[\exp\left(-\frac{2\pi i k \xi_{G+q}(j)t}{\gamma_q} \right) - 1 \right].
$$

(5.193)

The sum is taken over all permitted modes of propagation (Sect. 4.5.2).

The general expression for the phonon-scattered intensity can be obtained for a simple symmetric Laue case with a very small $|q|$. By following the procedure used in deriving (5.189), the integrated intensity is found to be proportional to

$$
I_G(q) \propto \frac{\hbar}{\omega_q} P_{q+q}^2 [n_\omega(q) + 1 \pm 1] \, |\chi_{G-O}|^2 \exp(-4\pi k_{O\,i}T_G)
$$

$$
\times \left([|E_O(1)|^2 \exp(-2\pi k \xi^i T_G) + |E_O(2)|^2 \exp(2\pi k \xi_{O+q}^i T_G)] \right.
$$

$$
\times \frac{\sinh(2\pi k \xi^r T_G/2)}{\xi^i} - \frac{\xi^r}{(\xi^r)^2 + (\xi^i)^2} \{E_2 + [1 - 2\cosh(2\pi k \xi^i T_G/2)]
$$

$$
\left. \times [E_2 \cos(2\pi k \xi^r T_G) + E_1 \sin(2\pi k \xi^r T_G)] \right),
$$

(5.194)

where $T_G = T/\gamma_G$, $\xi^r = \xi_{G+q}^r(1) - \xi_{G+q}^r(2)$, and $\xi^i = \xi_{G+q}^i(1) - \xi_{G+q}^i(2)$. Modes 1 and 2 for a singly polarized incident wave are considered. The quantities E_1 and E_2 are the real and the imaginary parts of $E_O(1)E_O^*(2)$. The interference between mode 1 and mode 2 appears as the sine and the cosine functions. The intensities of the diffuse scatterings at O and $g + q$ mentioned previously, for a very small $|q|$, depend on the thickness of the crystal and are roughly proportional to $|E_O|^2$ and $|E_G|^2$, respectively.

For a symmetric two-beam Bragg reflection in an infinitely thick crystal, there is only one permitted mode of propagation for a singly polarized incident wave (Sect. 4.5.2). The intensity of the phonon-scattered wave takes the simple form

$$I_G(q) \propto \frac{1}{4} P^2_{g+q} \chi^2_{(g+q)-O} A^+(g+q) A^-(g+q) \cdot \frac{E^2_O}{|\xi_{G+q}|^2}. \qquad (5.195)$$

Since $|\xi^r_{G+q}| < |\xi^i_{G+q}|$ in the total reflection region, range (II), the integrated intensity is proportional to $1/\xi^i_{G+q}$ for small $|q|$. This implies that the primary extinction still dominates the absorption because $1/\mu \gg 1/k\xi_{G+q}$. Thus, the intensity of the phonon-scattered waves is much reduced by the primary extinction.

If the angles of incidence lie in ranges (I) and (III) (Sect. 4.6) which are slightly off the total reflection region, the perturbation in ξ^i_O due to the elastic two-beam reflection becomes comparable with the perturbation due to the phonon scattering. It would be difficult to treat the interaction between phonons and x-rays quantitatively in these two ranges. Numerical calculation should therefore be employed.

We have discussed x-ray photon and phonon interaction, i.e., x-ray diffraction by vibrating crystals, following the treatment given in [5.34]. Similar treatments can be found in the work of *Schürmann* [5.36], *Haruta* [5.37], and *Kuriyama* and *Miyakawa* [5.38]. All these theories deal with either small vibration amplitudes or relatively large phonon wavelengths. In the following, we shall consider as a special case the interaction of x-rays with a highly coherent, monochromatic, intense phonon beam, generated by making use of the acoustoelectric effect [5.39, 40].

Based on the theory of *Köhler* et al. [5.41], the lattice displacement of vibration of this special kind can be expressed as

$$U(r, t) = u \exp(i\omega_q t - 2\pi i q \cdot r) + \text{c.c.}, \qquad (5.196)$$

where u is the amplitude, and q and ω_q are the wavevector and frequency of the phonon. The crystal vibration modifies the electric susceptibility in such a way that the χ in (4.19) is modified as

$$\chi_{\text{mod}}(r, t) = \sum_G \chi_G \exp(-2\pi i g_G \cdot r) \exp[2\pi i g_G \cdot U(r, t)], \qquad (5.197)$$

where g_G is the reciprocal lattice vector of the x-ray reflection G. The last exponential function can be expanded in a power series of $i(\omega_q t - q \cdot r)$. This leads (5.197) to an expression involving a Bessel function J_l of integer order l:

$$\chi_{mod}(r, t) = \sum_{G,l} \chi_G J_l(4\pi |g_G \cdot U|) \exp[i\omega_q t - 2\pi i(g_G + lq) \cdot r] . \quad (5.198)$$

By substituting χ_{mod} and the wavefield $E(r, t)$,

$$E(r, t) = \exp[i(\omega t - 2\pi K_O \cdot r)] \sum_{G,l} E_{G,l} \exp[i\omega_q t - 2\pi i(g_G + lq) \cdot r] ,$$
$$(5.199)$$

into (4.32), the fundamental equation of the wavefield can be written as

$$\frac{K_{G,l}^2 - k_l^2}{k_l^2} = \sum_{M,p} \chi_{G-M} J_{l-p}(4\pi |g_{G-M} \cdot U|) E_{M,p} \quad (5.200)$$

with

$$K_{G,l} = K_O + g_G + lq \quad \text{and} \quad (5.201)$$

$$K_l = (\omega + l\omega_q)/2\pi c . \quad (5.202)$$

Assuming the wavelength of vibration is relatively short such that $2k|\chi| \ll q$ and $2k^2|\chi| \ll qK$, the corresponding $2\xi_q$ defined in (5.3) is large. This condition makes the two-beam approximation valid (Sect. 5.1). Equation (5.200) can therefore be reduced to a two-beam form, like (5.5). Considering the case involving two strong wavefields, $E_{O,l=0}$ and $E_{G,l=0}$, the integrated reflection can be obtained following the same derivation given in Sects. 5.1.4, 5. For detailed information about the intensity without approximations, the numerical calculation described in Sect. 5.2 should be employed.

Experimental observations on the increase in x-ray diffracted intensity and in the peak width have been reported by *Köhler* et al. [5.41] for CdS, by *Carlson* et al. [5.42, 43] and *Ishibashi* et al. [5.44] for epitaxial GaAs, and by *Leroux* et al. [5.45] for InSb. It is found in [5.45] that the theory described above gives a fairly good account of the experimental results for x-ray Bragg reflection, while the theory still needs modification in order to interpret satisfactorily the diffracted x-ray intensities for transmission cases.

6. Case Studies

We have discussed in the preceding chapter three approaches to account for the intensities of multiple diffractions. The applicability of these three approaches depends, as in two-beam diffraction, on both perfection of the crystal quality and experimental conditions. Kinematical theory is valid for diffraction from small crystals and imperfect crystals. Cases involving weak reflections and short-wavelength radiation, according to the discussion given in Sect. 5.1, can also be treated kinematically. In these kinematical diffraction situations, the interaction among the diffracted beams in crystals is limited by the crystal imperfection, the lack of space, and the weak wavefield amplitudes, so that dynamical effects are not appreciable in the diffraction process. On the contrary, diffraction from perfect and large crystals, and from planes of strong reflections should be considered from the dynamical viewpoint, since the perfect lattice and strong wavefields provide favorable situations for wave interaction in crystals. The other approach, the quantum theory of diffraction, has not so far been frequently used, because in the usual diffraction experiments the kinematical and dynamical theories give a satisfactory account for experimental observations, except when the diffraction process involves to a great extent excitations of electrons, atoms and molecules, such as photoelectric transitions, phonon interactions, Compton scattering and so forth. In this chapter, only the kinematical and dynamical aspects of x-ray multiple diffraction for some representative cases are presented. Numerical calculation is indispensable for illustration and explanation of the diffraction mechanism, especially for the multi-beam dynamical diffraction, because analytical expressions accounting for the diffracted intensities are difficult to obtain in most cases (Sect. 5.2).

6.1 Bragg-Type Multiple Diffraction from Gallium Arsenide, Indium Arsenide and Indium Phosphide – Kinematical Interpretation

We first consider reflection-(Bragg-) type multiple diffraction from gallium arsenide single crystals. Figure 6.1 is a 45-degree asymmetric portion of the indexed (002) multiple diffraction pattern of GaAs for CuK_{α_1} radiation [6.1]. The experimental set-up used is similar to the one shown in Fig. 2.5. The an-

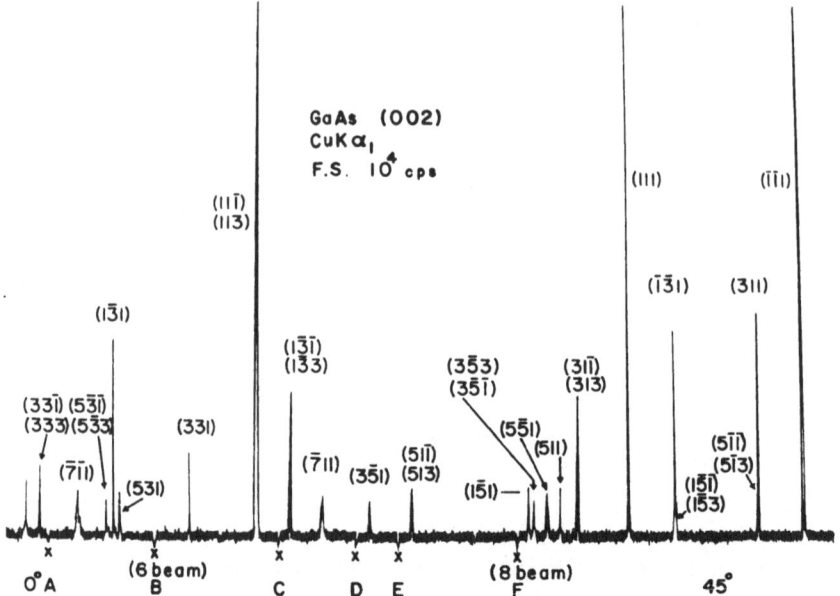

Fig. 6.1. Multiple diffraction pattern of GaAs (002) reflection for CuK_{α_1} (after *Chang* and *Post* [6.1])

gular divergence of the incident beam is about two minutes of arc, such that the doublets α_1 and α_2 of CuK radiation are separated. The primary reflection is (002), which is a very weak reflection according to the space group $F_{\bar{4}3m}$ [6.2]. Only the indices of the secondary reflections are indicated in Fig. 6.1. Those peaks with odd indices have moderate intensities because the secondary and coupling reflections involved are not weak (see Table 6.1 for the corresponding structure factors). Those reflections with even indices are rather weak. They are, however, detectable on closer inspection (Fig 6 2) In Fig. 6.2, various reflection "dips" are obtained, labeled A, B, C, D and E. Cases A, C, D and E are four-beam cases, and cases B and F are six-beam and eight-beam cases.

Gallium arsenide single crystals are usually grown as nearly perfect crystals. It is clear that the kinematical theory cannot be adopted to deal with strong diffractions from these crystals, since they are far from the category of ideal mosaic crystals. However, according to *Hirsch* and *Ramachandran* [6.3] and the discussion in Sect. 5.1, the kinematical theory can give reasonably good agreement between the calculated and the measured intensities for weak reflections. Therefore, (3.48), based on kinematical theory, is used for this calculation. The diffracted intensities for weak odd reflections, even reflections, and strong odd reflections, are listed respectively as groups (a), (b) and (c) in Table 6.2. The strong reflections are used for the purpose of comparison. Group (a) shows good agreement between calculated and measured

Table 6.1. Structure factors for the reflections of GaAs, InAs, InP at 25 °C (the + and − signs indicate that $h+k+l$ is $4n+1$, $4n-1$, respectively)

hkl	$\|F\|$		
	GaAs	InAs	InP
111 (+)	147.971	209.778	179.572
111 (−)	146.975	194.960	172.117
113 (+)	114.179	168.289	137.216
113 (−)	113.311	154.106	127.600
115 (+)⎫	80.865	126.299	105.159
333 (−)⎭	80.251	113.012	98.806
133 (+)	95.032	143.882	120.182
133 (−)	94.296	130.088	113.780
135 (+)	69.863	112.677	93.863
135 (−)	69.339	100.019	87.649
335 (+)	60.863	101.767	85.026
335 (−)	60.406	89.827	79.051
155 (+)	53.559	92.728	77.753
155 (−)	53.156	81.530	72.076
355 (+)	47.529	85.123	71.628
355 (−)	47.158	74.662	66.279
002	15.809	60.741	118.863
006⎫ 244⎭	6.538	34.679	63.256
024	6.538	42.491	81.839
022	174.219	239.243	182.361
004	142.769	200.174	151.332
044	103.526	153.000	116.049
026	89.927	136.989	104.385

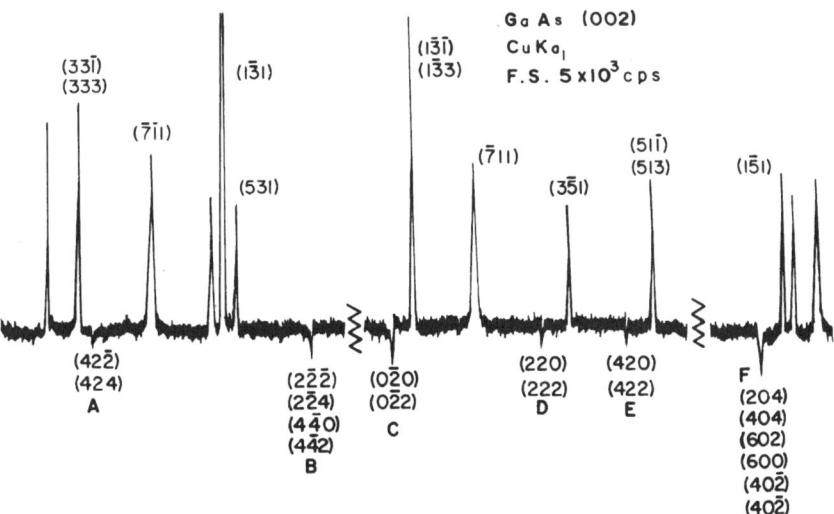

Fig. 6.2. Close view of Fig. 6.1 (after *Chang* and *Post* [6.1])

Table 6.2. Measured integrated intensities of the multiple diffractions of GaAs with (002) as the primary reflection for CuK_{α_1}

	hkl	N	ϕ [°]	$\Delta I/I_{002}$ (Integrated intensities)	
				Measured	Calculated
	$33\bar{1}/333$	4	0.50	7.8	7.82
(a)	$5\bar{3}\bar{1}/5\bar{3}3$	4	4.93	4.8	5.10
	$51\bar{1}/513$	4	25.05	5.4	5.08
	$42\bar{2}/424$	4	0.99	−0.8	−0.66
	6-beam[a]	6	8.23	−1.6	−0.99
(b)	$0\bar{2}0/0\bar{2}2$	4	16.45	−1.2	−0.82
	$220/222$	4	21.39	−1.1	−1.05
	$420/422$	4	24.14	−0.5	−0.33
	8-beam[b]	8	31.83	−2.6	−3.95
	$1\bar{3}1$	3	5.33	24.1	13.69
(c)	$11\bar{1}/113$	4	14.96	70.9	97.75
	111	3	39.25	119.8	172.04

[a] $(2\bar{2}\bar{2})\,(2\bar{2}4)\,(4\bar{4}0)\,(4\bar{4}2) + (000)\,(002)$
[b] $(204)\,(404)\,(602)\,(600)\,(40\bar{2})\,(20\bar{2}) + (000)\,(002)$

intensities for the weak reflections with odd indices. The agreement in even-indexed reflections varies from case to case. The six- and eight-beam cases, which involve strong interaction among diffracted beams such as $(20\bar{2})$, $(2\bar{2}4)$ and (404), are expected to give poorer agreement between the calculation and the measurement. The dynamical effect of beam interaction dominating the diffraction process results in the disagreement between theory and experiment for the strong reflections in group (c). If the beam divergence is now increased to, for example, 20 min of arc, the incident beam becomes less sensitive to the crystal perfection or the mosaic distribution than the beam of 2-min divergence in the previous case. Moreover, if the exponential function is used to replace the truncated two-term Taylor series for the reflection power P in (3.16), as described in Sect. 3.2, better agreement between the kinematical calculation and the measurements may be achieved.

Figures 6.3 − 5 depict (006) multiple diffraction patterns of GaAs, InAs and InP single crystals for both CuK_{α_1} and CuK_{α_2}, the beam divergence being 20′ [6.4]. The purpose of using a high-order reflection like (006) is to separate the α_1 peaks from the α_2 peaks. Because the incident beam possesses a large angular divergence, overlapping of several multiple diffraction peaks cannot be avoided. The overlapped peaks, shown in Figs. 6.3 − 5, are indexed with two or more sets of Miller indices. Those peaks labeled as "5-, 6- and 8-beam" are $(000)\,(006)\,(1\bar{1}1)\,(1\bar{1}5)\,(\bar{3}33)$, $(000)\,(006)\,(\bar{2}22)\,(\bar{2}24)\,(2\bar{2}2)\,(2\bar{2}4)$ and $(000)\,(006)\,(\bar{2}02)\,(200)\,(402)\,(\bar{2}04)\,(206)\,(404)$ reflections. Only the peak intensities of well-resolved multiple diffraction peaks are measured and subjected to calculation. The change of peak position according to the ratio of the

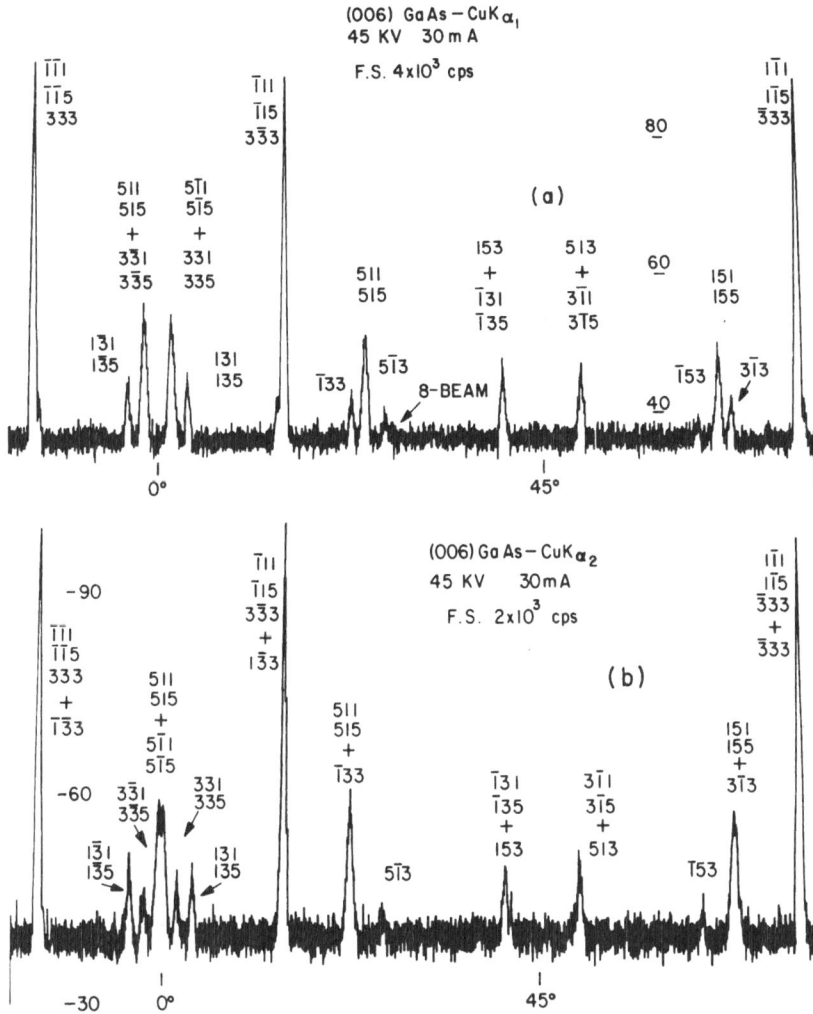

Fig. 6.3a, b. Multiple diffraction patterns of GaAs (006) reflection for (a) CuK_{α_1} and (b) CuK_{α_2} (after *Chang* [6.4])

wavelength to the lattice constant (2.25) can be visualized by comparing these six multiple diffraction patterns. The lattice constants for GaAs, InAs, and InP are 5.6539 Å, 6.0580 Å, and 5.8696 Å, respectively. The difference in the intensity background among these three crystals is due to the difference in the structure factors (Table 6.1). The peak intensities were measured with a slit of aperture 0.3° at the azimuthal position of each peak. The total counts for each peak were accumulated up to more than 10^4, making the error in counting statistics less than 1%.

Theoretical values for diffracted peak intensities can be calculated according to (3.106) of Sect. 3.7, in which the kinematical intensity problem is treat-

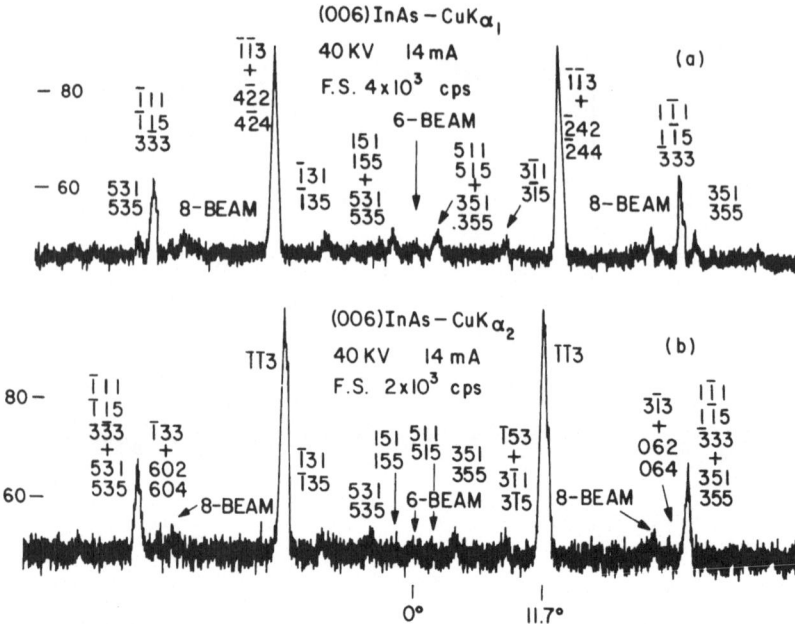

Fig. 6.4a, b. Multiple diffraction patterns of InAs (006) reflection for (a) CuK_{α_1} and (b) CuK_{α_2} (after *Chang* [6.4])

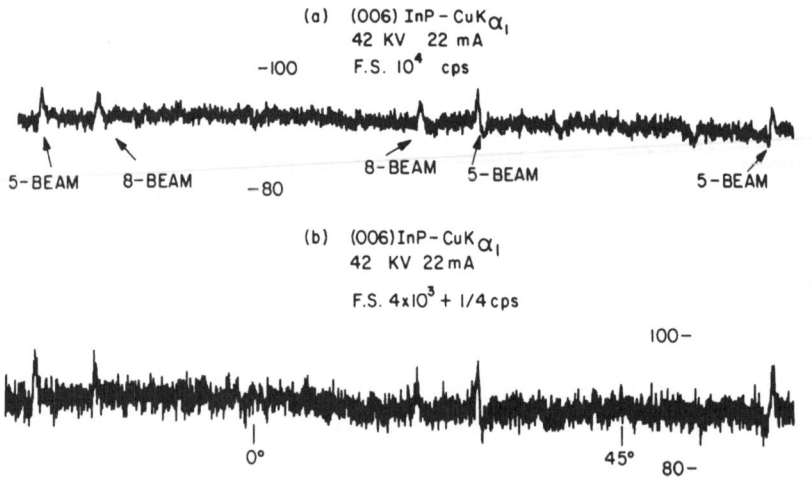

Fig. 6.5a, b. Multiple diffraction patterns of InP (006) reflection for (a) CuK_{α_1} and (b) CuK_{α_2} (after *Chang* [6.4])

ed as an eigenvalue problem. Similar to the classical dynamical formalism, the modes of attenuation and the excitation of mode can be specified and calculated (Sect. 3.7).

Because of the geometric relationship between the reciprocal lattice and the Ewald sphere, secondary reflections with their reciprocal lattice points lying on the equatorial plane circle are surface reflections (Sect. 2.1). In the present case, (006) is the primary reflection; therefore, secondary reflections (hkl) with $l = 3$ are surface reflections according to (2.3). The corresponding direction cosines γ are equal to zero. Equation (3.103) then has a singularity for these cases. To overcome this difficulty, an approximate solution can be obtained in the following way: Surface reflection can be treated as either a Bragg reflection with $\gamma < 0$ or a Laue transmission with $\gamma > 0$. The approximate value for peak intensity is the value interpolated from the calculated intensity curve versus γ, as γ approaches zero.

The cases (000) (006) and (000) (006) $(1\bar{3}1)$ $(1\bar{3}5)$ from GaAs and (000) (006) $(\bar{2}02)$ (200) (402) $(\bar{2}04)$ (206) (404) from InP for CuK_{α_1} are chosen to illustrate the calculations for two-, four- and eight-beam diffractions; (006) is a symmetric Bragg reflection. Table 6.3 gives the attenuation coefficients α, the number of permitted modes N_p, the excitations of modes Ex and the peak intensities of the Bragg reflection I_{Bragg}. The path lengths of the diffracted beams are chosen to be $1/\mu$, where μ is the linear absorption coefficient (Sect. 3.4). The values of $1/\mu$ with respect to CuK_{α_1} are 24.8 and 10.1 μm for GaAs and InP, respectively. The positive α's are approximately equal to μ/γ_L, where γ_L is the direction cosine of the transmitted beam L. The negative α's, approximately equal to μ/γ_M, are associated with the reflected beams. The small difference between α and μ/γ is due to the term ΣQ_{ij} in (3.103). The mode associated with the direct beam (000) is always the most excited mode.

Table 6.3. Kinematical calculations for 2-, 4- and 8-beam cases

	N	Laue	Bragg	α [mm^{-1}]	N_p	Ex [%]	I_{Bragg}
GaAs	2	000		49.276	1	100.00	
			006	− 49.276			0.27888×10^{-6}
GaAs	4	000		49.290		99.95	
		1$\bar{3}$1		73.942	2	0.05	
			006	− 49.281			0.28211×10^{-6}
			1$\bar{3}$5	− 73.933			0.24240×10^{-4}
InP	8	000		125.797		57.39	
		$\bar{2}$02		377.297	4	0.01	
		200		125.751		42.60	
		402		377.236		0.2×10^{-2}	
			006	− 125.729			0.78176×10^{-6}
			$\bar{2}$04	− 377.225			0.69310×10^{-5}
			206	− 125.750			0.24106×10^{-4}
			404	− 377.223			0.12261×10^{-4}

For GaAs, the attenuation α of this mode increases from 49.2755 to 49.2902 mm^{-1} as N changes from 2 to 4. The corresponding excitation decreases from 100% to 99.5%. The excitation of the mode with $\alpha = 73.942$ mm^{-1} is only about 0.05%. The (006) reflected intensity also changes from 0.2789×10^{-6} to 0.2821×10^{-6}.

In the four-beam, (000) (006) ($1\bar{3}1$) ($1\bar{3}5$), case, the ratio $I_{006}/I_{1\bar{3}5} = 1.16 \times 10^{-2}$ is not identical with $|F_{006}|^2/|F_{1\bar{3}5}|^2$, whose value is 0.88×10^{-2}. This signifies the fact that the interaction among the four diffracted beams modifies the reflected intensity of each two-beam reflection involved. For the eight-beam case from InP, two modes, $\alpha = 12.5797$ and 12.5751 cm^{-1}, are excited. The excitations for these two modes are 57.39% and 42.60%, respectively. Those modes with α very close to μ/γ_{000} play a more important role in the attenuation than those with α very different from μ/γ_{000}. Also, because of this attenuation, the reflected intensities are no longer proportional to $|F|^2$, for example, $I_{006} > I_{\bar{2}04}$, and $I_{206} > I_{404}$, although $|F_{006}| < |F_{\bar{2}04}|$ and $|F_{206}| < |F_{404}|$.

The measured and calculated peak intensities for the well-resolved multiple diffraction peaks are given in Table 6.4 for GaAs and CuK$_{\alpha_1}$. The terms ΔI and t_{eff} are the difference in intensity between the multiple diffraction peak and the (006) background, and the effective thickness by which the sample diffracts the incident beam O. t_{eff} is given by

$$I_{006} = P_M^T(t_{\text{eff}}) , \tag{6.1}$$

Table 6.4. The effective thicknesses and the calculated and observed (006) reflected intensities of multiple diffractions for CuK$_{\alpha_1}$ from GaAs

N	hkl	I_{006}		I/I^*		$\Delta I/I_{n=2}$		t_{eff} [μm]
		Obs.[a]	Calc. (10^{-6})	Obs.	Calc.	Obs.	Calc.	
2		27545	0.27888	0.928	0.999	0	0	67.95
3	$3\bar{1}3$	32432	0.27980	1.093	1.002	0.177	0.003	61.28
3	$\bar{1}53$[b]	29677	0.27914	1.00	1.00	0.077	0.001	65.70
4	151 155	38795	0.28284	1.307	1.013	0.408	0.014	67.54
4	$1\bar{3}1$ $1\bar{3}5$	33996	0.28211	1.146	1.011	0.234	0.012	67.09
4	131 135	33424	1.28221	1.126	1.011	0.213	0.012	67.05
5	$1\bar{1}1$ $1\bar{1}5$ $\bar{3}33$	69366	0.30934	2.337	1.108	1.518	0.109	60.11

[a] counts per minute
[b] the intensity of this reflection is used as unity for normalization

where the reflected power P_M^T is calculated according to (3.117). The values of t_{eff} in Table 6.4 are of the order of magnitude of about $3/\mu$. They increase when the multiple diffraction involves more transmission reflections and fewer Bragg and surface reflections, since transmission reflections tend to decrease absorption through Borrmann effects and Bragg reflections tend to shorten the length of beam penetration. The stronger the transmission, the lower the absorption, and the weaker the Bragg reflection, the longer the penetration path.

The ratio I/I^* in Table 6.4 shows a fair quantitative agreement between the experimental and calculated results for weak reflections. However, the calculated values of $\Delta I/I$ are about one or two orders of magnitude smaller than the observed values. This is because $\Delta I/I$ is very sensitive to small variations in diffracted intensity when I is very close to $I_{N=2}$, the two-beam diffracted intensity. Better agreement can be obtained by normalizing $\Delta I/I$ in the same way as I/I^*. The lack of quantitative agreement for $\Delta I/I$ can be directly understood from (3.4) and (3.103). For high-absorption cases, μ is much greater than Q_{ij} for GaAs crystals and CuK_{α_1} radiation. By taking this into account, the equations (3.4, 103) imply that α is almost equal to μ/γ and that the participation of secondary reflections will only give a third-order modification to the diffraction power of the primary reflection for these particular cases. As far as the diffraction in real crystals is concerned, the dynamical effects always accompany the diffraction process. The dynamical interaction among diffracted beams usually gives the two-beam reflected intensity a zeroth-order modification for strong reflections and a first- or second-order modification for medium or weak reflections. It is not surprising that the calculated ($\Delta I/I$)'s are much smaller than the observed ones.

6.2 Three-Beam Borrmann Diffraction – Dynamical Calculation

Two three-beam cases, (a) (000) (111) (11$\bar{1}$) and (b) (000) (111) (1$\bar{1}\bar{1}$), for germanium and CuK_{α_1} radiation are discussed here as examples. Case (a) was used to demonstrate anomalous transmission of x-rays at the three-beam diffraction point by *Borrmann* and *Hartwig* in 1965 [6.5]. Several investigations have been carried out since then [6.6 – 16]. This case involves a weak coupling (interaction) reflection, i.e., (111) − (11$\bar{1}$) = (002), which is a forbidden reflection. In order to reveal the influence of the coupling reflection on multibeam diffraction, case (b) is chosen here to compare with case (a). Case (b) differs from case (a) in having a strong coupling reflection (220) [6.9, 14, 17].

Figure 6.6a shows the relations between the reciprocal lattice vectors of the (111), (11$\bar{1}$) and (002) reflections. In order to reveal the detailed intensity distribution which reflects the dynamical effects, the divergent-beam technique,

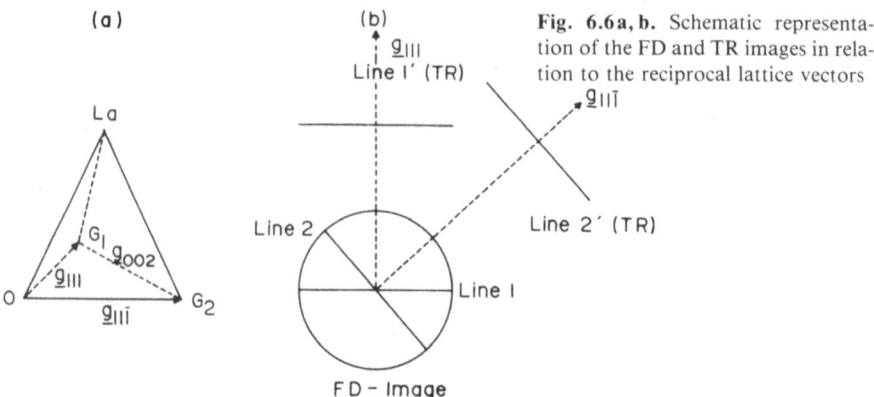

Fig. 6.6a, b. Schematic representation of the FD and TR images in relation to the reciprocal lattice vectors

mentioned in Sect. 2.2.2, is employed. Since the diffracted beam width is mainly affected by the vertical divergence (perpendicular to the plane of incidence containing the incident and reflected beams), and less affected by the horizontal divergence (Sect. 3.9), the images of the diffracted beams appear as lines perpendicular to the reciprocal lattice vectors g. For case (a) the diffraction lines 1 and 2 are the images of the forward diffracted (FD) (111) and (11$\bar{1}$) beams (Sect. 2.2.2). The intersection of lines 1 and 2 is the FD image of the three-beam diffraction. The transmitted reflected (TR) images are lines 1' and 2'. The separations between lines 1 and 1' and between lines 2 and 2' are proportional to the angle $2\theta_{111}$, for the (111) and (11$\bar{1}$) reflections, respectively (Fig. 6.6b). Figure 6.7 shows the experimental FD and TR images for this case [6.14]. Intensity enhancement at the exact three-beam diffraction position for both FD and TR images from a 0.05 cm thick germanium crystal for CuK_α radiation is readily visible [6.14].

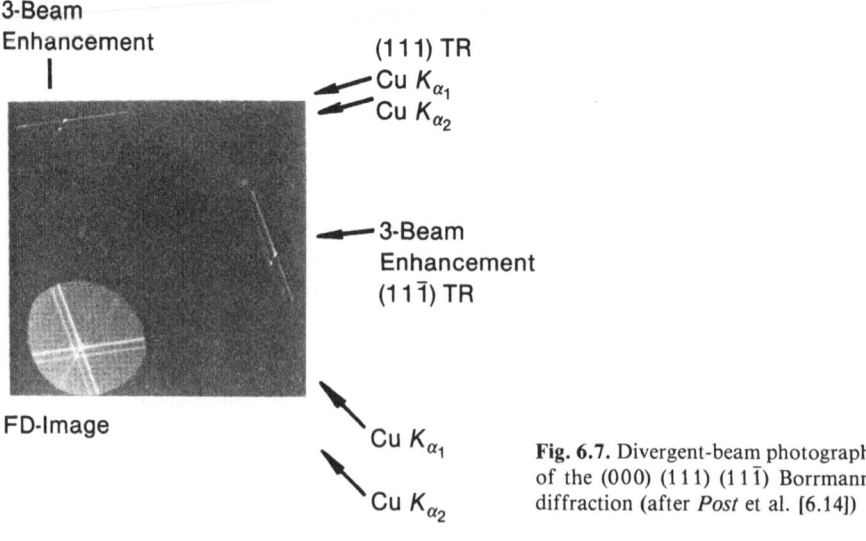

Fig. 6.7. Divergent-beam photograph of the (000) (111) (11$\bar{1}$) Borrmann diffraction (after *Post* et al. [6.14])

Table 6.5. Real and imaginary parts of the structure factors of Ge

hkl	F' (real)	F'' (imaginary)
000	245.60	7.345
111	145.62	5.107
311	113.54	4.909
331	94.73	4.694
002	0	0
220	173.78	7.045
004	142.93	6.749
222	1.09	0.048

To understand the mechanism of the dynamical diffraction, calculations based on Sect. 5.2 are carried out for the excitation of mode, the configuration of the dispersion surface, absorption, and the diffracted intensities. The results are shown in Figs. 6.8 – 11. The structure factors given in Table 6.5 are used.

There are six dispersion sheets corresponding to the two polarization states σ and π. The intersections of these six sheets of the dispersion surface with the plane of incidence for the (111) reflection (along line 1 of Fig. 6.6), in the vicinity of the three-beam diffraction position are shown in Fig. 6.8. The intersection of the dispersion surface with the plane of incidence for the $(11\bar{1})$ reflection is exactly the same as that shown in Fig. 6.8, because (111) and $(11\bar{1})$ are equivalent reflections in the three-beam case. Interchanging the vectors of the (111) and $(11\bar{1})$ reflections in Fig. 6.6 causes no modification of

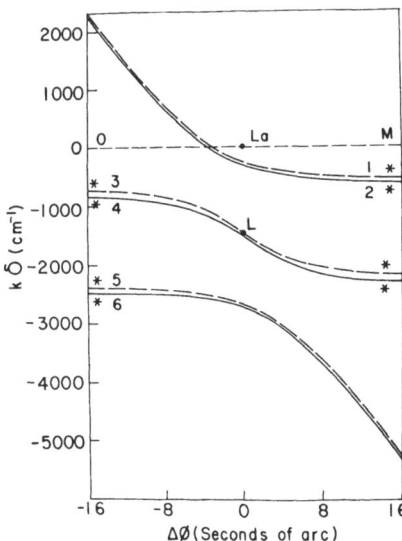

Fig. 6.8. Section of the dispersion surface of (000) (111) $(11\bar{1})$ Borrmann diffraction for CuK_{α_1} (case a)

the three-beam geometry. Therefore, only the dispersion sheets along the (111) line are discussed. The point La is the Laue point for the three-beam case and L is the Lorentz point. The horizontal line *OM* is the locus of the Laue point for the two-beam (111) reflection. The three-beam Laue point La, which indicates the exact three-beam diffraction position, is chosen as the origin of the figure. The ordinate stands for the accommodation, $k\delta$ in (4.86), with respect to the line *OM*. The direction of $k\delta$ is perpendicular to the entrance surface of the crystal. In this particular case, $k\delta$ is perpendicular to the plane containing the three reciprocal lattice points (000), (111) and (11$\bar{1}$) (Fig. 6.6a). The distance from the Laue point La to the (000) reciprocal lattice point O is $1/\lambda$, i.e., about $6.5 \times 10^7 \, \text{cm}^{-1}$ for $\text{Cu}K_{\alpha_1}$. The abscissa represents the azimuthal angle $\Delta\phi$ of the rotation of the crystal around the $\langle 111 \rangle$ direction. In the two-beam region, i.e., when $\Delta\phi$ is not near the three-beam point ($\Delta\phi = 0$), only four sheets are excited. They are marked with asterisks. The dispersion sheets are grouped in pairs due to the two polarizations. When $\Delta\phi$ is very far from the exact three-beam diffraction point, the separation between the pair of modes 1 and 2 and the pair of modes 3 and 4 for positive $\Delta\phi$ is expected to approach the separation of the dispersion sheets in the two-beam case. This separation s_G can be expressed, according to (4.54), as

$$s_G = \frac{r_e \lambda}{\pi V} \frac{C|F_H|}{|\cos\theta_G|}, \tag{6.2}$$

where the terms C, r_e, V and F have been defined in Sect. 4.1.

The relative excitation of mode, i.e., $\text{Ex}_G(j)$ in (4.228a), along the (111) line for the (111) diffracted beam is shown in Fig. 6.9. Since the polarizations of σ and π give almost the same results in calculations, the two lines of the same pair of modes have been combined into a single line. In the two-beam regions, modes 3, 4, 5 and 6 for negative $\Delta\phi$ (for example, $\Delta\phi = -60''$) and modes 1, 2, 3 and 4 for positive $\Delta\phi$ (for example, $\Delta\phi = 60''$) are strongly ex-

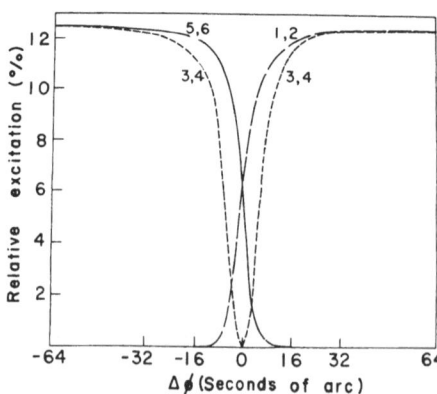

Fig. 6.9. Relative excitations of modes along the (111)

cited. Each of these modes, which are almost equally excited, has an excitation of 12.5% for the (1 1 1) reflection. This value can be obtained by dividing 100% by the number of beams (i.e., two-beam) and by the number of modes (i.e., 4 modes). In addition, the sum of the excitations over the four excited modes on each side of $\Delta\phi$ and over the two reflections (1 1 1) and (1 1 $\bar{1}$) are equal to unity, i.e., 8×0.125. When the crystal setting is varied from negative $\Delta\phi$ to positive $\Delta\phi$, the excitation of modes 3, 4 and 5, 6 decreases monotonically while the excitation of modes 1 and 2 increases before reaching the exact three-beam position, $\Delta\phi = 0$. At the three-beam point, modes 1, 2, 5, 6 have the same excitation of 6.125%, and modes 3 and 4 have zero excitation. As $\Delta\phi$ increases, the excitations of modes 1, 2 and 3, 4 increase and saturate at the value of 12.5%, while those of modes 5 and 6 are reduced towards zero. It is interesting to note that the asymmetry of the excitation curves of modes 1, 2 and 5, 6 and the symmetry of that for modes 3 and 4 with respect to $\Delta\phi = 0$ are consistent with the symmetry and asymmetry of the corresponding dispersion sheets about the Lorentz point.

The linear absorption coefficients of the three pairs of modes of propagation along the (1 1 1) reflection line are shown in Fig. 6.10. According to *Ewald* [6.18], the closer the tie point on the dispersion sheet to the Laue point, the lower the absorption. At the three-beam point, modes 1 and 2 have the lowest absorption coefficients, 17 and 28 cm^{-1}, respectively. The absorption coeffi-

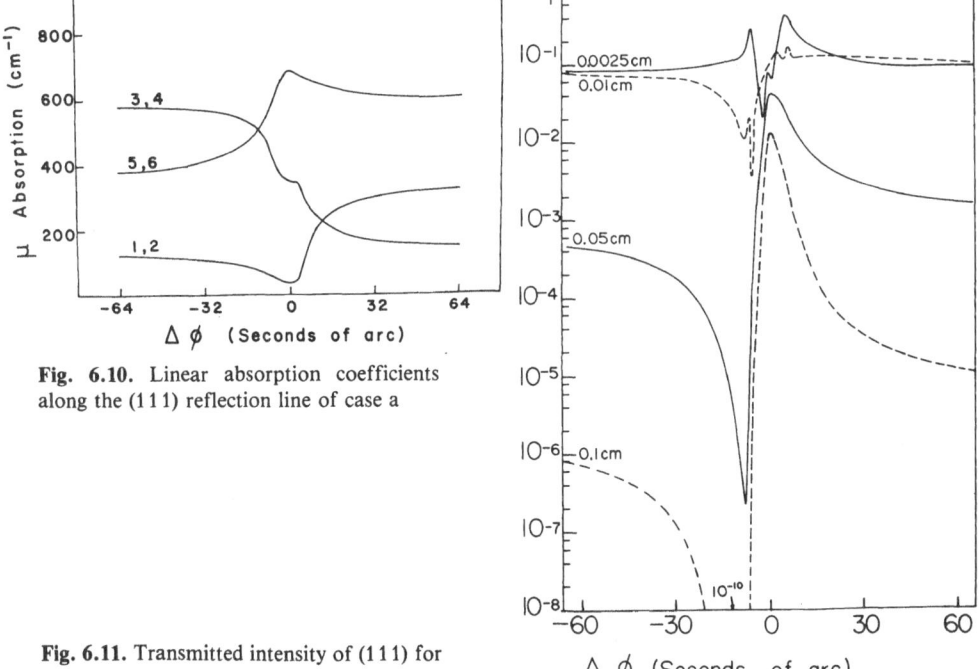

Fig. 6.10. Linear absorption coefficients along the (1 1 1) reflection line of case a

Fig. 6.11. Transmitted intensity of (1 1 1) for case a

cients for the rest of the modes are 356, 357, 672 and 683 cm^{-1} for modes 3, 4, 5 and 6. The lowest absorption is about twenty times smaller than the normal absorption coefficient, 352 cm^{-1}, in the one-beam case. Away from the three-beam point, the absorption of modes 1, 2 for negative $\Delta\phi$ and of modes 3, 4 for positive $\Delta\phi$ approaches the value of the minimum absorption coefficient for two-beam diffraction, i.e., 106 cm^{-1}, which is related to (4.195). That the sum of the absorption coefficients over all the modes is a constant, 352 cm^{-1}, for CuK_{α_1} for any crystal setting is in agreement with (4.193). This is due to the fact that the sum of the imaginary eigenvalues is independent of the angles $\Delta\theta$ and $\Delta\phi$. This fact provides another check on the dynamical calculation.

Figure 6.11 shows the (1 1 1) transmitted intensity for crystal thicknesses of 0.0025, 0.01, 0.05 and 0.1 cm. The intensity line profile for a 0.05-cm-thick crystal gives good agreement with the experimental result shown in Fig. 6.7 for the same crystal thickness. The asymmetry in the diffracted intensity is clearly seen. When the crystal thickness decreases to 0.0025 cm, this reso-nance-like intensity profile changes to an oscillation form. This change is a dynamical effect, since in a very thick crystal, say $\mu t > 10$, only the modes with lowest absorption, i.e., modes 1, 2, can survive after penetrating through the thick crystal. In Figs. 6.9, 10, modes 1 and 2 show asymmetrical behaviors for absorption and excitation over the angle $\Delta\phi$. The transmitted intensity evidently exhibits a similar asymmetry. For thin crystals, all the modes can survive after undergoing absorption in the crystal. The interaction among these modes within the crystal produces sub-peaks for the transmitted beam. It should be noted that the intensity minimum in Fig. 6.11 for the 0.05-cm-thick crystal appears for all wavelengths in the divergent-beam experiment. A deficient line (or 'eclipse' region [6.13]) is therefore formed on the divergent-beam photographs.

The calculated dispersion surface, the excitation of modes, the absorption and the transmitted intensity are given in Figs. 6.12 – 15 for case (b), in which a strong interaction reflection (2 2 0) is involved. The difference between case (a) and case (b) can be directly visualized from the difference in the dispersion surface, or more analytically, from the equation of dispersion (4.41). For sim-plicity, let us consider the equation at the exact three-beam diffraction posi-tion for a singly polarized beam for both cases:

$$Z^3 - (|F_{G_1}|^2 + |F_{G_2}|^2 + |F_{G_1-G_2}|^2) + (F_{-G_1}F_{G_2}F_{G_1-G_2}$$
$$+ F_{G_1}F_{-G_2}F_{G_2-G_1}) = 0 , \tag{6.3}$$

where

$$Z = F_O - 2\varepsilon_O \zeta , \tag{6.4}$$

with

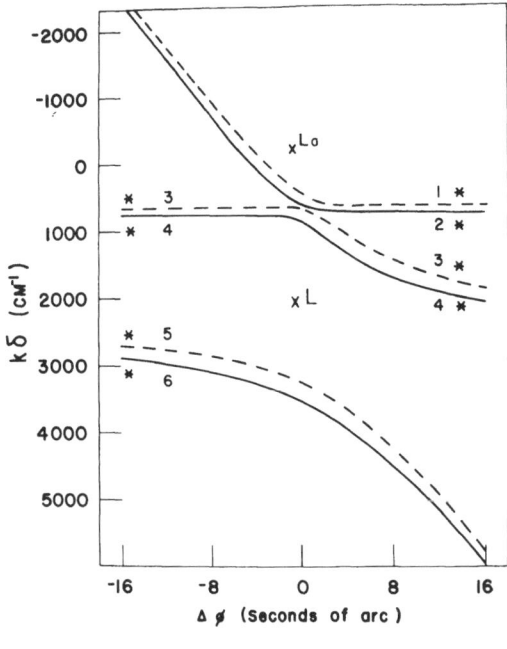

Fig. 6.12. Section of the dispersion surface along the (1 1 1) line for case b

$$\zeta = \frac{1}{\Gamma} = \frac{\pi V}{r_e \lambda^2}. \tag{6.5}$$

For centrosymmetric crystals like germanium,

$$F_{-G_1} F_{G_2} F_{G_1-G_2} + F_{G_1} F_{-G_2} F_{G_2-G_1} = 2 |F_{G_1} F_{G_2} F_{G_1-G_2}| \cos \Psi, \tag{6.6}$$

where Ψ is the phase angle of the product $F_{-G_1} F_{G_2} F_{G_1-G_2}$. This equation determines three values of Z corresponding to three dispersion sheets. The Lorentz point corresponds to $Z = 0$. It has been shown [6.19] that the location of the three dispersion sheets depends on the sign of $\cos \Psi$. For case (b), $\cos \Psi$ is positive. The dispersion sheets of modes 3 and 4 lie between the Laue and the Lorentz points. The phase effect on the dispersion surface will be discussed in Chap. 7. Figure 6.12 shows the same section for case (b) as for case (a) in Fig. 6.8. Two pairs, modes 1, 2 and 3, 4, of the dispersion sheets are very close to the Laue point at the exact three-beam point, $\Delta\phi = 0$. In contrast to this case, (a) involves a forbidden reflection (0 0 2) with $F_{002} = 0$. This leads to $Z = 0$ for (6.3). The dispersion sheets of modes 3 and 4 therefore pass through the Lorentz point (Fig. 6.8).

The linear absorption coefficients for the six modes in case (b) are shown in Fig. 6.13. Because of the closeness of the four dispersion sheets for modes 1, 2, 3 and 4 to the Laue point, the absorption coefficients of these modes do not differ very significantly from one another near the three-beam point. In the region of relatively large $|\Delta\phi|$, two pairs of modes approach the maxi-

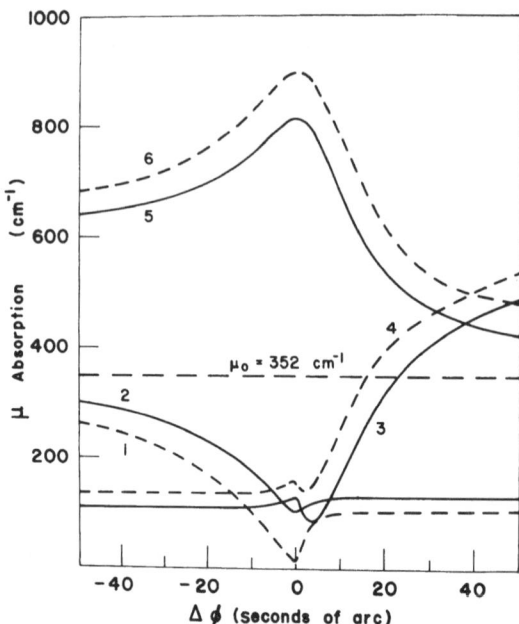

Fig. 6.13. Linear absorption coefficients along the (1 1 1) line for case b

mum and the minimum absorptions of the two-beam (1 1 1) reflection according to (4.194, 195), i.e., 600 and 106 cm^{-1}, respectively. This fact again reveals the two-beam characteristic for large $|\Delta\phi|$. The remaining modes, i.e., 1 and 2 for $\Delta\phi < 0$ and 5 and 6 for $\Delta\phi > 0$, approach asymptotically the normal absorption $\mu_0 = 352$ cm^{-1} as $|\Delta\phi|$ increases. The absorption of mode 1 at the three-beam point is equal to the minimum two-beam absorption of the (2 2 0) reflection, $\mu \simeq 4.2$ cm^{-1} [6.8, 9, 12, 17]. Determination of the direction of the Poynting vectors normal to the dispersion sheets at $\Delta\phi = 0$ shows that the direction of the Poynting vector of mode 1 is the same as that of the mode with $\mu = 4.2$ cm^{-1} in the two-beam (2 2 0) reflection case [6.9, 17]. This mode therefore behaves like the wave propagating in the (2 2 0) planes in the simple two-beam case.

The excitation of mode, defined in Sect. 4.8 as the magnitude of the resultant Poynting vector, is shown for all the modes of propagation in Fig. 6.14. The relative excitation of modes along the line for the (1 1 1) reflection (Fig. 6.15) shows a similar behavior. The mode with the lowest absorption coefficient has a zero excitation. The rest of the modes, which are strongly excited, have an absorption coefficient greater than $\mu_{min}^{(111)} = 106$ cm^{-1}. In comparison with the excitation and the minimum absorption in case (a), the effect of anomalous transmission in case (b) is not as pronounced as in case (a) at the three-beam position. For large $|\Delta\phi|$, two-beam characteristics remain, i.e., modes 1, 2, 3 and 4 for positive $\Delta\phi$ and modes 3, 4, 5 and 6 for negative $\Delta\phi$ are excited with excitation approximately equal to 25% each (Fig. 6.14).

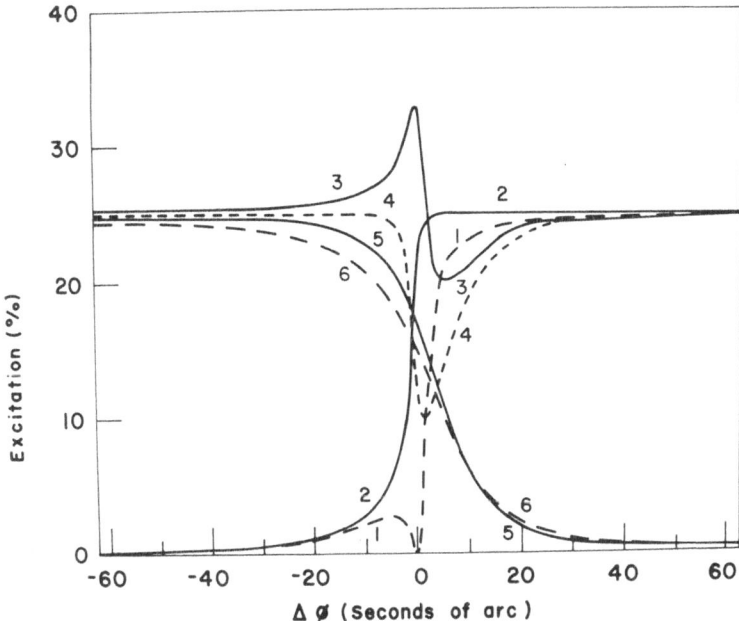

Fig. 6.14. Excitations of modes along the (1 1 1) line for case b

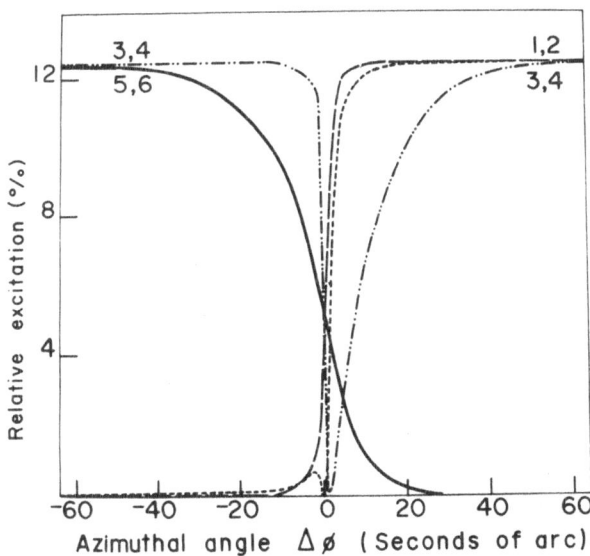

Fig. 6.15. Relative excitation along the (1 1 1) reflection line for case b

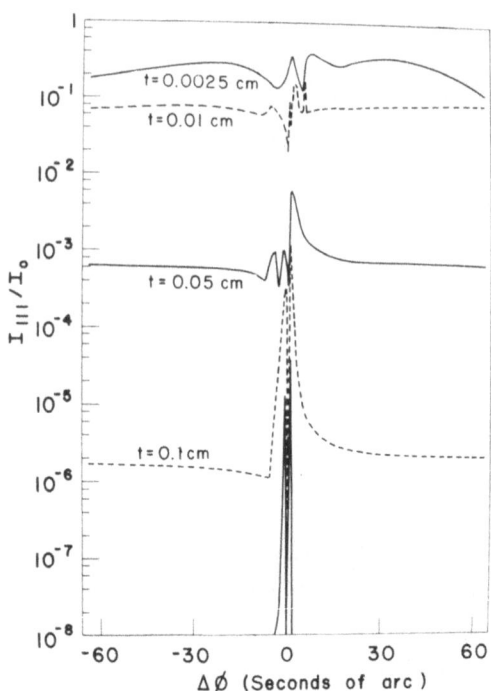

g. **6.16.** Transmitted intensity of 11) for case b

The (111) transmitted intensities of case (b) for the thicknesses 0.0025, 0.01, 0.05 and 0.1 cm are plotted against $\Delta\phi$ in Fig. 6.16. Intensity oscillation appears even for thickness up to 0.05 cm. For the thickness 0.1 cm, two pronounced peaks, spaced one or two seconds apart, appear on both sides of $\Delta\phi = 0$. This is due to the fact that mode 1 dominates the diffraction for such a thick crystal. The appearance of the excitation of this mode resembles the intensity curve near the three-beam point.

The experimental intensity distribution of the (000), (111) and ($\bar{1}\bar{1}$1) reflected beams for case (b) are shown in Fig. 6.17 for a 0.05 cm thick crystal.

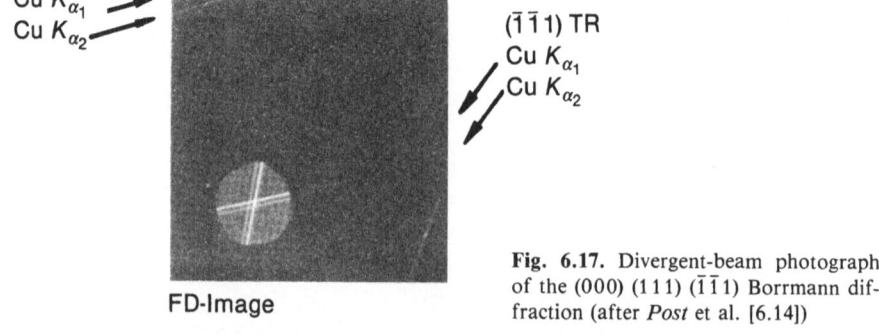

Fig. 6.17. Divergent-beam photograph of the (000) (111) ($\bar{1}\bar{1}$1) Borrmann diffraction (after *Post* et al. [6.14])

The anomalous transmissions of all the beams are less pronounced than in case (a). The anomalous transmission and absorption can be easily visualized by considering the intensity of the wavefield for the modes of propagation, as mentioned in Sect. 4.9. Referring to (4.187, 188), the photoelectric absorption is directly proportional to the intensity of the wavefield I_F of (4.238) provided that there is no thermal vibration, i.e., $\Omega = 1$ in (4.187, 196). If the maximum values of I_F appear at the atomic sites where all electrons are concentrated, anomalous absorption takes place. If the minimum I_F's coincide with the atomic sites, the absorption is zero.

For simplicity, the quantity γ_μ is defined for point atoms to connect the absorption (4.187) with the wavefield (4.238):

$$\gamma_\mu = \frac{\mu(j)}{\mu_0} = \frac{\sum\limits_{m \neq l} E_{Gm}(j)^* \cdot E_{Gl}(j) \exp[-2\pi i(g_m - g_l) \cdot r]}{|E_O^e|^2}, \qquad (6.7)$$

where μ_0 is the ordinary linear absorption coefficient. The vector r indicates the atomic positions in a crystal unit cell. For germanium, the atoms are located at $(1/8, 1/8, 1/8)$ and $(7/8, 7/8, 7/8)$ plus face-centered equivalent positions. There are eight atoms per unit cell. For the one-beam, (000), case, $|E_{000}| = |E_O^e|$ and $\gamma_\mu = 1$. The linear absorption coefficient is equal to μ_0, which is consistent with (4.45). μ_0 is $352\ \text{cm}^{-1}$ for germanium and CuK_{α_1} radiation. For the two-beam, (000) and (111), case, $|E_{000}| = |E_{111}| = |E_O^e|/\sqrt{2}$ (4.116), and

$$\gamma_\mu = 1 \pm C \cos(2\pi g_{111} \cdot r), \qquad (6.8)$$

where the polarization factors C are 1 and $\cos 2\theta_{111}$ for the σ and π waves, respectively. For the lowest absorbing mode, $C = 1$, the wavefield intensity at $(1/8, 1/8, 1/8)$ is proportional to $[1 + \cos(3\pi/4)]$. The minimum $\gamma_\mu^{\min}$ is therefore equal to 0.293. The maximum value can similarly be determined to be $\gamma_\mu^{\max} = 1.707$. The corresponding linear absorption coefficients are 103 and $601\ \text{cm}^{-1}$, respectively, at the temperature 0 K. The values of γ_μ at all the eight atomic sites are the same for the lowest absorbing mode. Referring to Table 6.6, γ_μ takes the following form for the lowest absorbing mode 3 in case (a):

$$\gamma_\mu = 1 - \frac{0.963}{\sqrt{2}} \{\cos[2\pi(x+y+z)] - \cos[2\pi(x+y-z)]\}$$

$$- \frac{0.852}{2} \cos(4\pi z), \qquad (6.9)$$

where the polarization vectors, σ and π, involved in the scalar products of (6.7) are defined in Fig. 6.18, and

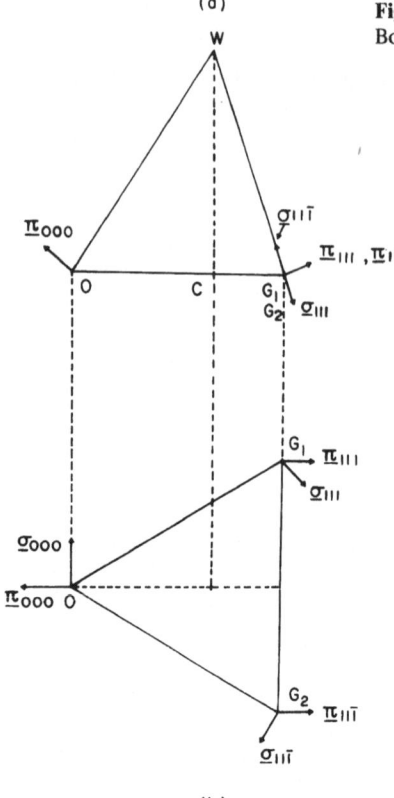

(a)

Fig. 6.18. Polarization vector for the three-beam Borrmann diffraction (case a)

(b)

$$\pi_{000} \cdot \pi_{111} = \pi_{000} \cdot \pi_{11\bar{1}} = -\frac{\cos 2\theta_{111}}{\cos \theta_{002}} = m_1 \,,$$

$$\pi_{000} \cdot \sigma_{11\bar{1}} = -\pi_{000} \cdot \sigma_{11\bar{1}} - \sqrt{1 - m_1^2} \sin \theta_{002} - m_2 \,,$$

$$\sigma_{000} \cdot \sigma_{111} = \sigma_{000} \cdot \sigma_{11\bar{1}} = -\cos \theta_{002} = m_3 = -0.963 \,,$$

$$\sigma_{111} \cdot \sigma_{11\bar{1}} = -0.852 \,.$$

(6.10)

The angles θ_{111} and θ_{002} are the Bragg angles of the (111) and (002) reflections. The wavefield-amplitude ratios and the eigenvalues (Sect. 4.5.1) are listed in Table 6.6 [6.7], where the parameters ξ and η are defined as

$$\xi = \frac{m_2}{2\sqrt{m_1^2 + m_2^2}}, \qquad \eta = \frac{m_1}{2\sqrt{m_1^2 + m_2^2}} \,.$$

(6.11)

The minimum values for γ_μ and μ are therefore equal to 0.037 and 13 cm^{-1}, respectively. This minimum absorption $\mu_{\min}$ can also be obtained, according to (4.194, 195), from the imaginary part of the eigenvalues given in Table 6.6:

Table 6.6. Ratios of the wavefield amplitudes and the parameter y' of the three-beam, (000) (111) (11$\bar{1}$), Borrmann diffraction at the exact three-beam point

Mode	1	2	3	4	5	6
y'	0	0	$\sqrt{2}\,m_3\chi_{111}$	$-\sqrt{2}\,m_3\chi_{111}$	$\sqrt{2(m_1^2+m_2^2)}\,\chi_{111}$	$-\sqrt{2(m_1^2+m_2^2)}\,\chi_{111}$
$E_{\sigma O}$	0	0	$\dfrac{1}{\sqrt{2}}$	$\dfrac{1}{\sqrt{2}}$	0	0
$E_{\sigma G_1}$	0	$\sqrt{2}\,\eta$	$-\tfrac{1}{2}$	$\tfrac{1}{2}$	ζ	$-\zeta$
$E_{\sigma G_2}$	0	$\sqrt{2}\,\eta$	$\tfrac{1}{2}$	$-\tfrac{1}{2}$	ζ	$-\zeta$
$E_{\pi O}$	0	0	0	0	$\dfrac{1}{\sqrt{2}}$	$\dfrac{1}{\sqrt{2}}$
$E_{\pi G_1}$	$\dfrac{1}{\sqrt{2}}$	$\sqrt{2}\,\xi$	0	0	$-\eta$	η
$E_{\pi G_2}$	$\dfrac{1}{\sqrt{2}}$	$-\sqrt{2}\,\xi$	0	0	η	$-\eta$

$$\mu_{\min} = \mu_0 - \frac{2\pi}{\lambda}\, y_i'(3) . \tag{6.12}$$

With $\chi_{111}^i\,(T = 0\,\text{K}) = 6.1176 \times 10^{-7}$ in Table 6.6, $\mu = 13\,\text{cm}^{-1}$.

The calculated and experimental values for $\mu_{\min}$ from various investigators are listed in Table 6.7. The variation of the minimum absorption for the transition from one-beam to two-beam and from two-beam to three-beam diffraction is illustrated in Fig. 6.19a along the $\langle 110 \rangle$ direction. The corresponding atomic positions projected on the $(1\bar{1}0)$ plane are shown in Fig. 6.19b. The anomalous transmission in the three-beam diffraction, case (a), can be directly visualized for the minimum absorbing modes, for which the atomic sites coincide with the minimum absorption.

Table 6.7. Minimum linear absorption coefficient of the (000) (111) (11$\bar{1}$) Borrmann diffraction of germanium for CuK_{α_1} at the exact three-beam point

$\mu_{\min}\,[\text{cm}^{-1}]$	References
15.4	*Borrmann* and *Hartwig* [6.5]
17.0	*Joko* and *Fukuhara* [6.7]
18.7	*Hildebrandt* [6.6], *Penning* [6.11]
19.0	*Uebach* and *Hildebrandt* [6.10]
18.4	*Feldman* and *Post* [6.13]
17.9	*Balter* et al. [6.12]

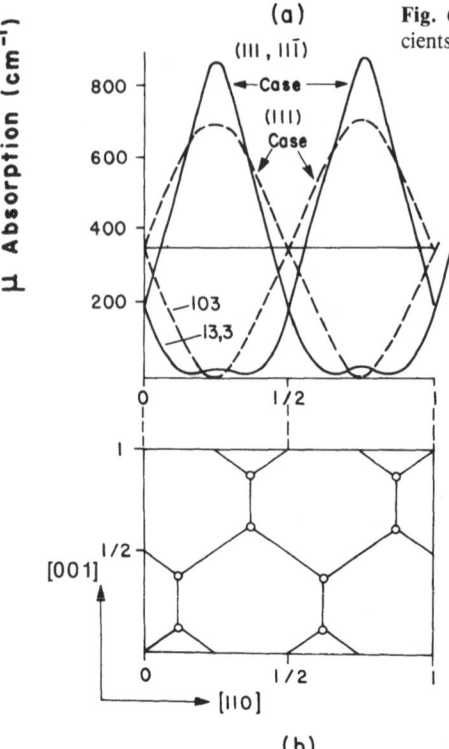

Fig. 6.19a, b. Calculated linear absorption coefficients along ⟨110⟩ versus the atomic sites for case a

6.3 Simultaneous Four-Beam Borrmann Diffraction

A four-beam case, (000) (1$\bar{1}$1) (20$\bar{2}$) (3$\bar{1}\bar{1}$), in germanium, for CuK_{α_1} radiation, is used to illustrate the four-beam dynamical interaction in single crystals. This case was first studied by *Katsnelson* et al. [6.20 – 23] and *Post* et al. [6.24], and involves two strong reflections, (1$\bar{1}$1) and (20$\bar{2}$), and a medium strong reflection (3$\bar{1}\bar{1}$). Their structure factors are given in Table 6.5. Although (3$\bar{1}\bar{1}$) is a relatively weak reflection with respect to (1$\bar{1}$1) and (20$\bar{2}$), one cannot ignore this reflection in the dynamical calculation, since the coupling among (1$\bar{1}$1), (20$\bar{2}$) and (3$\bar{1}\bar{1}$), i.e., 1$\bar{1}$1 – 3$\bar{1}\bar{1}$ = $\bar{2}$02 and 20$\bar{2}$ – 3$\bar{1}\bar{1}$ = $\bar{1}$1$\bar{1}$, are strong reflections. Following *Post* et al. [6.24], we discuss the calculations for the dispersion surface, absorption, excitation and diffracted intensities.

The four reciprocal lattice points (000), (1$\bar{1}$1), (20$\bar{2}$), (3$\bar{1}\bar{1}$), form a rectangle. The reciprocal lattice vector $g_{3\bar{1}\bar{1}}$ is the diameter of the reflection circle circumscribing the four points. The zone axis of these four reflection planes is ⟨121⟩. The diffraction-line images are, as mentioned in the three-beam cases, perpendicular to the reciprocal lattice vectors. The calculations are therefore performed for angular settings of the crystal along these diffraction lines by fixing $\Delta\theta$ for one reflection and varying $\Delta\phi$. The dispersion surface along the reflection lines is shown in Fig. 6.20. The coordinates are the same as in the three-beam cases discussed in the previous section.

The dispersion sheets along the $(20\bar{2})$ and $(1\bar{1}1)$ lines, shown in Fig. 6.20a and b, are similar to those of the two three-beam cases discussed in the previous section. Asymmetries of these sheets about $\Delta\phi = 0$ are seen. However, the section of the dispersion surface along $(3\bar{1}1)$ shown in Fig. 6.20c is symmetric about the three-beam point. This is obviously due to the geometrical relation between the $(3\bar{1}\bar{1})$, $(1\bar{1}1)$ and $(20\bar{2})$ reflection lines. The $(3\bar{1}\bar{1})$ line lies inbetween the $(1\bar{1}1)$ and $(20\bar{2})$ lines. There is no difference between the rotation around the reciprocal lattice vector $g_{3\bar{1}\bar{1}}$ from $-\Delta\phi$ to $+\Delta\phi$ and the rotation from $+\Delta\phi$ to $-\Delta\phi$. The symmetry of the $(3\bar{1}\bar{1})$ section is therefore preserved. Since this case involves all the reflections having positive values for the structure factors (including all the coupling reflections), the dispersion equation at $\Delta\phi = 0$ can be written in the form

$$AZ^4 - BZ^3 + CZ^2 - DZ + E = 0, \qquad (6.13)$$

where all the coefficients A, B, C, D and E are positive if the real part of Z is considered. The two pairs of dispersion sheets for modes 3, 4 and modes 5, 6 therefore lie between the Laue point and the Lorentz point. It should be noted that the dispersion sheet at $\Delta\phi = 0$ cannot be situated above the Laue point, since the modulus of the wavevector in the crystal is smaller than that in vacuum at this exact N-beam point.

The relative excitation of mode is shown in Fig. 6.21. The eight modes are equally excited at the exact four-beam diffraction position ($\Delta\phi = 0$) for each reflection. In the two-beam region (far away from $\Delta\phi = 0$), those modes with the dispersion sheets close to the Laue point are excited. They are marked with asterisks in Fig. 6.20. Again the excitations along the $3\bar{1}\bar{1}$ reflection line are symmetric about $\Delta\phi = 0$. Figure 6.22 shows the absorption coefficient for each mode along each reflection line. At the four-beam point, the minimum value of the absorption coefficient is $6.2\,\text{cm}^{-1}$, in contrast to the value in the one-beam case, $352\,\text{cm}^{-1}$, and the minimum two-beam values 14.4, 107.3, $116.8\,\text{cm}^{-1}$ for $(20\bar{2})$, $(1\bar{1}1)$, $(3\bar{1}\bar{1})$ reflections, respectively. Asymmetry in absorption with respect to $\Delta\phi = 0$ is expected due to the asymmetry of the excitation curves.

The transmitted intensities of the diffracted beams are shown in Fig. 6.23 for various thicknesses. The forward diffracted intensities for $t = 0.025$ cm are also shown. The oscillations of the intensity curves for thin crystals are due to the cross terms (oscillation terms) in (4.207, 211) of Sect. 4.7. For thick crystals, the averages of the cross terms over the crystal thicknesses are null. Well-defined intensity asymmetries are therefore seen for thick crystals.

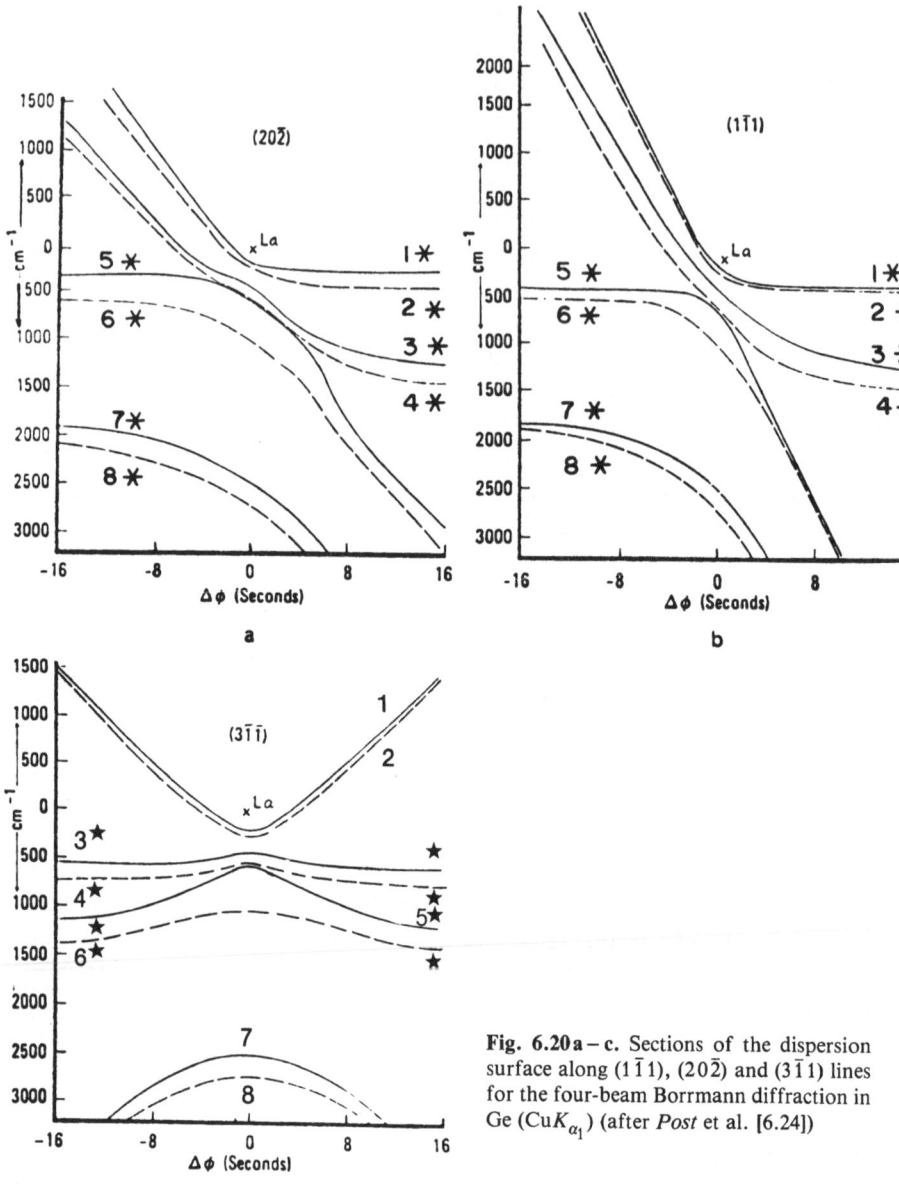

Fig. 6.20 a – c. Sections of the dispersion surface along (1 $\bar{1}$ 1), (20 $\bar{2}$) and (3 $\bar{1}$ $\bar{1}$) lines for the four-beam Borrmann diffraction in Ge (CuK_{α_1}) (after *Post* et al. [6.24])

The FD and TR images of CuK_{α_1} and CuK_{α_2} are shown in Fig. 6.24 for a 0.05-cm-thick germanium crystal [6.24]. Since the incident beam is polychromatic, four-beam diffraction images for both continuous and characteristic radiations appear as a continuous line on the FD photograph (Fig. 6.24a). Intensity enhancements occur at the intersection points. An extinction line is also observed near the four-beam point. The extinction line corresponds to the

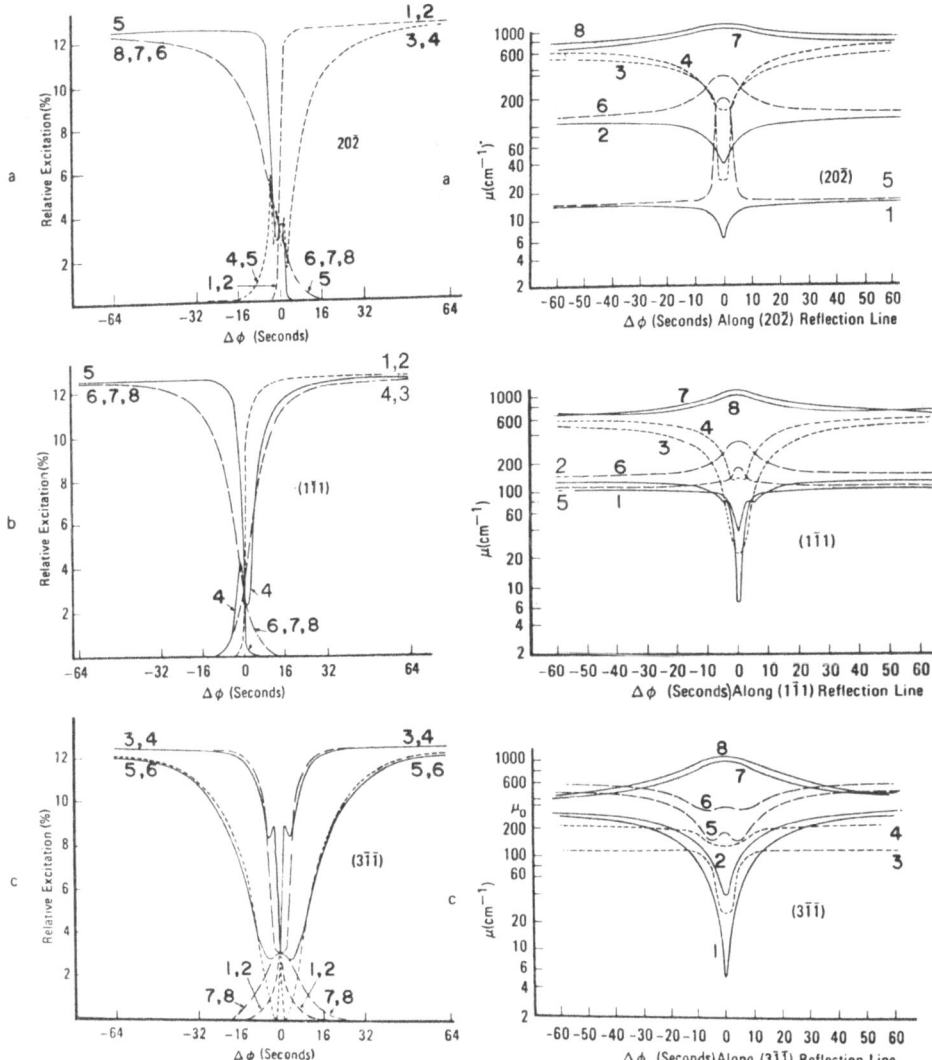

Fig. 6.21a – c. Relative excitations of modes for Fig. 6.20 (after *Post* et al. [6.24])

Fig. 6.22a – c. Linear absorption coefficients for Fig. 6.20 (after *Post* et al. [6.24])

dips of the $(20\bar{2})$ and $(1\bar{1}1)$ transmitted intensities shown in Fig. 6.23. This extinction is extended to all wavelengths.

In Figs. 6.24b, c, d, the TR lines are shown for both CuK_{α_1} and CuK_{α_2}. The discontinuities in the $(20\bar{2})$ CuK_{α_1} and K_{α_2} lines of Fig. 6.24 corresponding to the dips in the calculated intensities are visible. For the $(1\bar{1}1)$ and $(3\bar{1}\bar{1})$ reflection lines, strong enhancements at the four-beam positions are also

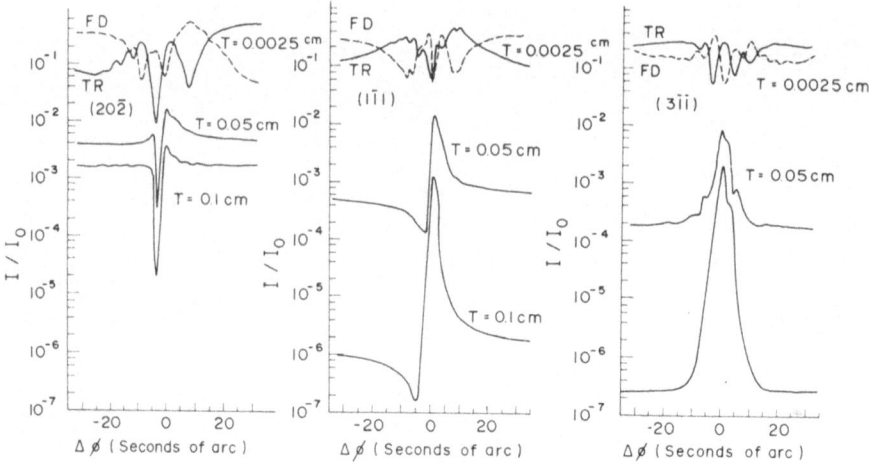

Fig. 6.23. Transmitted intensities for Fig. 6.20 (after *Post* et al. [6.24])

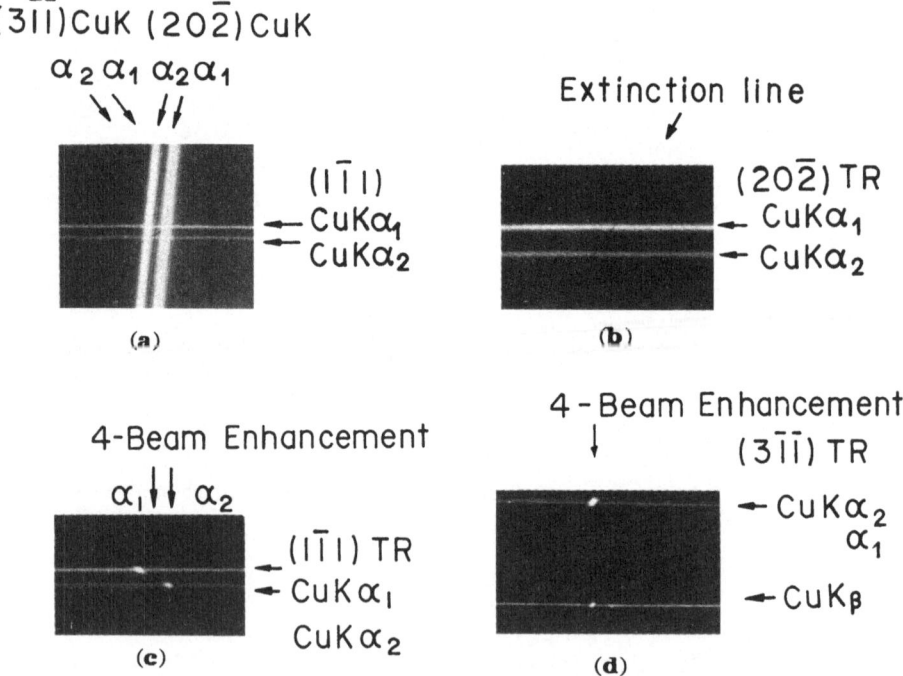

Fig. 6.24a – d. Divergent-beam photographs for the four-beam Borrmann diffraction (after *Post* et al. [6.24])

Table 6.8. Separations of sheets along the $(20\bar{2})$ line $[\text{cm}^{-1}]$

$\Delta\phi$ [s]	A 1 to 4	B 5 to 8	$\dfrac{(A+B)}{2}$
16	1157	1803	1480
32	1311	1659	1485
64	1396	1575	1485.5
128	1441	1530	1485.5
256	1463	1509	1486

clearly seen in Figs. 6.24c and d. The observation of the enhancement agrees qualitatively with the results of the calculations shown in Figs. 6.23b and c.

The long-range effect of the presence of the secondary reflection on the two-beam primary reflection is worth mentioning. This effect appears as a distortion on the two-beam dispersion surface and as a modification of the excitation of two-beam diffraction. These modifications are shown in Tables 6.8 and 6.9. Table 6.8 lists the separation between the sheets of the dispersion surface along the $(20\bar{2})$ line. The individual values listed under A and B for modes 1, 2, 3, 4 and 5, 6, 7, 8 approach very slowly the two-beam value, $1486\ \text{cm}^{-1}$. Table 6.9 gives the average absorption coefficients of the modes which are strongly related to the two-beam reflections of $(1\bar{1}1)$, $(3\bar{1}\bar{1})$ and $(20\bar{2})$. In the two-beam region ($\Delta\phi$ is very far from $\Delta\phi = 0$), the average absorption over the strongly excited modes is equal to μ_0 ($= 352\ \text{cm}^{-1}$). In the vicinity of the four-beam position, the average absorption approaches μ_0 as $|\Delta\phi|$ increases. At $\Delta\phi = 10'$, the influence of the four-beam diffraction on the absorption of the two-beam diffractions is still perceptible. This influence is stronger for the strong reflection. It is this long-range effect that is responsible for the inaccuracy in measuring the two-beam diffracted intensity.

Moreover, the omission of weak reflections in N-beam dynamical calculation is unjustified for x-rays, in contrast to the electron case where such omissions are commonly adopted in N-beam electron diffractions. Table 6.10 il-

Table 6.9. Average absorption coefficients of modes 'active' in two-beam regions near the four-beam point $[\text{cm}^{-1}]$

$\Delta\phi$ [s]	Reflection line		
	$(1\bar{1}1)$	$(3\bar{1}\bar{1})$	$(20\bar{2})$
128	336.9	349.8	332.1
300	345.6	351.7	343.6
600	348.9	352.0	347.8

Table 6.10. Effects of omission of the $(3\bar{1}\bar{1})$ reflection from the four-beam, (000) $(1\bar{1}1)$ $(20\bar{2})$ $(3\bar{1}\bar{1})$, calculations at the exact N-beam point

Mode	Absorption coefficients [cm^{-1}]		Excitation [%]	
	4-beam	'3-beam'	4-beam	'3-beam'
1	6.2	20	12.5	28.6
2	37	77	12.5	28.5
3	25	132	12.5	1.1
4	133	234	12.5	2.4
5	177	776	12.5	20.5
6	324	875	12.5	18.8
7	1000		12.5	
8	1115		12.5	

lustrates the influence of the omission of the $(3\bar{1}\bar{1})$ reflection on the excitation and absorption coefficients at the exact four-beam diffraction position. It is clear from this table that the omission of the $(3\bar{1}\bar{1})$ reflection in the four-beam case leads to serious errors in the calculated results.

Similar investigations on the multi-beam Borrmann effect using synchrotron radiation have been reported [6.25, 26].

6.4 Three-Beam Bragg-Laue and Bragg-Bragg Diffraction

Dynamical calculations for Bragg-type multiple diffraction have very seldom been reported, because of the difficulty associated with the total reflection in Bragg cases. In the literature, only a few articles [6.27 – 30] deal with the intensities in this type of diffraction. The configurations of the dispersion surface at the exact N-beam diffraction position [6.29] and for a four-beam case in which the crystal is regarded as a non-absorbing material [6.30] have only been briefly discussed. In order to make comparisons with Borrmann diffractions, calculations are preserved here on the dispersion surface, the absorption, the excitation of modes and the intensity distribution of the diffracted beams for Bragg-type multiple diffraction. Although results for these items cannot be found in the literature, the dynamical calculation can be carried out by following the procedures given in Sect. 5.2.

Dynamical calculations for two three-beam Bragg-type multiple diffractions are considered. Case A is the (000) (111) $(\bar{1}\bar{1}1)$ Bragg-Laue diffraction of germanium for CuK_{α_1} radiation, in which the primary reflection (111) is a symmetric Bragg and the secondary reflection $(\bar{1}\bar{1}1)$ a Laue transmission. The other case, case B, is the (000) (111) (220) Bragg-Bragg diffraction for the same crystal and radiation, in which the primary reflection (111) is a symmetric Bragg and the secondary reflection $(2\,2\,0)$ is an inclined Bragg reflection. The coupling reflection, $(2\,2\,0)$ $(= 1\,1\,1 - \bar{1}\,\bar{1}\,1)$, of the Bragg-Laue case is the secondary Bragg reflection of the Bragg-Bragg case, and vice versa. The purpose of choosing these two cases for illustration is to reveal the effect that interchanging the secondary and the coupling reflections has on the dynamical diffractions, and the connection between the two cases [6.31].

The crystal surface involved in this particular Bragg-Laue case is parallel to the $(1\,1\,1)$ planes so that $(1\,1\,1)$ is a symmetric Bragg reflection. Since this case involves one Bragg reflection, i.e., $(1\,1\,1)$, and two transmissions, $(0\,0\,0)$ and $(\bar{1}\,\bar{1}\,1)$, the boundary conditions (5.119) are composed of a (4×4) matrix A, a (2×2) matrix B, a (2×2) null matrix C and a (2×2) unit matrix D. The number of permitted modes, according to (4.181), is $2 \times (3 - 1) = 4$.

Before discussing these three-beam Bragg-type multiple diffractions, the dispersion surface, the absorption, the excitation, and the reflected intensity of the two-beam $(1\,1\,1)$ Bragg reflection ought to be mentioned. The dispersion surface of the $(1\,1\,1)$ reflection in germanium for CuK_{α_1} radiation is shown in Fig. 6.25a. The horizontal axis is the angular deviation, $\Delta\theta$, from the Bragg angle of the $(1\,1\,1)$ reflection. The ordinate is proportional to the accommodation $k\delta$. The Laue point La is the origin, and L is the Lorentz point of the two-beam case. Four dispersion sheets for modes 1, 2, 3 and 4 are shown. The region between $\Delta\theta = 5''$ and $\Delta\theta = 20''$ is the total-reflection range, denoted as range II. In comparison with the dispersion surface in the two-beam transmission case (see Fig. 6.25b), there are no dispersion sheets within range II, because there are no intersections between the crystal surface normals (which are parallel to the ordinate) and the dispersion surface. Mathematically, however, there exist solutions to the dispersion equation in the range of total reflection [6.33, 34]. To distinguish the dispersion surface in this range from that of the transmission case, the former is called the 'solution' dispersion surface [6.32]. Intersection between the 'solution' dispersion sheets is common in the total-reflection region [6.32], while there is no such intersection in the transmission case. Figures 6.25c, d, and e shows the absorption coefficients, the excitations of modes, and the diffracted intensities for the symmetric $(1\,1\,1)$ reflection, respectively [6.32].

The dispersion surface and the absorption coefficient of the two-beam $(1\,1\,1)$ case are modified by the presence of the additional $(\bar{1}\,\bar{1}\,1)$ reflection in case A. Section planes perpendicular to Fig. 6.25a and parallel to the ordinate of this figure at the Laue Point $(\Delta\theta = 0)$, Lorentz point $(\Delta\theta = 13'')$ and the position with $\Delta\theta = 25''$ are used to reveal these modifications. These three angular positions, $\Delta\theta = 0$, $13''$ and $25''$, are the angular settings of the crystal in ranges III, II and I, respectively. The dispersion sheets and the linear absorption coefficients for the possible modes of wave propagation at these three positions, shown in Figs. 6.26, 27, illustrate the transition from range I to range II and from range II to range III.

The dispersion sheets of the six possible modes for $\Delta\theta = 0$ are shown in Fig. 6.26a. The permitted modes are modes 3, 4, 5 and 6. The forbidden dispersion sheets are marked with 'x'. For large $|\Delta\phi|$, the dispersion sheets approach the two-beam values, ± 3275 cm^{-1} for $k\delta$, which can be seen directly from Fig. 6.26a. In the vicinity of $\Delta\phi = 0$, modes 3 and 4 vary almost linearly with $\Delta\phi$ from 3275 to -3275 cm^{-1}. This inclined line (plane) can be regarded as the intersection of the dispersion surface of the two-beam ($\bar{1}\bar{1}1$) reflection with the (111) plane. The accommodations of modes 5 and 6 decrease very slowly as $\Delta\phi$ increases from large negative angular positions and drops abruptly along the inclined line near $\Delta\phi = 0$. The dispersion sheets of all the permitted modes are below the Laue point at $\Delta\phi = 0$ so that the corresponding absorption coefficients are positive (Fig. 6.25c). This fact is consistent with the discussion given in Sect. 4.5.2. The presence of a dispersion surface in

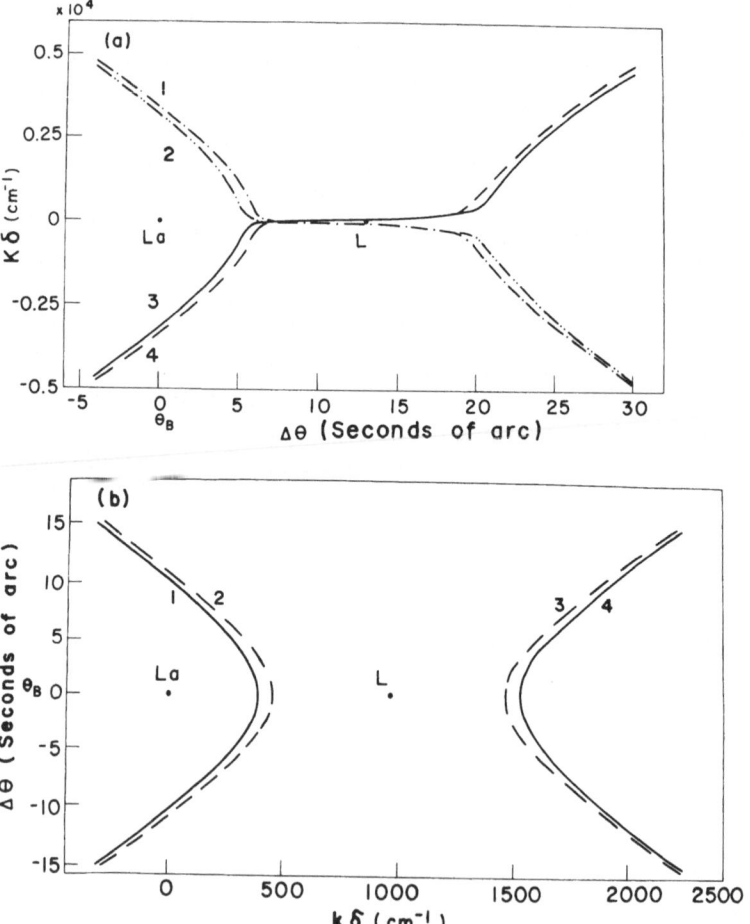

Fig. 6.25a, b. Figure caption see opposite page

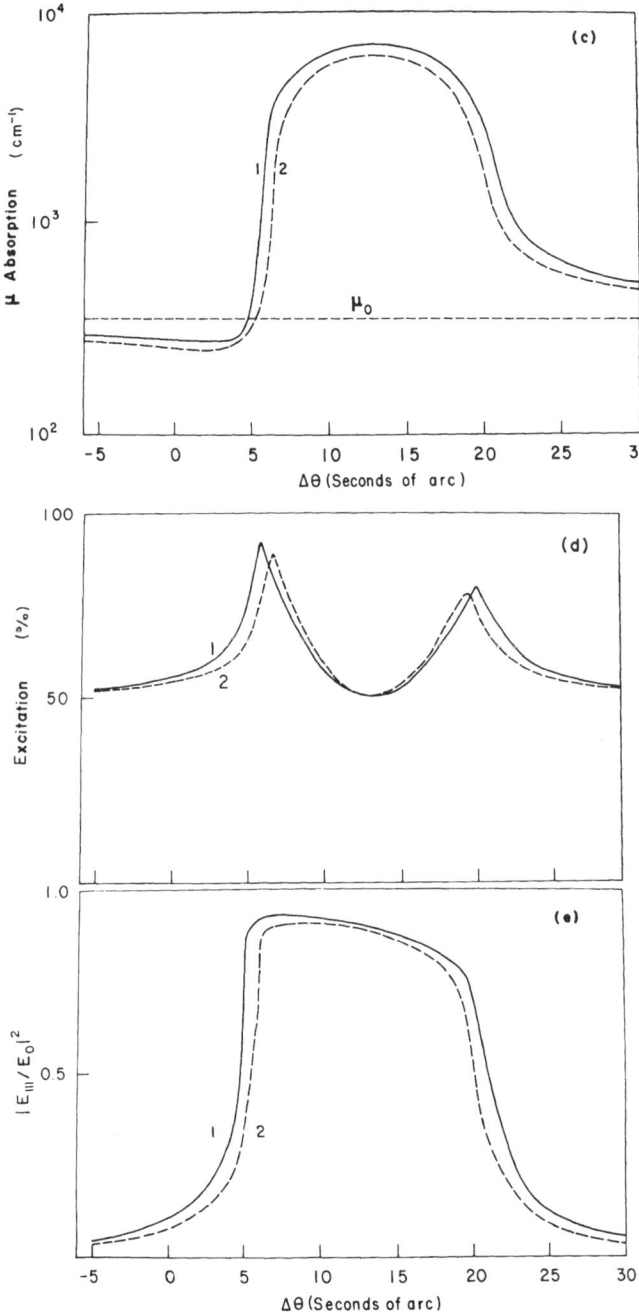

Fig. 6.25. (a) Solution dispersion surface of (111) Bragg reflection for CuK_{α_1}; (b) dispersion surface of transmission; (c) linear absorption coefficient; (d) excitation of modes; (e) reflectivity $|E_{111}/E_{000}|^2$ for the Bragg reflection (after *Chang* [6.32])

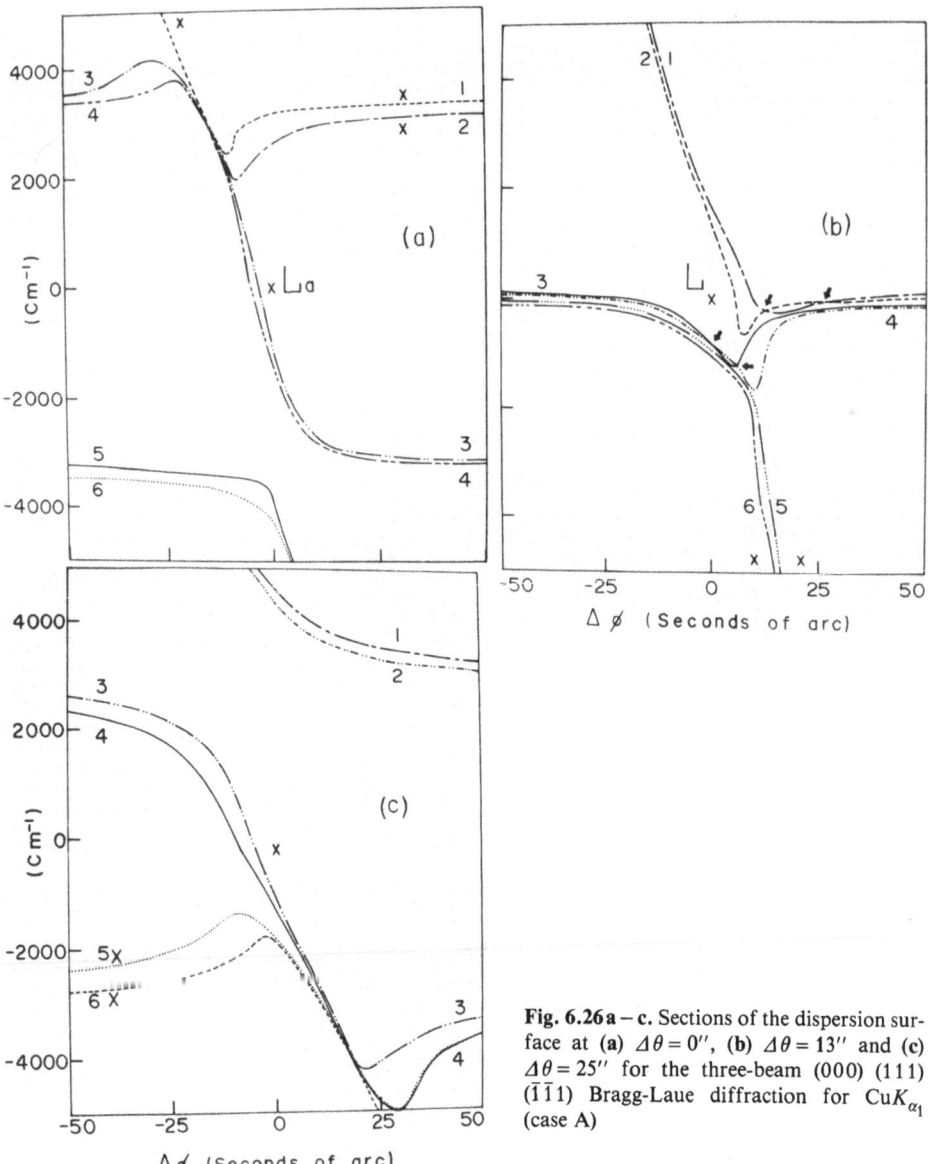

Fig. 6.26 a–c. Sections of the dispersion surface at **(a)** $\Delta\theta = 0''$, **(b)** $\Delta\theta = 13''$ and **(c)** $\Delta\theta = 25''$ for the three-beam (000) (111) ($\bar{1}\bar{1}1$) Bragg-Laue diffraction for CuK_{α_1} (case A)

the upper-left corner of Fig. 6.26a indicates the intersection of the (111) and ($\bar{1}\bar{1}1$) two-beam dispersion sheets. This complication leads to further modification of the x-ray refractive index within the crystal. Thus, the linear absorption coefficients vary accordingly. Figure 6.27a shows the linear absorption coefficients for the four allowed modes. Modes 3, 4, 5 and 6 maintain their two-beam characteristics for large $|\Delta\phi|$, i.e., the absorption coefficients at these positions are smaller than the ordinary value, 352 cm^{-1}. The asymptot-

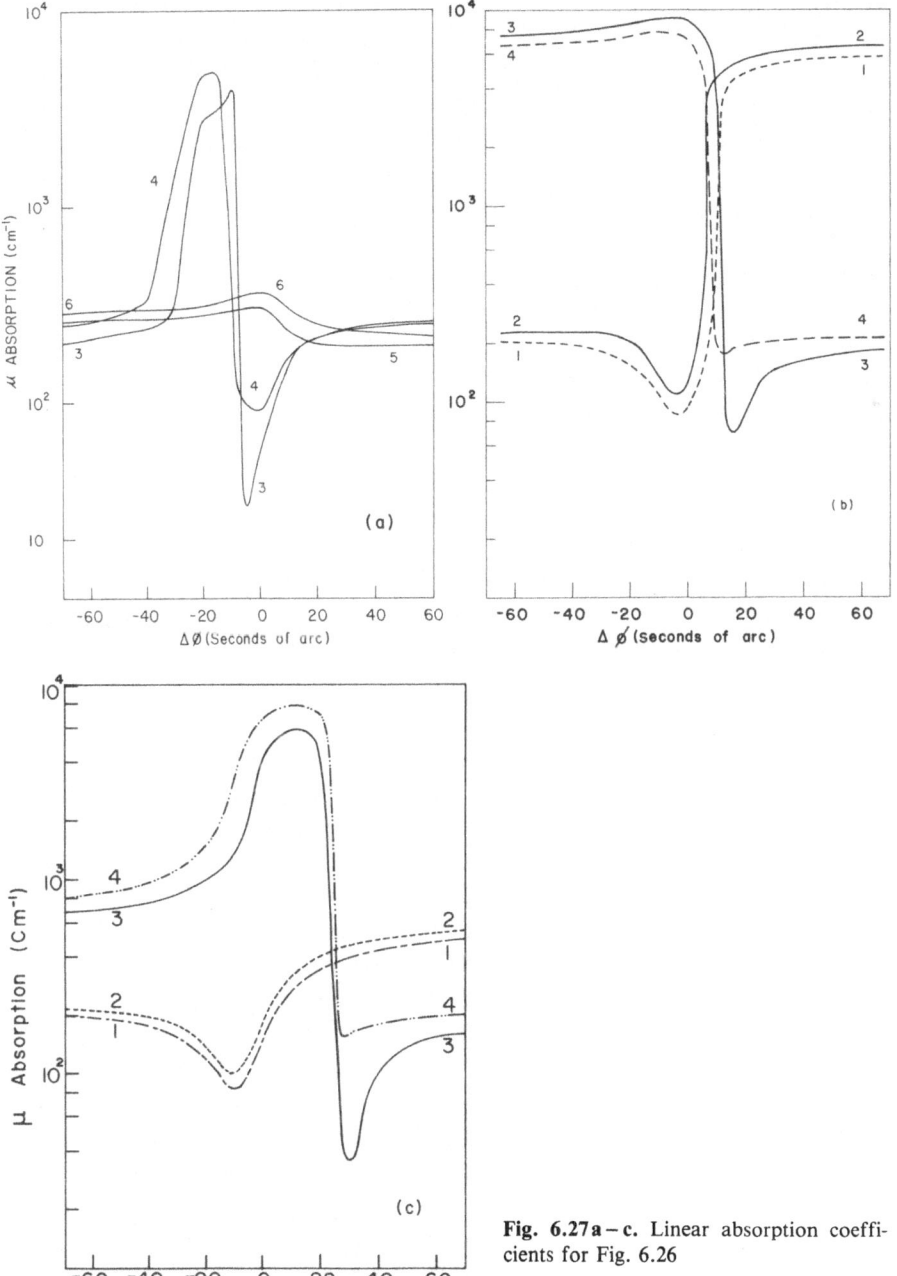

Fig. 6.27a–c. Linear absorption coefficients for Fig. 6.26

ic value is about 257 cm^{-1} (Fig. 6.25c). The absorption of modes 3 and 4 is a maximum near $\Delta\phi = -15''$ due to the interaction between the two pairs of dispersion sheets of modes 1 and 2 and modes 3 and 4. It should be noted that although modes 1 and 2 are forbidden modes, their presence in the dispersion relation (4.41) still affects the eigenvectors, i.e., the amplitude ratio of the wavefields. The interaction between the allowed and the forbidden modes should therefore be considered.

The dispersion sheets of modes 3, 4 and 5, 6 in the vicinity of $\Delta\phi = 0$ resemble the hyperbola in the two-beam transmission case. Modes 3 and 4 are the low absorption modes because their dispersion sheets are very close to the Laue point. Modes 5 and 6 are the high absorption modes. This implies the influence of the transmission $(\bar{1}\bar{1}1)$ reflection on the $(1\,1\,1)$ Bragg reflection. The linear absorption coefficients of modes 3 and 4 are therefore lowered via the anomalous transmission (the Borrmann effect) of the $(\bar{1}\bar{1}1)$ reflection, while the absorptions of modes 5 and 6 show their maxima close to $\Delta\phi = 2''$. The absorption here is very similar to the absorption of modes 5 and 6 shown in Fig. 6.13 in the three-beam Borrmann diffraction discussed in Sect. 6.2.

The cross section (Fig. 6.26b) of the dispersion surface along the $(1\,1\,1)$ reflection line at the Lorentz point ($\Delta\theta = 13''$), the center of range II, shows that the dispersion sheets of modes 3, 4, 5 and 6 approach asymptotically the two-beam value for $k\delta$, 0 cm^{-1}, in the regions far away from the three-beam diffraction point ($\Delta\phi = 0$). At this particular setting, modes 1, 2, 3 and 4 are the allowed modes. Intersections of the dispersion sheets between the two pairs of modes 1 and 2, and 3 and 4 are marked with arrows in Fig. 6.26b. Since modes 1 and 3 are associated with the σ-polarized wavefields and modes 2 and 4 with the π-polarized ones, these intersections indicate the polarization dependence of the degeneracy of the eigenvalues in (4.41). This fact is one of the characteristics of the total reflection in two-beam Bragg reflections [6.32]. It is worth mentioning that at $\Delta\theta = 13''$ we are dealing with the 'solution' dispersion sheets resulting from the $(1\,1\,1)$ total reflection plus the actual dispersion sheets from the $(\bar{1}\bar{1}1)$ transmission reflection. In the two-beam regions, modes 1, 2, 3 and 4 for positive $\Delta\phi$ and modes 3, 4, 5 and 6 for negative $\Delta\phi$ approach the solution dispersion sheets of the two-beam $(1\,1\,1)$ reflection shown in Fig. 6.26b. The actual dispersion sheets associated with the $(\bar{1}\bar{1}1)$ reflection are those labeled 1 and 2 for negative $\Delta\phi$ and those labeled 5 and 6 for positive $\Delta\phi$ in the same figure.

The linear absorption coefficients of the four allowed modes are shown in Fig. 6.27b. In the two-beam regions, only modes 3 and 4 for large negative $\Delta\phi$ and modes 1 and 2 for large positive $\Delta\phi$ have values close to the two-beam maximum absorption coefficient, i.e., about 6990 cm^{-1} (Sect. 4.6 and Fig. 6.25c). These four modes in their respective angular settings are expected to be strongly excited by the incident wave to preserve the two-beam characteristics. This argument agrees with the calculated excitation of modes shown in

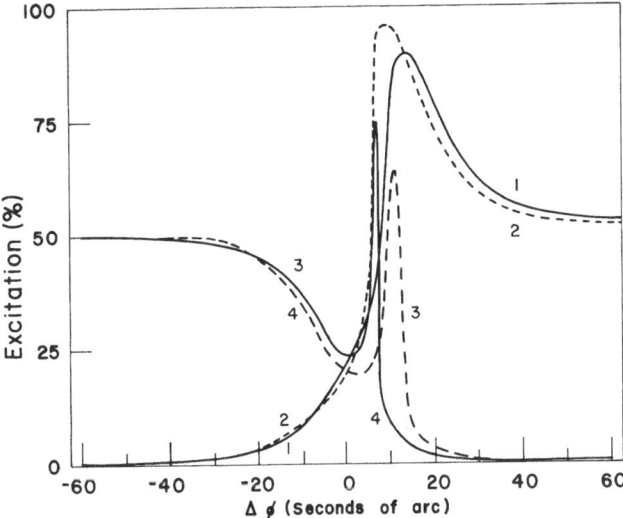

Fig. 6.28. Excitation of modes for Fig. 6.26b

Fig. 6.28. At the other settings, i.e., negative $\Delta\phi$ for modes 1 and 2 and positive $\Delta\phi$ for modes 3 and 4, the linear absorption coefficients approach the minimum value, 257 cm^{-1}, which was previously mentioned in the discussion of Fig. 6.27a. The absorption dips appearing in Fig. 6.27b are also attributed to the anomalous transmission near the Lorentz point.

Figure 6.26c shows the dispersion sheets for $\Delta\theta = 25''$ in range I. This figure is an inversion of Fig. 6.26a about the Lorentz point. This inversion is also seen in Fig. 6.25a for the two-beam (1 1 1) reflection. Modes 1, 2, 3 and 4 are the permitted modes for $\Delta\theta = 25''$. The corresponding absorption coefficients, shown in Fig. 6.27c, behave similarly to those shown in Fig. 6.27a for $\Delta\theta = 0$, except that the asymptotic two-beam absorption of modes 1 and 2 for $\Delta\phi > 0$ and of modes 3 and 4 for $\Delta\phi < 0$ is greater than 352 cm^{-1}. This is consistent with the absorption of the two-beam (1 1 1) reflection in range I (Fig. 6.25c). The other asymptotic value in the two-beam regions is about 257 cm^{-1}.

Since the reflected intensity has its maximum at the Lorentz point, we shall consider the excitation of mode at this point for illustration. Modes 1 and 2 for large positive $\Delta\phi$ and modes 3 and 4 for large negative $\Delta\phi$ are equally excited with an excitation of 50% per mode. The minimum excitation, about 25%, for modes 3 and 4 occurs at $\Delta\phi = 0$, because most energy is reflected from the crystal at this point. Consequently, the crystal is weakly excited. Actually, the four allowed modes at this position are equally excited. Maximum excitations occur in the range $\Delta\phi = 10''$ to $\Delta\phi = 15''$. These increases in excitation are caused by the interaction between the (1 1 1) and the ($\bar{1}\bar{1}1$) reflections (Fig. 6.28).

It should be noted that the expression (4.228) for the calculation of the excitation of mode, which takes no account of the directional dependence of the energy flow, provides a measure of the density of the wavefields associated with the modes of propagation. The maximum value, the upper limit, of the excitation is 200%. If one considers the Poynting vectors for the diffracted waves, the following relation holds for all crystal settings:

$$S_{000} + S_{111} + S_{\bar{1}\bar{1}1} = S_0, \tag{6.14}$$

where S is defined in Sect. 4.4 and S_0 is the Poynting vector of the incident wave.

These calculations are function of $\Delta\theta$ and $\Delta\phi$. The calculated reflected intensity $|E_{111}|^2 / |E_0|^2$, or I_{111}/I_0, versus $\Delta\phi$ and $\Delta\theta$ is the intensity distribution shown in Fig. 6.29. The direction of the reflection lines (111) and $(\bar{1}\bar{1}1)$ are indicated. The integrated intensity as a function of $\Delta\phi$ can be obtained by

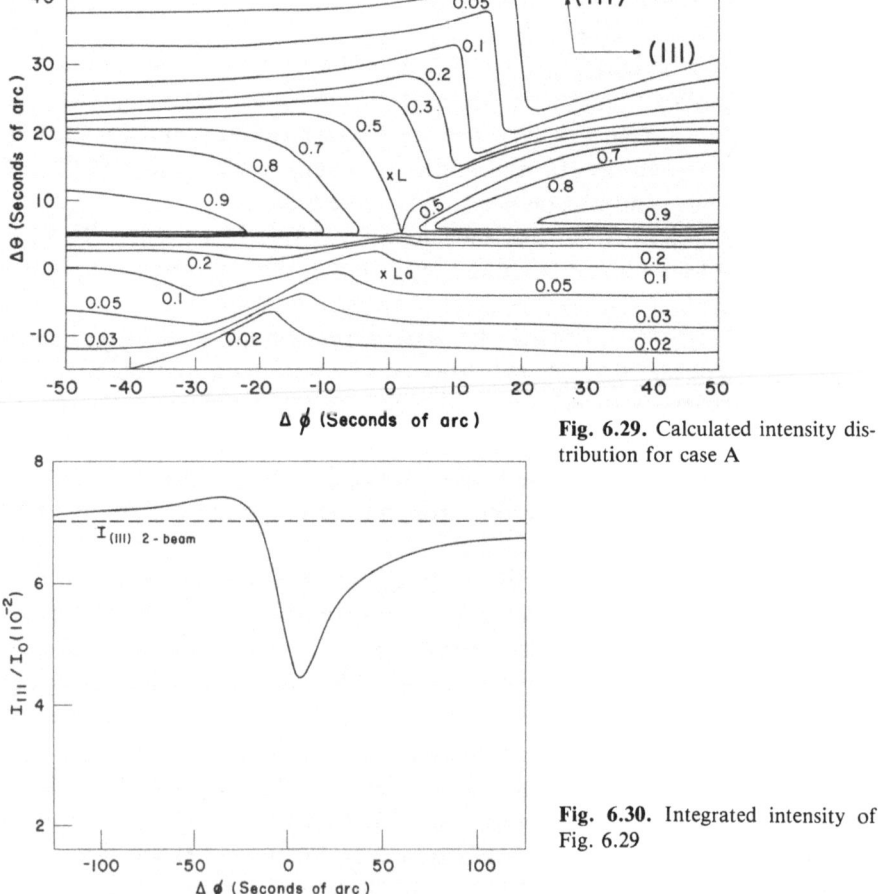

Fig. 6.29. Calculated intensity distribution for case A

Fig. 6.30. Integrated intensity of Fig. 6.29

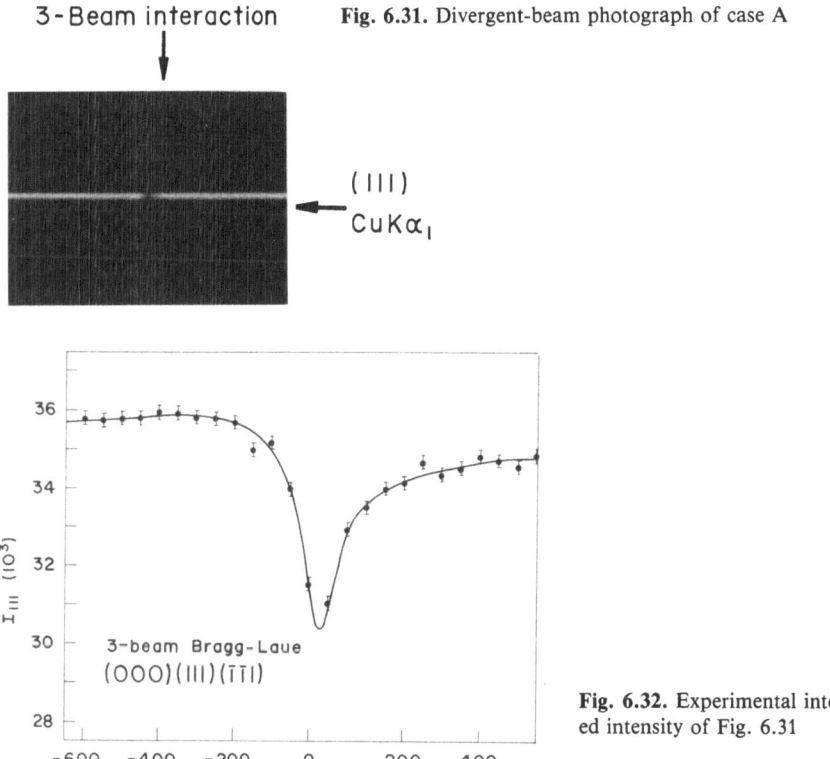

3-Beam interaction

Fig. 6.31. Divergent-beam photograph of case A

(111)
CuKα₁

3-beam Bragg-Laue
(000)(111)(1̄1̄1)

Fig. 6.32. Experimental integrated intensity of Fig. 6.31

integrating the reflected intensity over $\Delta\theta$. The asymmetry of the integrated intensity is shown in Fig. 6.30 with the integration interval $\Delta\theta$ equal to one minute of arc. More rigorous calculations may be carried out by integrating the intensity over $\Delta\phi$. Here, the interval $\Delta\phi$ corresponds to the experimental conditions, i.e., the beam divergence. When a well-collimated incident beam is used, single integration over $\Delta\theta$ for the integrated intensity is sufficient to provide qualitative information about the multiple diffracted intensities. In Fig. 6.30, the maximum intensity near $\Delta\phi = -30''$ is mainly due to the fact that $|E_{111}|^2/|E_0|^2 = 0.9$ near $\Delta\phi = -30''$ and $\Delta\theta = 5''$, while the minimum intensity at $\Delta\phi = 0''$ results from the smaller intensity distribution (Fig. 6.29) in the vicinities of La and L. Figures 6.29 and 30 can be compared, respectively, with the photograph, Fig. 6.31, obtained with the divergent-beam method mentioned in Sect. 2.2.2, and with the integrated intensity line profile, Fig. 6.32, obtained with the collimated-beam arrangement (Sect. 2.2.1). The agreement between the calculated and the experimental results can be seen.

The dispersion surface at $\Delta\theta = 13''$ for case B, shown in Fig. 6.33, can be obtained by applying an inversion operation about L plus a mirror reflection

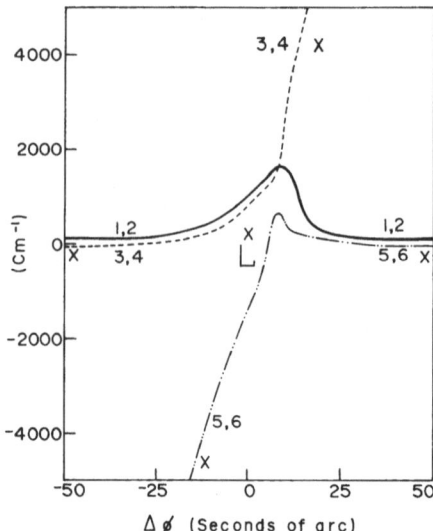

Fig. 6.33. Section of the dispersion surface at $\Delta\theta = 13''$ of the Ge (000) (111) (220) Bragg-Bragg diffraction for CuK_{α_1} (case B)

about $\Delta\phi = 0$ in Fig. 6.26b. The three pairs of dispersion sheets are represented by three lines. The number of permitted modes is $2(3-2) = 2$, because there are two Bragg reflections, (111) and (220), involved in this case. Only modes 1 and 2 are allowed.

It is not surprising that Fig. 6.33 is related to Fig. 6.26b by inversion and mirror symmetry, since cases A and B form a reciprocal pair according to the reciprocity theorem [6.35]. Figure 6.34 shows the wavevectors of both cases, which are related to the reciprocity theorem, i.e.,

$$K_{000}(B) = -K_{111}(A) \,,$$

$$K_{220}(B) = -K_{\bar{1}\bar{1}1}(A) \,, \tag{6.15}$$

$$K_{111}(B) = -K_{000}(A) \,.$$

Because of this reciprocity, the allowed modes at $\Delta\phi = 0$ in case A become forbidden in case B and the only forbidden pair, modes 5 and 6, in case A are the allowed modes 1 and 2 in case B.

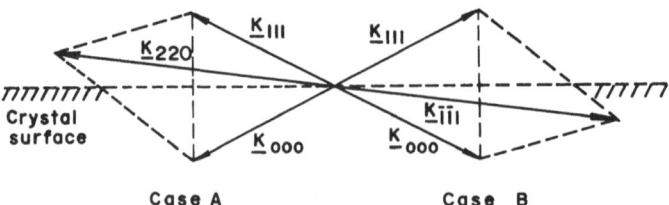

Fig. 6.34. Geometrical relation between the Bragg-Laue case A and the Bragg-Bragg case B

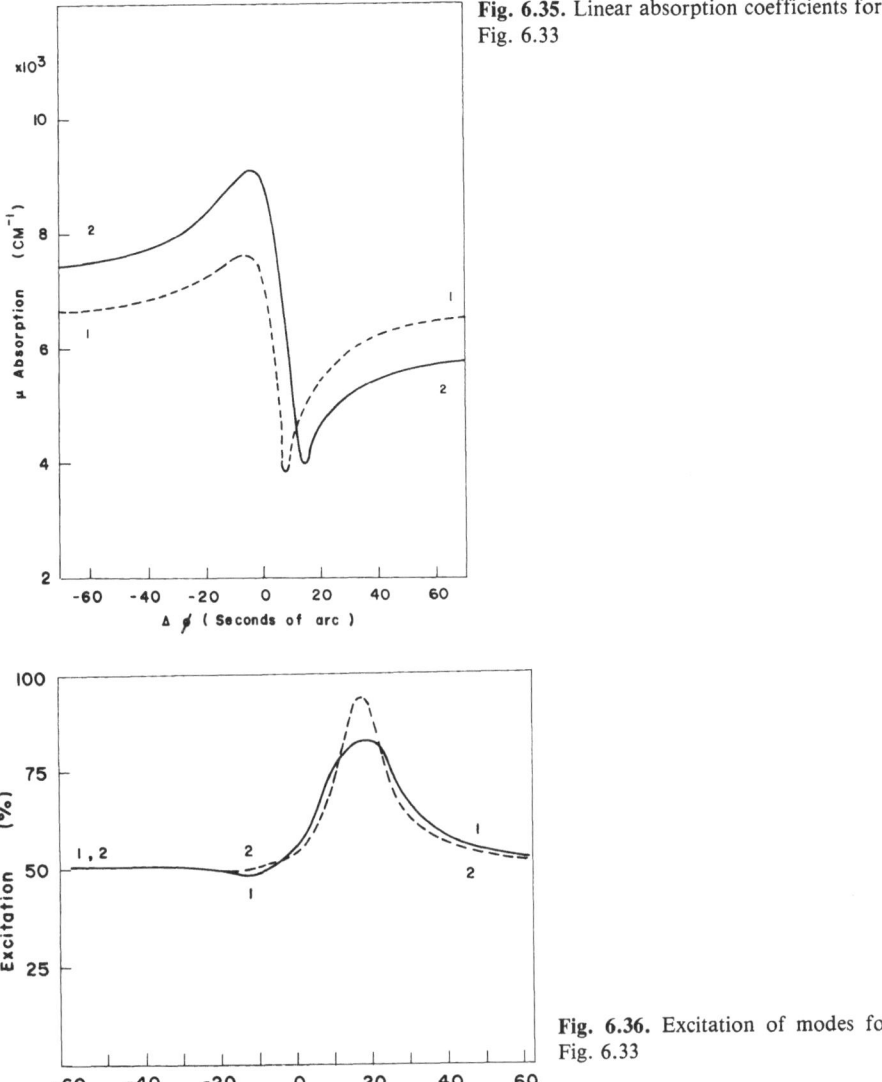

Fig. 6.35. Linear absorption coefficients for Fig. 6.33

Fig. 6.36. Excitation of modes for Fig. 6.33

The absorption coefficients and the excitations of modes for $\Delta\theta = 13''$ are shown in Figs. 6.35 and 36. The absorption coefficients for case B resemble the upper half of Fig. 6.27b for case A. Both modes 1 and 2 approach the two-beam absorption value, 6990 cm^{-1}, for large $|\Delta\phi|$. The excitations of modes shown in Fig. 6.36 also have features similar to those in the upper half of Fig. 6.28. By considering the effective absorption defined in (4.234), it is very easy to prove that the effective absorption coefficients of both cases are equal. The intensity distribution and the integrated intensity of the (1 1 1) reflection

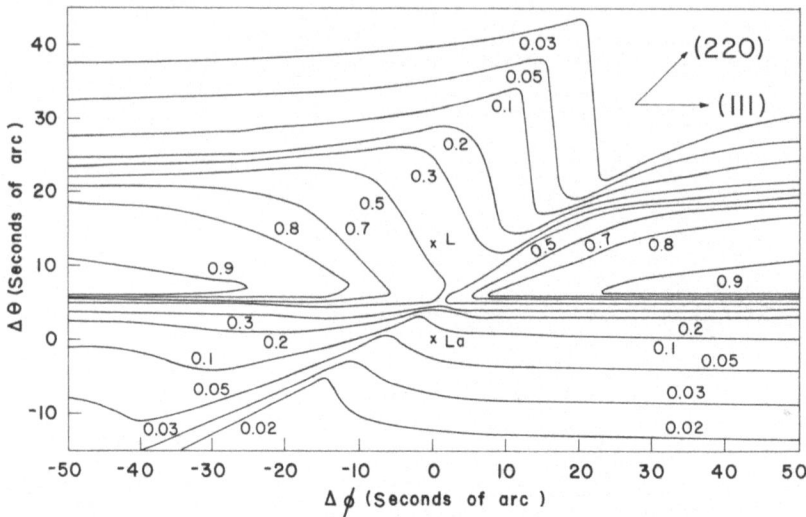

Fig. 6.37. Calculated intensity distribution of case B

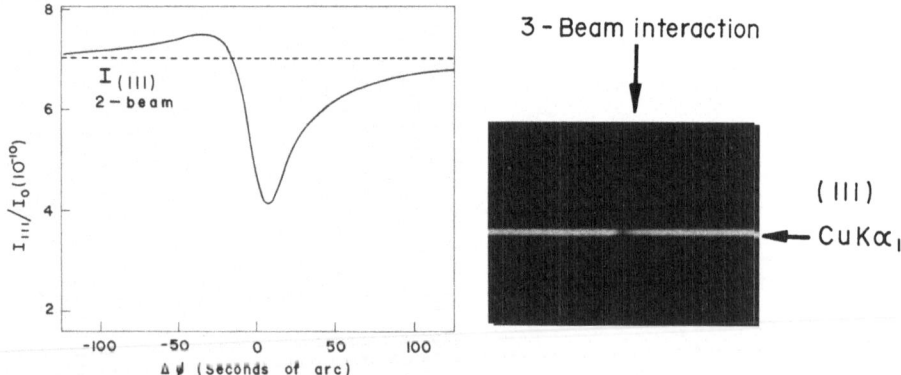

Fig. 6.38. Calculated integrated intensity of Fig. 6.37

Fig. 6.39. Divergent-beam photograph of Fig. 6.37

should be the same. Figures 6.37 – 40 show the calculated and the experimental results. They are the same as those shown in Figs. 6.29 – 32 for case A, respectively. This interesting characteristic of the two cases is undoubtedly due to the reciprocity theorem.

6.5 Four-Beam Bragg-Laue Diffraction

The four-beam case, (000) (004) ($\bar{1}\bar{1}$1) ($\bar{1}\bar{1}$3), in germanium for CuK_{α_1} consists of a symmetric (004) Bragg reflection, an inclined Bragg reflection ($\bar{1}\bar{1}$3)

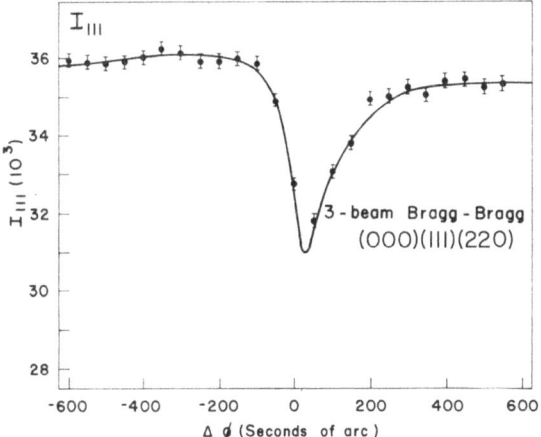

Fig. 6.40. Observed integrated intensity of Fig. 6.39

and a Laue transmission ($\bar{1}\bar{1}1$). There are two Bragg reflections involved in this case. The number of permitted modes is therefore four, according to (4.181).

The center of the total reflection region for the (004) reflection is at a deviation $\Delta\theta = 6$ seconds of arc from the exact Bragg angle of (004). For simplicity, let us consider only the cross section of the dispersion surface perpendicular to the (004) plane of incidence at $\Delta\theta = 6''$. The eight dispersion sheets of the possible modes are shown in Fig. 6.41. Only modes 1, 2, 3 and 4 are per-

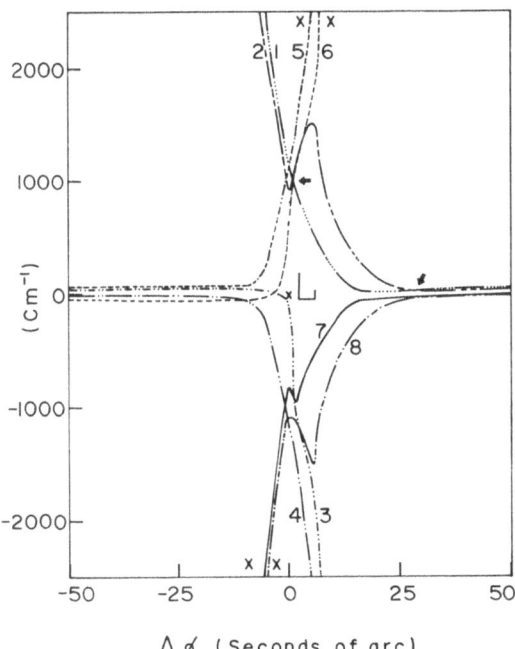

Fig. 6.41. Section of the dispersion surface at $\Delta\theta = 6''$ for the Ge (000) (004) ($\bar{1}\bar{1}3$) ($\bar{1}\bar{1}1$) Bragg diffraction of CuK_{α_1}

mitted. The point with $k\delta = 0$ at $\Delta\phi = 0$ is the Lorentz point. The two pairs of dispersion sheets of modes 1 and 2 and modes 7 and 8 are symmetric about $k\delta = 0$. Similarly, the dispersion sheets of modes 3 and 4 are symmetrically related to those of modes 5 and 6. This symmetry about $k\delta = 0$ is associated with the symmetry of the dispersion surface of the two-beam symmetric Bragg reflection. The same symmetry can be seen in Fig. 6.25a for the $(1\,1\,1)$ reflection. The intersections of the dispersion sheets of the permitted modes are marked with arrows; the forbidden modes are marked with 'x'. Modes 1 and 2 approach the two-beam value, $k\delta \simeq 0\ \text{cm}^{-1}$, for large positive $\Delta\phi$, while modes 3 and 4 approach the same value for large negative $\Delta\theta$. Modes 1 and 2 in the region of large negative $\Delta\phi$ and modes 3 and 4 in the region of large positive $\Delta\phi$ approach the same inclined line (plane) which represents the dispersion sheets of the two-beam $(\bar{1}\,\bar{1}\,1)$ transmission reflection. Further distortions near the Lorentz point are caused by the presence of the inclined $(\bar{1}\,\bar{1}\,3)$ Bragg reflection.

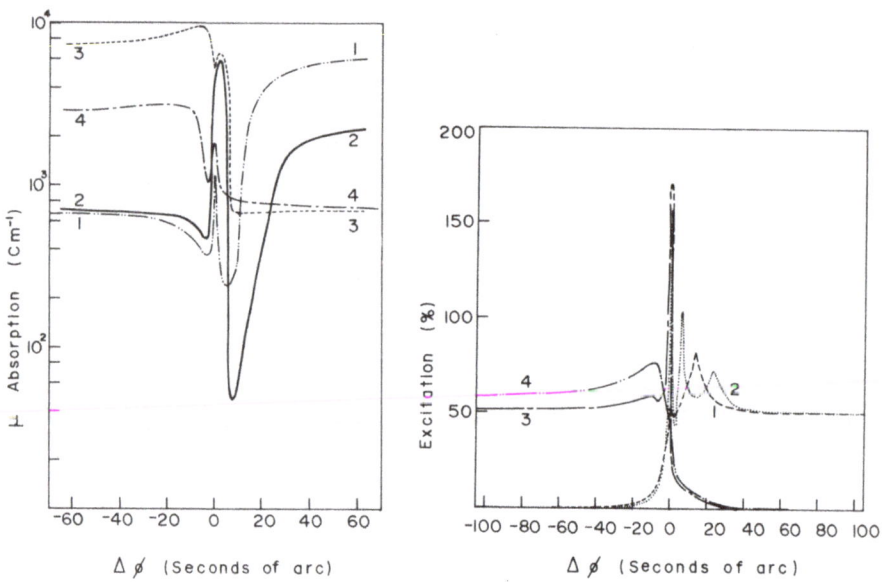

Fig. 6.42. Linear absorption coefficients for Fig. 6.41

Fig. 6.43. Excitation of modes for Fig. 6.41

Figure 6.42 shows the linear absorption coefficients of the four permitted modes along the same direction [the $(0\,0\,4)$ reflection line] as in Fig. 6.41. Modes 1 and 2 behave similarly; the corresponding absorption coefficients for large negative $\Delta\phi$ have the maximum two-beam value of the $(1\,1\,1)$ reflection. Modes 1 and 2 for positive $\Delta\phi$ and modes 3 and 4 for large negative $\Delta\phi$ approach the maximum two-beam values, 2550 and 6650 cm^{-1}, in the range of

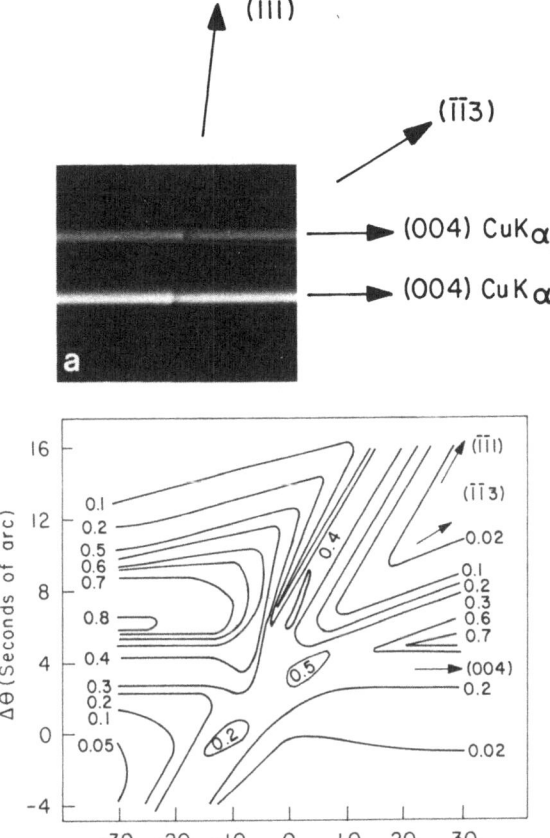

Fig. 6.44. (a) Observed and (b) calculated intensity distribution I_{004} of the four-beam case

total reflection of the (004) reflection. The dips close to $\Delta\phi = 0$ are due to the anomalous transmission of the $(\bar{1}\bar{1}1)$ reflection. The sharp peaks at $\Delta\phi = 0$ result from the total reflection of the inclined $(\bar{1}\bar{1}3)$ reflection.

The excitation of modes, shown in Fig. 6.43, is similar to that in the three-beam Bragg-Laue case (Fig. 6.28). The complex features at $\Delta\phi = 0$ and in the range $\Delta\phi = 0$ to $\Delta\phi = 30''$ are due to the interaction between the dispersion sheets of the two Bragg reflections (004) and $(\bar{1}\bar{1}3)$.

The experimental and calculated intensity distributions of the (004) reflection around the four-beam diffraction position ($\Delta\theta = 0$, $\Delta\phi = 0$) are shown in Fig. 6.44a and b, respectively. The influence of the $(\bar{1}\bar{1}1)$ and $(\bar{1}\bar{1}3)$ reflections on the (004) reflected intensity is clearly seen in the photograph (Fig. 6.44a). The observed intensity distribution agrees qualitatively with the calculated one (Fig. 6.44a and b).

The integrated intensity obtained experimentally using the collimated-beam method is shown in Fig. 6.45a. Only the indices of the secondary reflec-

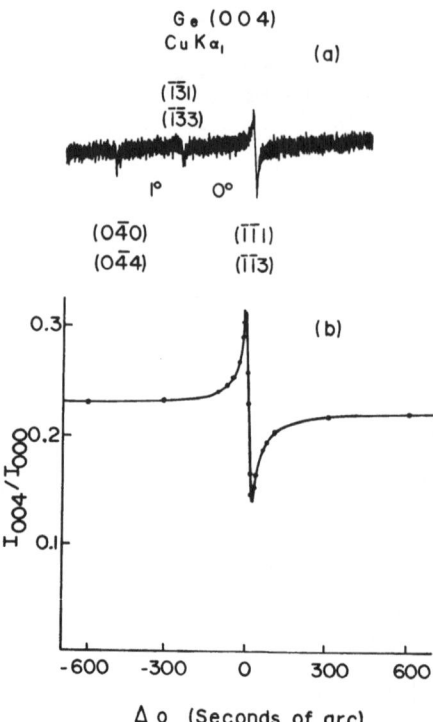

G$_\bullet$ (0 0 4)
Cu Kα_1 (a)

($\bar{1}3\bar{1}$)
($\bar{1}\bar{3}3$)

1°   0°

($0\bar{4}0$)   ($\bar{1}\bar{1}1$)
($0\bar{4}4$)   ($\bar{1}\bar{1}3$)

Fig. 6.45. (a) Observed and (b) calculated integrated intensity I_{004} for Fig. 6.44

Δ o (Seconds of arc)

tions are labeled. The resonance-like profile can also be calculated by integrating the diffracted intensity over $\Delta\theta$. The calculated integrated intensity shown in Fig. 6.45b reveals the same line profile as that experimentally observed.

7. Applications

Multiple diffraction of x-rays, as mentioned in Chap. 1, provides three-dimensional information about the structure of matter. Its applications to the study of materials are many. In this chapter, applications to the determination of x-ray reflection phases, lattice constants, lattice mismatches in layered materials, and its potential in solving problems related to surface physics are discussed. Moreover, for a given crystal with a known structure, multiple diffraction can be used, in turn, to determine precisely the wavelength of the radiation. This makes possible the spectroscopic characterization of radiation sources. This application is also included. Finally, at the end of the chapter, further possible development and applications of the methods using multiple diffraction are discussed.

7.1 Experimental Determination of X-Ray Reflection Phases; Application to Crystal Structure Determination

7.1.1 General Consideration of the X-Ray Phase Problem

Determination of a crystal structure requires information about the relative positions of the atoms and molecules in the crystal unit cell. From the definition of the structure factor (4.22), this information is deposited in the amplitude and the phase of the structure factor. Intensity measurement from a simple two-beam reflection determines the square of the amplitude of the corresponding structure factor. The information about the phase of the structure factor is lost since the intensity is proportional to the multiplication of the structure factor with its complex conjugate. More rigorously, this fact can be easily proved by considering the dispersion equation of a two-beam case. Setting $F_{G_2} = F_{G_1 - G_2} = 0$ in (6.3) and neglecting the virtual solution $Z = 0$, we obtain the two-beam dispersion equation:

$$Z^2 - |F_{G_1}|^2 = 0 . \tag{7.1}$$

The positions of the excited tie points related to Z are independent of the phase of F_{G_1} [7.1]. The diffracted intensity, which depends on the dispersion surface, is therefore not sensitive to the phase variation in F_{G_1}. This difficulty in determining the phase of F_{G_1}, the reflection phase, constitutes the so-called x-ray phase problem in crystallography and x-ray optics.

The phase of the structure factor F_{G_1} depends not only on the positions of the atoms and molecules in a crystal unit cell but also on the choice of origin for the cell. However, the phase of the product of a group of structure factors is an invariant when the reciprocal lattice vectors of the reflections form a polygon in the reciprocal space. This is the only phase that is physically significant.

Attempts to solve this phase problem have been carried out in two principal directions, one applying mathematical techniques and the other utilizing existing physical phenomena to extract the phase information from the intensities of x-ray diffractions. In the mathematical approaches, a huge collection of two-beam intensity measurements are analyzed using the principles of mathematics to determine the phases on a trial and error basis. One of the most frequently used methods is the direct method [7.2 – 8], which is based on the assumption that the atoms in a crystal are represented by identical, spherically symmetric electron densities which do not overlap, and that the electron density is everywhere positive. A certain phase relationship, often called the Σ_1 and Σ_2 relation [7.2], can be derived from this assumption [7.4] with calculable probabilities [7.7, 9, 10] which depend on the structure factors, or, more precisely, the normalized structure factors $E(G)$ defined as

$$E^2(G) = \frac{|F_G|^2}{\varepsilon_G \sum\limits_{j=1}^{n} f_j^2} , \qquad (7.2)$$

where F is the corresponding structure factor. The real atoms are replaced by point atoms at rest. The quantity ε_G is an integer which is generally equal to 1 but may assume other values for special sets of reflections in certain space groups. The summation is taken over all the atoms in a unit cell. In practice, a few phases are assigned arbitrarily to form the starting phases. Approximate phases are then developed via the phase relations, Σ_1 and Σ_2.

To improve the approximate phases obtained from the Σ_2 and Σ_1 relationships, the tangent formula [7.11] is used:

$$\tan \Psi_G = \frac{\sum Q'_{GH}\sin(\Psi_{G-H}+\Psi_H)}{\sum Q'_{GH}\cos(\Psi_{G-H}+\Psi_H)} = \frac{T_G}{B_G} , \qquad (7.3)$$

where

$$Q'_{GH} = \frac{w_G w_H |E_H||E_{G-H}|}{1-|U_G|^2} , \qquad (7.4)$$

with the weighting factors

$$w_G = \tan\left[\sigma_3 \sigma_2^{-3/2} |E_G| \sqrt{(T_G^2 + B_G^2)}\right] , \qquad (7.5)$$

the unitary structure factor

$$|U_G| = \frac{|F_G|}{\sum\limits_{j=1}^{n} f_j}, \quad \text{and} \tag{7.6}$$

$$\sigma_l = \sum\limits_{j=1}^{n} Z_j^l. \tag{7.7}$$

The quantity Z_j is the atomic number of the jth atom.

The plausible phase set is the one with the highest absolute figure of merit FOM and the lowest ψ_0 and residual (RESID) values [7.5, 11, 12].

The physical approaches include the technique of replacing light atoms by a heavy atom in a crystal and the use of anomalous scattering and multiple x-ray diffraction. In the replacement method, it is easy to determine the position of the heavy atom which serves as the reference for the other atoms. The difference in the diffraction patterns between the natural and the heavy-atom crystals reveals the phase information [7.13 – 16]. In anomalous scattering, the difference in the real and imaginary parts of the atomic scattering factors of the atoms for x-ray wavelengths above and below an absorption edge also provides a means for phase determination [7.17, 18]. In the multiple diffraction of x-rays, the interaction between the wavefields should, in principle, be closely related to the reflection phases.

7.1.2 Reflection Phases and Multiple Diffraction

The reflection phases of visible light, in contrast with x-rays, are easily determined experimentally by the interference of two coherent beams coming from a single source or a pair of sources. Optical holography is a well-known example where the phase contrast of two coherent beams, an incident beam and a reference beam, is used.

The coherent dynamical interaction in x-ray multiple diffraction has long been considered to provide clues to phase determination [7.19]. X-ray multiple diffraction is analogous to optical holography, since one of the diffracted x-ray beams can be treated as a reference for the other beam. The relative phase difference between the two beams modifies the diffracted intensity of the reference beam. Phase information can therefore be extracted from the intensity variation on the reference beam.

Lipscomb [7.19], in 1949, investigated the possibility of using the intensity of multiple diffraction for phase determination and concluded that it is very difficult to extract phase information from the intensity measurements. A similar investigation has also been performed on quartz in 1957 by *Williamson* and *Fankuchen* [7.20].

In contrast to x-ray diffraction, a theoretical and experimental investigation based on the dynamical diffraction of electrons was carried out by *Kambe* and *Miyake* [7.21 – 23]. In that study, the dynamical theory of electron diffraction of *Bethe* [7.24] was adopted for three-beam electron diffraction for which one weak diffraction was involved. The dispersion surface for three-beam electron diffraction was discussed theoretically in relation to the reflection phases. Analytical expressions for diffracted intensities as a function of the phases were found, under the assumption that one of the three reflections is very weak, so that the dispersion equation can be solved as in two-beam cases. Experimental evidence for the phase dependence of the diffracted intensities of the Kikuchi lines was observed for three-beam electron diffraction in a graphite crystal. The intensity distribution of Kikuchi lines at the three-beam diffraction position is a key fact for phase determination. However, this investigation is limited to three-beam electron diffractions involving at least one weak reflection.

Hart and *Lang* [7.25] in 1962 adopted the theoretical considerations given by *Kambe* [6.23] to a three-beam x-ray diffraction in a wedge-shaped crystal. The variation in crystal thickness causes the generation of interference fringes, the so-called Pendellösung fringes [7.26]. It was found that the periodicity of Pendellösung fringes between the intensity maxima and minima depends on the reflection phases. In 1977 *Post* [7.27] considered the position of dispersion surfaces as a function of phases for three-beam x-ray diffractions. Intensity distributions for two three-beam transmission cases in an Al_2O_3 crystal were obtained experimentally using the divergent-beam technique (Sect. 2.2.2). This consideration is similar to that given by *Kambe* and *Miyake*, except that *Post*'s method is more general than *Kambe*'s and can be applied to cases involving strong reflections. However, the effect of crystal thickness on the transmitted intensities was not considered. A similar investigation has also been carried out by *Jagodzinski* [7.28] using four- and five-beam diffractions.

In 1981 *Chapman* et al. [7.29] investigated the phase dependence of two overlapped three-beam Umweg diffractions, which were treated as a four-beam Bragg-type reflection. The complexity of the phases involved in the four-beam case complicates their determination. The sense of crystal rotation in multiple diffraction experiments, which plays an important role in phase determination, was not considered.

In the same year, *Chang* [7.30] reconsidered the relative position of the dispersion surface in connection with the reflection phases for three-beam Bragg-type diffractions, both Umweg and Aufhellung, and included explicitly the effect of crystal rotations on the diffracted intensities in the phase determination. A simple relation between the signs of the phases, the crystal rotations, and the intensity distribution was obtained. The application of this relation to a central symmetric crystal was realized in conjunction with the conventional direct methods.

Other investigations along this line of research have also been since reported [7.31 – 33].

7.1.3 Experimental Methods for Phase Determination Using Multiple Diffraction

The details of proposed experimental methods for phase determination, using multiple diffraction of x-rays, are given in this section. The justification and applicability of these methods are also discussed.

a) Three-beam Borrmann diffraction. The dispersion surface, as discussed in Sect. 6.2, contains the information about the phases of the triple product of the structure factors. From (6.3, 6), we see that the phase angle Ψ is either zero or π for a centrosymmetric crystal. We define the sign

$$S_{P,T} = \cos \Psi, \qquad (7.8)$$

where Ψ is the phase angle of the triple product of the structure factors, $F_{-G_1} F_{G_2} F_{G_1-G_2}$. $S_{P,T}$ is either positive or negative. As mentioned previously, the angle Ψ is independent of the choice of the origin of the crystal unit cell. According to *Post* [7.27], at the exact three-beam diffraction position, the signs of Z are $(- - +)$ for a positive $S_{P,T}$ and $(- + +)$ for a negative $S_{P,T}$. The direction of positive Z is towards the crystal interior. The intersection of the dispersion surface with the plane of incidence of the G_1 reflection (the plane containing the reciprocal lattice points O and G_1 and the three-beam Laue point) is shown schematically in Fig. 7.1a. The relation between the dispersion sheets and the angular deviation $\Delta\theta$ from the Bragg angle θ_{G_1} of the G_1 reflection is revealed. The points La and L are the Laue and Lorentz points, respectively. Branches 1 and 2 are the distorted two-beam dispersion sheets for the G_1 reflection. For a positive $S_{P,T}$, the third dispersion sheets

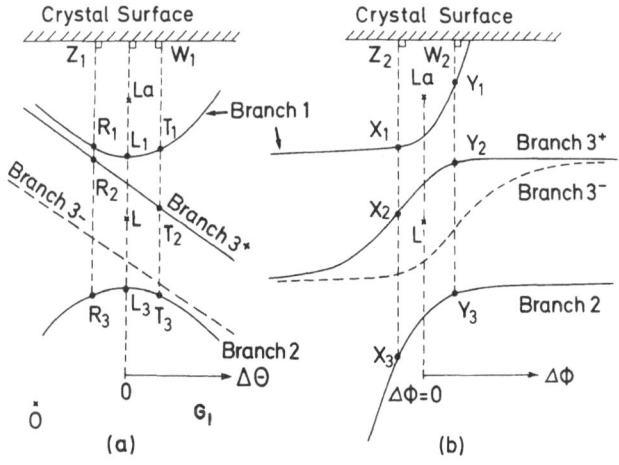

Fig. 7.1. (a) Schematic representation of the dispersion surface of a three-beam Borrmann diffraction (after *Chang* [7.34]); (b) a side view of (a)

(branch 3^+) is situated between L and branch 1. For a negative $S_{P,T}$, the third branch 3^- lies between L and branch 2.

The excitations of the tie points on the dispersion sheets are determined, as mentioned in Chap. 4, by the line normal to the crystal surface from a given entrance point on the incident wavefront. For a positive $S_{P,T}$ and $\Delta\theta < 0$, three tie points, R_1, R_2 and R_3 (see Fig. 7.1), are excited. Since the Poynting vectors at R_1 and R_2 perpendicular to the corresponding dispersion sheet point along the direction of the incident beam, i.e., along LaO, the two modes associated with R_1 and R_2 are more strongly excited than the mode at R_3. For $\Delta\theta > 0$, the tie points T_2 and T_3 are excited more strongly than T_1. Similar situations occur at the crystal settings which are off the exact three-beam diffraction point ($\Delta\phi = 0$). The angle ϕ is the azimuthal rotation angle of the secondary reciprocal lattice point G_2 around OG_1. The projection of Fig. 7.1a on the plane perpendicular to and bisecting OG_1 is shown in Fig. 7.1b. For $\Delta\phi < 0$, tie points X_1 and X_2 are more strongly excited than X_3. For $\Delta\phi > 0$, Y_2 and Y_3 are more strongly excited than Y_1. For a negative $S_{P,T}$, the excitation of the tie points is the same as that for a positive $S_{P,T}$, except that the dispersion sheet 3^- replaces the branch 3^+.

In the three-beam Pendellösung experiment reported by *Hart* and *Lang* [7.25], the three-beam Borrmann diffraction, (000) $(2\bar{2}0)$ $(3\bar{1}\bar{1})$[1], from a wedge-shaped germanium single crystal was investigated using AgK_{α_1} radiation. For this three-beam case, $S_{P,T}$ is positive. The Pendellösung fringe spacing of two-beam diffraction is inversely proportional to the separation between the dispersion sheets. When the three-beam case with $S_{P,T} > 0$ comes into play, the fringe spacing of the direct (000) beam at $\Delta\phi < 0$ depends on the distance X_1X_2 of the two strongly excited tie points. (If the experiment involves an inclined crystal entrance surface, the lines Z_2X_3 and W_2Y_3 in Fig. 7.1b are tilted accordingly.) At $\Delta\phi > 0$, the fringe spacing is inversely proportional to Y_2Y_3. When the crystal surface is, for example, cut in such a way that lines Z_2X_3 and W_2Y_3 are parallel to the asymptotic line passing through L for branches 1 and 2 of Fig. 7.1b, then

$$|X_1X_2| < |L_1L_2| < |Y_2Y_3|.$$

The distance L_1L_2 is the separation between branches 1 and 2 for the two-beam G_1 reflection. The inequality also holds for other inclined crystal surfaces. The $(2\bar{2}0)$ Pendellösung fringe spacing increases for $\Delta\phi < 0$ and decreases for $\Delta\phi > 0$. For a negative $S_{P,T}$ the situation is reversed.

A more detailed investigation has recently been carried out by *Hoier* and *Aanestad* [7.35] for the three-beam (000) $(2\bar{2}0)$ $(0\bar{2}2)$ Pendellösung of a silicon single crystal for MoK_α radiation. The calculated Pendellösung fringes

---

[1] This diffraction is the 4-beam case discussed in Sect. 6.3

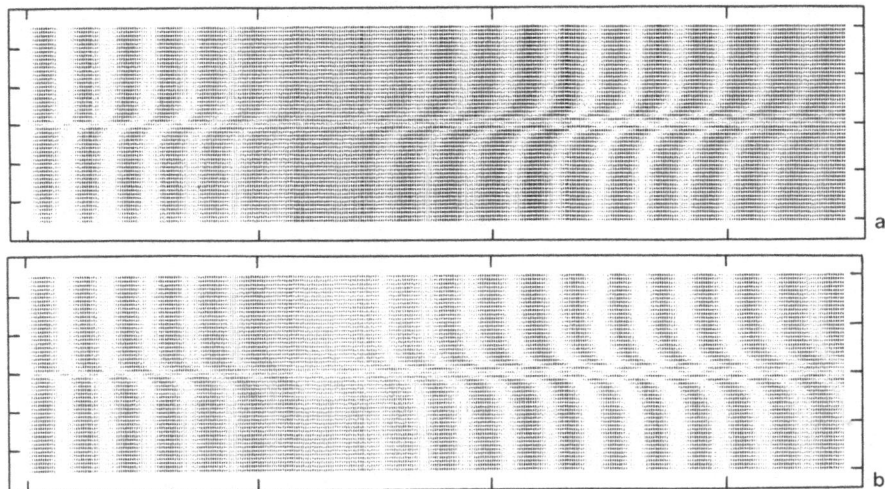

Fig. 7.2a, b. Calculated Pendellösung fringe contrast for the phase angles (**a**) $\Psi = 0°$ and (**b**) $\Psi = 180°$ (courtesy of *Hoier* and *Aanestad* [7.35])

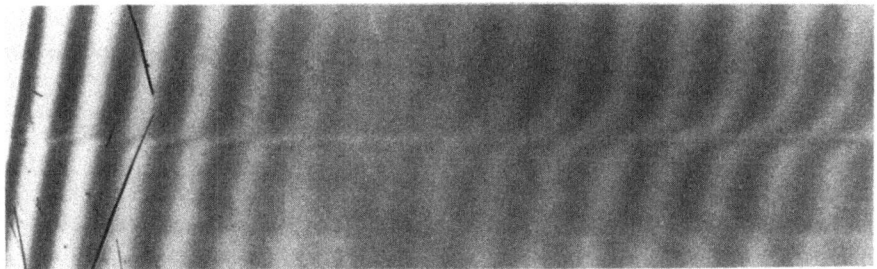

Fig. 7.3. The observed $(2\bar{2}0)$ Pendellösung fringe contrast (courtesy of *Hoier* and *Aanestad* [7.35])

for the positive $S_{P,T}$ are given in Fig. 7.2. [The reflections $(2\bar{2}0)$, $(0\bar{2}2)$, and the coupling $(20\bar{2})$ have positive structure factors.] For a negative $S_{P,T}$ the calculated fringe image is reversed. The horizontal axis then represents the decrease in crystal thickness. The experimental result is shown in Fig. 7.3, which can be compared with the calculation.

This method is impractical in the real crystal-structure determination because it needs a crystal wedge, which, however, is difficult to prepare for crystals to be used in crystal-structure determination.

Alternatively, the divergent-beam technique, discussed in Sect. 2.2.2, was used by *Post* [7.27] to extract phase information from the intensity distribution for a plane-parallel crystal plate in the vicinity of a three-beam Borrmann diffraction point. In Post's experiment, two three-beam cases, (000) $(1\bar{1}2)$ $(1\bar{2}0)/(012)$ and (000) $(1\bar{1}2)$ $(0\bar{1}\bar{2})/(104)$, of an aluminium oxide (corundum) single crystal for CuK_{α_1} were investigated. The indices after the slashes indicate the coupling reflections. The former three-beam case has a

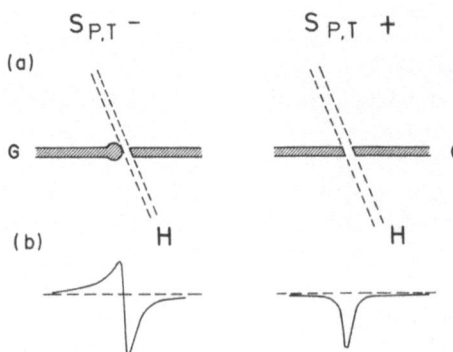

Fig. 7.4a, b. Schematic representations of (a) the intensity distribution of three-beam Borrmann diffractions with a negative $S_{P,T}$ and a positive $S_{P,T}$. (b) the line profiles along the diffracted lines

positive $S_{P,T}$, the latter a negative $S_{P,T}$. An asymmetric intensity distribution, similar to that shown in Fig. 6.7, was observed near the three-beam point for negative $S_{P,T}$. For the case with a positive $S_{P,T}$, no intensity asymmetry could be detected. For illustration, a schematic representation of the diffraction lines of these two cases is given in Fig. 7.4a. The asymmetry can be seen clearly by drawing the intensity line profile along the diffraction lines. These are also shown in the same figure (Fig. 7.4b).

According to *Ewald* and *Heno* [7.1], the modes with dispersion sheets closest to the Laue point have the lowest absorption in transmission cases. By considering the absorption and the excitation of modes, as discussed in Chap. 6, the three-beam transmitted intensity for a negative $S_{P,T}$ shows the intensity asymmetry about the three-beam point. No intensity asymmetry is expected for the case with a positive $S_{P,T}$. This is in agreement with the experimental results shown in Fig. 7.4. However, the transmitted intensity near the three-beam point, according to (5.85), depends not only on the term $\mathrm{Re}\{\chi_{GO,HO}\chi_{OG,OH}\}$ but also on the sine and hyperbolic sine functions. The real function can be written in the alternative form

$$\mathrm{Re}\{\chi_{GO,HO}\chi_{OG,OH}\} = \mathrm{Re}\left\{P^2|\chi_{G-O}|^2 - \frac{\alpha}{2}P\Gamma^3 T_p \right.$$
$$\left. + \frac{\alpha^2}{2}\chi_{H-O}\chi_{O-H}\chi_{G-H}\chi_{H-G}\right\}, \qquad (7.9)$$

where

$$T_p = F_{-G}F_H F_{G-H} + F_G F_{-H}F_{H-G}, \qquad (7.10)$$

which is phase dependent. P is the polarization factor of the two-beam G reflection and Γ is a constant defined in (6.5). For centrosymmetric crystals,

$$T_p = 2|F_G F_{-H}F_{H-G}|\cos\Psi. \qquad (7.11)$$

The quantity α defined in (5.9), depending on the rotation of the crystal, can be expressed in terms of the azimuthal angle $\Delta\phi$ [7.36]:

$$\alpha = -\frac{P_3}{k\,\Gamma \sin\psi \cos\theta_G \tan\varDelta\phi}, \tag{7.12}$$

where P_3 is the polarization factor of the three-beam diffraction, i.e., $d_1 d_1' + d_3 d_3'$ for the π polarization and d_2^2 for the σ polarization (5.9, 20). The angle ψ is defined in (5.2). The sine and hyperbolic sine functions in (5.34) have a thickness dependence. For a thick crystal, the diffracted intensity is described by (5.86). There is, however, no definite connection between the phase $\varPsi$ and the intensity asymmetry. For example, in the particular case of the three-beam, (000) (111) (11$\bar{1}$)/(002), diffraction, the intensity asymmetry is observed (Fig. 6.7) even for $S_{\mathrm{P,T}} = 0$, because $F_{002} = 0$. For a thin crystal, the sine function in (5.48) dominates. The intensity asymmetry may be smeared out by this sinusoidal variation. A recent report [7.37] on the calculated intensity also shows the thickness dependence of the intensity asymmetry. It can therefore be concluded that because of this thickness effect the use of three-beam Borrmann diffraction is not well suited for phase determination.

b) Two overlapped three-beam Umweg reflections and N-beam ($N > 3$) diffraction. The two overlapped three-beam Umweg reflections (000) (222) (1$\bar{1}\bar{1}$) and (000) (2$\bar{2}$2) (1$\bar{1}$1) of silicon (centrosymmetric) for $\mathrm{Cr}K_{\alpha_1}$ radiation were treated as a four-beam case by *Chapman* et al. [7.29] to demonstrate the effect of a phase change of any one of the reflections on the line profile of the multiple diffraction. This case ought properly to be treated as a five-beam case, since a three-beam, (000) (222) (22$\bar{2}$), case appears between the two three-beam cases mentioned. The coupling reflection between (22$\bar{2}$) and (222) is the (004) reflection. According to (3.20), the reflected intensity of this three-beam case is very weak. If this additional three-beam case is neglected because of its weak intensity, the corresponding dispersion relation for this assumed four-beam case is

$$\begin{vmatrix} Z & F_{\bar{2}\bar{2}2} & F_{\bar{1}11} & F_{\bar{1}1\bar{1}} \\ F_{222} & Z & F_{133} & F_{131} \\ F_{1\bar{1}\bar{1}} & F_{\bar{1}\bar{3}\bar{3}} & Z & F_{00\bar{2}} \\ F_{1\bar{1}1} & F_{\bar{1}\bar{3}\bar{1}} & F_{002} & Z \end{vmatrix} = 0. \tag{7.13}$$

This equation involves many phase triplets and quartets. Those quartets involving symmetry-related reflections can be treated as doublets, since the sign relation between the symmetry-related reflections can be determined from the space group. For example, the phase-sign quartet $S_{\bar{1}1\bar{1}}S_{1\bar{1}1}S_{131}S_{\bar{1}\bar{3}\bar{3}}$ is equal to $S_{131}S_{\bar{1}\bar{3}\bar{3}}$ with $S_{\bar{1}1\bar{1}}S_{1\bar{1}1} = +1$. If we neglect the doublets, including the symmetry-related quartets, and set $F_{002} = 0$ [(002) is a forbidden reflection], (7.13) still involves at least two phase triplets, $S_{\bar{2}\bar{2}2}S_{1\bar{1}\bar{1}}S_{133}$ and $S_{\bar{2}\bar{2}2}S_{1\bar{1}1}S_{131}$.

The former is positive and the latter negative. It is difficult to tell how the asymmetry of the 222 reflection intensity would be influenced by these two phase triplets. The same argument can be applied to N-beam cases with $N > 3$. In addition, some N-beam cases with $N > 3$ do not show the intensity asymmetry at the N-beam points. For example, in the four-beam case, (000) (004) (022) (0$\bar{2}$2), for germanium and CuK_{α_1} radiation, the (004) reflected intensity at the four-beam point is symmetric (see Fig. 7.5). In comparison with the case (000) (004) ($\bar{1}\bar{1}$1) ($\bar{1}\bar{1}$3) discussed in Sect. 6.5 (Fig. 6.45), it is very hard to extract the phase information from the former 4-beam case. However, when the N-beam cases ($N > 3$) involve many symmetry-related reflections such that only a three-beam interaction dominates the whole diffraction process, these N-beam cases can be treated as three-beam cases for phase determination using the method described below.

c) Three-beam Bragg reflection. The discussion follows the treatment of [7.30]. The intensity of a Bragg-type multiple diffraction, as described in (5.100, 102), is less affected by the crystal thickness than is the intensity of the transmission-type multiple diffraction. Moreover, the invariant triplet-phase relationship is also preserved in the Bragg-type diffraction. Three-beam Bragg multiple diffraction, either Umweganregung [7.38] or Aufhellung [7.39], should therefore be suitable for phase determination. As discussed previously, the intensity asymmetry of multiple diffraction provides the phase information. This asymmetry is further enhanced in the Bragg-type multiple diffraction. This fact can be easily understood by examining the square of the wavefield amplitude of the reflected wave in a simple Bragg reflection. Figure 6.25e shows the ratio $|E_G|^2/|E_O|^2$ for the (111) reflection of germanium and CuK_{α_1} radiation. The asymmetry appearing in this figure is due to the difference in absorption and the phase difference of 180° between the two dispersion sheets. More rigorously, this phase dependence of the asymmetry can be directly seen from the intensity I_F of the wavefields (standing waves discussed in Sect. 4.9). For a phase angle Ψ_G equal to 180°, the intensity I_F at or near the crystal surface for mode 1 is expressed in (4.250). For $\Psi_G = 0°$, the corresponding I_F takes the form

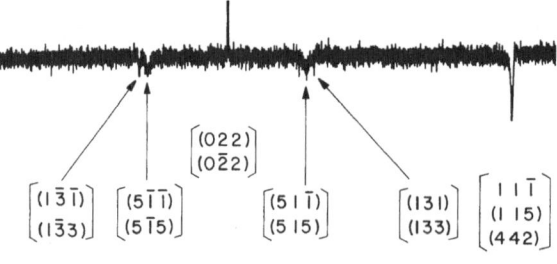

Ge   (004) CuK_{α_1}
F.S. 2×10^4 cps

Fig. 7.5. Observed multiple diffraction peak for the four-beam Bragg-Laue diffraction

$$I_F(1) = \tfrac{1}{2}[1 - |X| \exp(-v)]^2 \qquad\qquad (7.14a)$$

for range I,

$$I_F(1) = \tfrac{1}{2}[1 + |X| \exp(-v)]^2 \qquad\qquad (7.14b)$$

for range III, and

$$I_F(1) = \tfrac{1}{2}(1 + |X|^2 + 2|X| \cos v) \qquad\qquad (7.14c)$$

for range II. The expression (7.14a) for $\Psi_G = 0°$ in range I is the same as the one in (4.250b) for $\Psi_G = 180°$ in range III. Similarly, (7.14b) for $\Psi_G = 0°$ in range III is the same as (4.250a) for $\Psi_G = 180°$ in range I. A reversal in the line profile for I_F is therefore expected in two cases, one with $\Psi_G = 0°$ and the other with $\Psi_G = 180°$. Figure 7.6 shows this reversal for the (111) Bragg reflection of germanium for CuK_{α_1} [7.40]. The asymmetry of I_F clearly indicates the phase dependence of the line profile. Unfortunately, the reflected intensities of two-beam cases reveal no such feature.

The phase information is, however, preserved in the dispersion surface of the three-beam Bragg-type multiple diffraction. As discussed in Sect. 6.4, the dispersion surface in the Bragg case is different from that in the transmission case due to the presence of the total reflection. The way in which the dispersion sheets associate with the triplet phase in the Bragg case is therefore different from that in the transmission case. To illustrate the phase dependence of the dispersion surface, two Umweganregung reflections, (i) (000) (222) (1$\bar{1}\bar{1}$) with $S_{P,T}$ ($= S_{\bar{2}\bar{2}\bar{2}} S_{1\bar{1}\bar{1}} S_{133}$) > 0 and (ii) (000) (222) (113) with $S_{P,T}$ ($= S_{\bar{2}\bar{2}\bar{2}} S_{113} S_{11\bar{1}}$) < 0, are considered for germanium and CuK_{α_1} radiation.

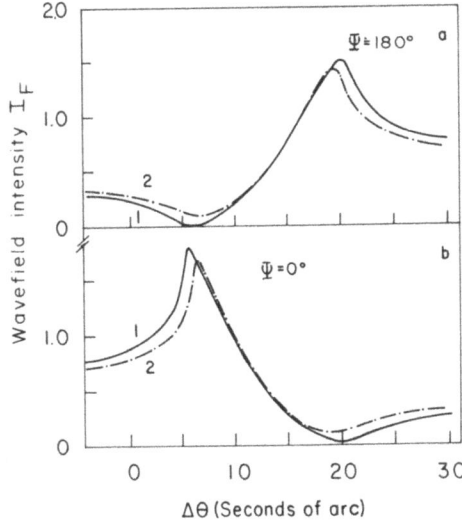

Fig. 7.6a, b. Calculated wavefield intensities of Ge two-beam (111) Bragg reflection (CuK_{α_1}) for the phase angles (a) $\Psi = 180°$ and (b) $\Psi = 0°$ (after *Chang* [7.40])

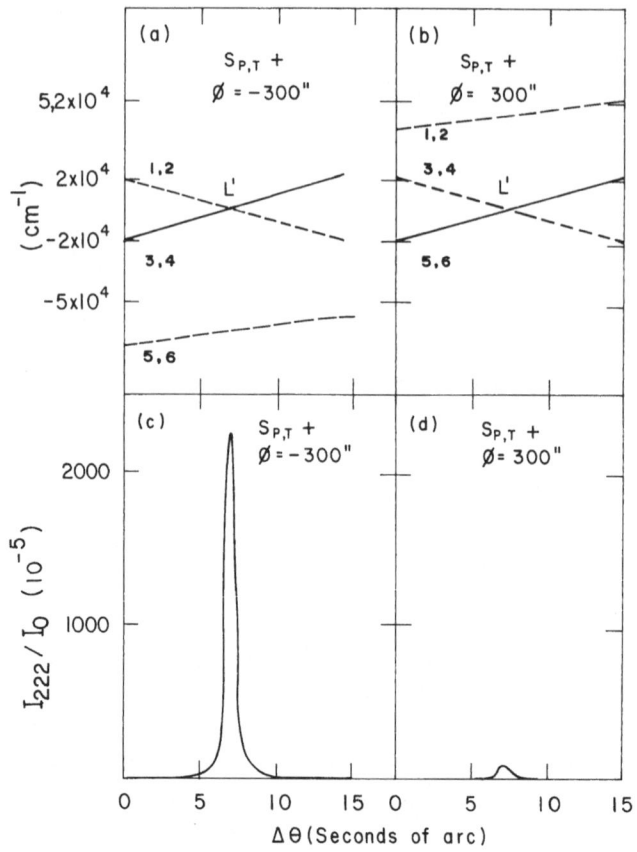

Fig. 7.7. Dispersion sheets and the calculated intensity I_{222} at $\Delta\phi = 300''$ and $-300''$ for a phase angle $\Psi = 0°$ in the Ge (000) (222) ($\bar{1}\bar{1}1$) Bragg diffraction (CuK_{α_1})

The calculated cross sections of the dispersion sheets at the azimuthal angles $\Delta\phi = -300''$ and $\Delta\phi = 300''$ for case (i) are shown in Fig. 7.7a and b. The point L' is the projection of the Lorentz point on the cross sections. The position $\Delta\theta = 7''$, corresponding to the point L', is the center of the range of total reflection. The dispersion sheets of modes 3 and 4 at $\Delta\phi = -300''$ and those of modes 5 and 6 at $\Delta\phi = 300''$ are strongly excited, with an excitation of more than 95%. They are indicated by the solid curves. In both Figs. 7.7a and b, the relative amplitude of the (222) reflected wave of the upper dispersion sheet is greater than that of the lower sheet (here we only consider the wavefield-amplitude ratios without taking the boundary conditions into account). In Fig. 7.7a, the upper dispersion sheets near the point L' correspond to modes 1, 2, 3 and 4. These four modes are therefore active and the strongly excited modes 3 and 4 thus contribute to the reflected intensity which is shown in Fig. 7.7c, while modes 1 and 2, associated with the uppermost dispersion sheets at $\Delta\phi = 300''$, are effective. The most strongly excited modes 5 and 6 do not then contribute to the reflected intensity. The reflected intensity shown in Fig. 7.7d is much weaker than that at $\Delta\phi = -300''$. Hence, the asymmetry on the reflected intensity appears.

The dispersion sheets at $\Delta\phi = 300''$ and $\Delta\phi = -300''$ shown in Fig. 7.7a and b are interchanged for $\Delta\phi = -300''$ and $\Delta\phi = 300''$ in case (ii) with a negative $S_{P,T}$. Accordingly, the reflected intensity at $\Delta\phi = 300''$ is much stronger than that at $\Delta\phi = -300''$. The asymmetry of the reflected intensity versus $\Delta\phi$ in this case is the reversal of that in case (i).

This intensity asymmetry with respect to $\Delta\phi$ also depends on how the positive and negative $\Delta\phi$'s are defined. In view of the geometrical representation of multiple diffraction in the reciprocal space (Fig. 2.15), multiple diffraction occurs at both incoming (IN) and outgoing (OUT) positions. At the incoming position, the secondary reciprocal lattice point P is initially outside the Ewald sphere. The crystal rotation tends to bring the point P towards the Ewald sphere. If P is initially inside the Ewald sphere, the rotation brings P towards the point where it leaves the sphere, an outgoing situation. These calculations are carried out for the incoming situation. For the outgoing situation, the asymmetry of the calculated intensity is reversed because of the interchange of $\Delta\phi = 300''$ and $\Delta\phi = -300''$. This indicates that the diffraction at the outgoing position for a positive $S_{P,T}$ can be treated as if the diffraction were at the incoming position for a negative $S_{P,T}$.

Physically, the difference in crystal rotation between the incoming and outgoing situations introduces a phase difference of 180°. This is similar to the phase change due to reflection in the optics of light.

The line profiles of the integrated intensities over $\Delta\theta$ for both cases (i) and (ii) are shown in Fig. 7.8. From these line profiles, the following relation is obtained for phase determination [7.30]:

$$S_{P,E} = S_L \cdot S_R , \tag{7.15}$$

where $S_{P,E}$ is the experimentally determined sign of the phase triplet. S_L is the sign defined from the line profiles which are given in Fig. 7.9 for both Aufhellung and Umweganregung three-beam diffractions (assume that the line profiles are recorded from the right side to the left side on a chart recorder). S_R is determined from the sense of crystal rotation, either incoming or outgoing. By considering (2.26), S_R is defined as the sign of the derivative,

$$S_R = S\left(-\frac{\partial(1/\lambda)}{\partial\phi}\right) . \tag{7.16}$$

S_R is therefore positive for the incoming and negative for the outgoing situations.

A portion (Fig. 7.10) of the multiple diffraction pattern (Fig. 2.6) shows the details of the line profiles of cases (i) and (ii) at their IN and OUT posi-

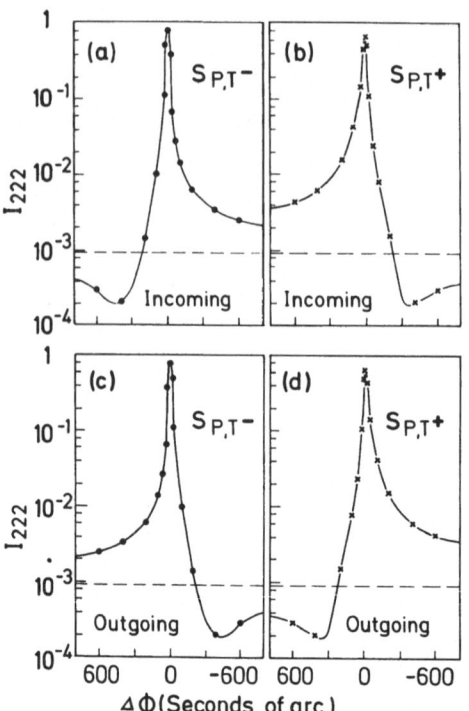

Fig. 7.8a–d. Calculated integrated intensity I_{222} profiles for the three-beam Umweg (000) (222) (113) for incoming (a) and outgoing (c) situations, and for the (000) (222) (11$\bar{1}$) case for incoming (b) and outgoing (d) situations (after *Chang* [7.34])

		LINE PROFILE		S_L
		STRONG REFLECTION (CONVERGENT BEAM)	WEAK REFLECTION (DIVERGENT BEAM)	
UMWEG		⋏	⋏	−
		⋏	⋏	+
AUFHE.		⋎	⋎	−
		⋎	⋎	+

Fig. 7.9. Definition of S_L from line profiles (after *Chang* [7.30, 34])

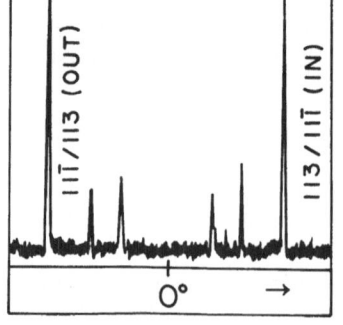
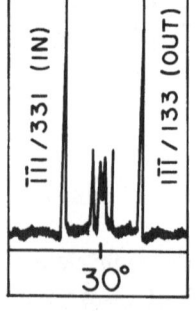

Fig. 7.10. Close view of the multiple diffraction pattern of Ge (222) reflection for CuK_{α_1} (after *Chang* [7.30])

Table 7.1. Experimentally determined phases

Reflection		S_R	S_L	$S_{P,E}$	$S_{P,T}$
Secondary	Coupling				
$11\bar{1}$	113	$-$	$+$	$-$	$-$
113	$11\bar{1}$	$+$	$-$	$-$	$-$
$\bar{1}\bar{1}1$	331	$+$	$+$	$+$	$+$
$1\bar{1}\bar{1}$	133	$-$	$-$	$+$	$+$

tions. A summary of the experimentally determined phases, using (7.15), is listed in Table 7.1. The agreement between $S_{P,T}$ and $S_{P,E}$ supports the correctness of (7.15).

In practice, multiple diffraction patterns can be easily obtained by making a ψ-scan (see Sect. 2.3.2) on a four-circle single-crystal diffractometer, and the sign $S_{P,E}$ can be determined in a straightforward manner from (7.15).

It is worth mentioning that the observation of the intensity asymmetry was first reported by *Renninger* in 1937 [7.38], and is re-mentioned in the article of *Cole* et al. [7.41]. Equation (7.15) would have been discovered much earlier if the sense of the crystal rotation had been considered in discussing the phase problem.

An analytic expression for the diffracted intensity of three-beam Bragg reflection near the three-beam point was given recently by *Juretschke* [7.36] to account for (7.15). By recalling (5.100, 102) and assuming either $z_i < \sqrt{|b|} \sqrt{Re\{\chi_{GO,HO}\chi_{OG,OH}\}}$ or $z_i > \sqrt{|b|} \sqrt{Re\{\chi_{GO,HO}\chi_{OG,OH}\}}$, the integrated reflection I_{GH} of the three-beam diffraction relative to the ordinary two-beam reflection I_G can be written as

$$\frac{I_{GH}}{I_G} = \frac{R_G(\text{three-beam})}{R_G(\text{two-beam})} \propto \left[1 + \left(\frac{2}{x}\right)\cos\Psi + \frac{1}{x^2}\right]^{(1/2 \text{ or } 1)} , \tag{7.17}$$

where

$$x = \left(\frac{P}{P_3}\right)\left(\frac{F^r_{G-O}}{F^r_{H-O}F^r_{G-H}}\right)\tan\Delta\phi . \tag{7.18}$$

The terms P and P_3 are the polarization factors of the two-beam G reflection and the three-beam case. The phase angle Ψ is defined as

$$e^{-2i\Psi} = \frac{F_{O-G}F_{G-H}F_{H-O}}{F_{G-O}F_{H-G}F_{O-H}} . \tag{7.19}$$

The intensity asymmetry therefore depends on the sign of the second term of (7.17) in such a way that

$$S_L \left(\frac{I_{GH}}{I_G} \right) = S_P (\cos \Psi) \cdot S_R \left(\frac{2}{x} \right). \tag{7.20}$$

S stands for the sign. This expression is exactly the same as (7.15).

7.1.4 Determination of Centrosymmetric Crystal Structures in Practice

The equation (7.15) determines experimentally the sign of the phase triplet for centrosymmetric crystals [7.42, 43]. It is, however, essential in crystal-structure determination to have information about the phase of each individual reflection. To determine the individual phase requires several triplet phase relations, like

$$S_P = \text{Sign}\,(F_{-G} F_H F_{G-H}) = S(-G)S(H)S(G-H), \tag{7.21}$$

where the sign $S(-G) = \cos \Psi_{-G}$. The angle Ψ_{-G} is the phase angle of the structure factor F_{-G}. In practice, for crystals with high symmetries, a convenient way of obtaining the required phase relations is to have several three-beam diffraction line profiles from a common primary reflection G_1. The secondary and coupling reflections involved in these three-beam cases should be symmetry related or repeatedly appear in the phase relations (7.21) so that one of the following two sets of phase relations can be formed [7.42]:

(i) Set A:

$$S_1 = S(-G_1)S(G_2)S(G_1-G_2), \tag{7.22a}$$

$$S_2 = S(-G_1)S(G_3)S(\{G_2\}), \tag{7.22b}$$

$$S_3 = S(-G_1)S(\{G_3\})S(\{G_1-G_2\}); \tag{7.22c}$$

(ii) Set B:

$$S_1 = S(-G_1)S(G_2)S(G_1-G_2), \tag{7.22d}$$

$$S_2 = S(-G_1)S(G_3)S(\{G_2\}), \tag{7.22e}$$

$$S_3 = S(-G_1)S(G_4)S(\{G_4\}). \tag{7.22f}$$

G_4 serves as an auxiliary reflection. $\{G\}$ represents a symmetry-related reflection of G. $\{G_2\}$ is equal to $G_1 - G_3$ in (7.22b) and $\{G_4\}$ is equal to $G_1 - G_4$ in (7.22f). $\{G_1 - G_2\}$ in (7.22c) is equal to $G_1 - G_3$. The phase relations between the symmetry-related reflections can be easily determined from the space group of the crystal. S_1, S_2 and S_3 are the experimentally obtained signs from (7.15). For crystals of low symmetry, more phase relations are needed to link the secondary and coupling reflections. In other words, more three-beam line profiles are required. Utilizing either set A or set B of these relations, two groups of individual phases $S(-G_1)$, $S(G_2)$, $S(G_3)$ and $S(G_1-G_2)$ are obtained. These two groups are, however, equivalent and are related to each other either by a translation of the origin of the unit cell or by the related symmetry operations. For a complicated crystal system, the number of equivalent groups of individual phases is different. By fixing the origin of the unit cell according to the lattice symmetry (space group), the corresponding individual phases can be used as a starting phase set for the direct methods, to develop more known phases. After the correct phases are developed, the structure can be determined by following routine procedures to perform Fourier transformations and refinements.

For illustration, the structure determination of a centrosymmetric $Cs_{10}Ga_6Se_{14}$ crystal, using the method proposed by *Chang* [7.30] together with the ordinary direct methods, is given here in detail. CsGaSe, which belongs to the space group $C2/m$, is a monoclinic crystal with lattice parameters $a = 18.2337$ Å, $b = 12.8895$ Å, $c = 9.668$ Å and $\beta = 108.2°$, where β is the angle between the a and c axes. There are two molecules per unit cell. This structure has recently been determined [7.44], using the *Patterson* method [7.13] with great effort, after a failure using the MULTAN program [7.45] of the direct methods. However, the structure was redetermined with ease by *Han* and *Chang* [7.43] using the experimentally determined phases in conjunction with the direct methods.

In *Han* and *Chang*'s experiment, a CsGaSe crystal of size $0.1 \times 0.1 \times 0.1$ mm^3 in a glass capillary is mounted on a P1 Syntex four-circle single-crystal diffractometer. A fine focus x-ray tube and a graphite monochromator provide an incident beam with 6 min of arc angular divergence. The intensities of two-beam reflections are collected automatically as is usual in conventional crystal-structure determination. MoK_α radiation is employed. The ψ-scan (Sect. 2.3.2), the rotation about the reciprocal lattice vector of a preselected primary reflection, is used to generate multiple diffractions using CuK_{α_1} radiation. The scanning speed is 0.05 degree per minute. The intensity of the primary reflection is monitored by a scintillation counter and recorded on a rolling paper chart.

In order to optimize the experimental conditions and obtain well-defined and useful three-beam diffraction profiles, the following facts are considered:

1) Choice of wavelength. The number of multiple diffractions, as described in Sect. 2.1.4, depends on the wavelength used and the size of the crystal unit cell. The use of a long wavelength can resolve the overlapping of multiple diffraction peaks. This provides clear diffraction profiles for phase determination. For this purpose, CuK_α radiation is used.

2) Choice of primary reflection G_1. Weak reflections are more suitable than strong ones as the primary reflection, since the former provide a better signal-to-noise ratio for multiple diffraction peaks than the latter. This can be seen from Figs. 6.32 and 7.10, in which the intensity asymmetry of the three-beam (000) (2̄2̄2̄) (1̄1̄1) case, with the weak (2̄2̄2̄) as the primary reflection, is clearly observed, while the three-beam, (000) (111) (1̄1̄1), case with the strong (111) reflection as the primary reflection gives only a smeared asymmetric profile.

The primary reflection, on the other hand, should provide enough three-beam cases with strong secondary and coupling reflections. The triplet phase relations in these three-beam cases ought to link the reflections involved, either the secondary or the coupling reflections, so that the individual phases can be determined. In practice, the data collection for CsGaSe shows 1907 independent two-beam reflections for $\sin \theta/\lambda$ between 0 and 1.15, namely, those corresponding to Miller indices (hkl) in the ranges: $0 < h < 19, 0 < k < 14$ and $-10 < l < 10$. Among these, 82 reflections are strong, with intensities greater than 1500 counts per second (cps), and 491 reflections are weak, with intensities less than 300 cps. Treating these strong reflections as the secondary reflections G_2 and the coupling reflections G_3, the corresponding primary reflections can be found by the addition $G_2 + G_3$. Remember that the secondary and the coupling reflections in the phase relation (7.21) are G_2 and $G_1 - G_2$, respectively. The primary reflection G_1 is therefore equal to $G_2 + (G_1 - G_2)$. Twenty weak reflections generated by this addition appear frequently in the results of additions. The reflections with their frequencies of appearance are listed in Table 7.2.

The reflection 3̄11, from Table 7.2, is the reflection most suitable as the primary reflection because of its high frequency of occurrence. This signifies that the number of useful three-beam reflections generated by using (3̄11) as the primary reflection is a maximum. Alternatively, the same primary reflection can be obtained by the same additions $G_2 + G_3$ of the last 30 reflections (the most linking and active reflection in the triplets) of the convergence map. The map lists these reflections, which have large E values and are frequently involved in (7.21) [7.11].

Table 7.2. Frequency of 20 weak reflections appearing in the phase relations

hkl	Frequency	hkl	Frequency
443̄	20	02̄2̄	5
022	5	130	22
221	26	110	22
312̄	18	312	12
3̄11	34	513̄	24
133	10	223	19
220	13	44̄1̄	14
42̄1̄	13	443	8
532	17	402	7
001	22		

As described previously, Bragg-type multiple diffractions are more suitable than Laue-type for phase determination, and one prefers to put the crystal in a position of more Bragg-like than Laue-like multiple diffraction. This can be easily achieved by slightly lowering the crystal so that the main diffraction of the beam is a Bragg reflection.

According to the geometry (Sect. 2.3) of the rotation about the direction $\langle \bar{3}11 \rangle$, the multiple diffraction should repeat every 360°, with a mirror at every 180°. In order to record all the possible diffraction peaks, the multiple diffraction pattern of $(\bar{3}11)$ for CuK_α is obtained covering an azimuthal angle of 180°. A portion of this pattern is shown in Fig. 7.11. Not all the diffracted peaks are used for phase determination. Only those with well-defined profile asymmetry are selected. They are marked with asterisks. All the peaks are labeled $G_2/(G_1 - G_2)$ with the indices of the secondary reflection G_2 and the coupling reflection $G_1 - G_2$. Fifteen multiple diffraction profiles are selected to provide information about the signs S_L (Fig. 7.9). The signs of the crystal rotation S_R, are determined during the indexing of the peaks, utilizing the procedure described in Sect. 2.3.2. Table 7.3 gives the indices of the three-beam diffractions selected, the signs S_L and S_R, and the sign of the product of S_L and S_R, which is also the sign of the triplet phase. As an example, the peak at the azimuthal angle $\phi = 295.2°$ has a positive S_L, according to Fig. 7.9, and is in an incoming situation, i.e., S_R is positive. The sign of the triplet phase, S_P, is therefore positive. From these triplet phase relations and the relations

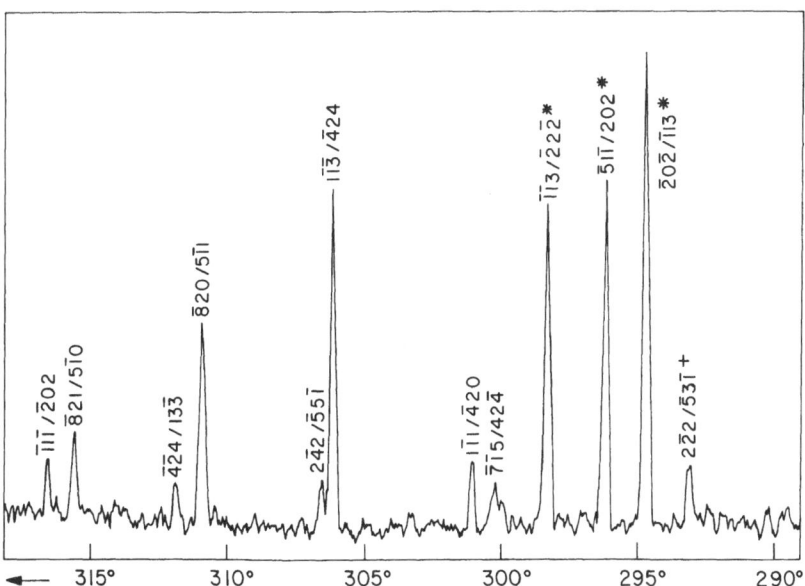

Fig. 7.11. Multiple diffraction pattern of $Cs_{10}Ga_6Se_{14}$ $(\bar{3}11)$ reflection for CuK_α (after *Han* and *Chang* [7.43])

Table 7.3. Useful triplet phase products obtained from $(\bar{3}11)$ multiple diffraction pattern of $Cs_{10}Ga_6Se_{14}$

ϕ [°]	S_R In or out	Secondary reflection	Coupling reflection	S_L	$(S_L \cdot S_R)$
260.8	out (−)	$\bar{4}24$	$11\bar{3}$	+	−
271.6	out (−)	$\bar{1}13$	$\bar{2}0\bar{2}$	−	+
293.6	out (−)	$2\bar{2}2$	$53\bar{1}$	−	+
295.2	in (+)	$\bar{2}0\bar{2}$	$\bar{1}13$	+	+
296.6	in (+)	$\bar{5}1\bar{1}$	202	+	+
298.6	out (−)	$\bar{1}\bar{1}3$	$\bar{2}\bar{2}\bar{2}$	+	−
320.8	in (+)	$\bar{4}20$	$1\bar{1}1$	+	+
323.5	in (+)	$\bar{2}\bar{2}2$	$\bar{1}\bar{1}3$	−	−
328.4	in (+)	$53\bar{1}$	$2\bar{2}2$	+	+
332.3	out (−)	$\bar{2}\bar{2}2$	$\bar{1}3\bar{1}$	−	+
349.0	in (+)	133	$\bar{4}\bar{2}4$	+	+
32.7	out (−)	$\bar{1}3\bar{1}$	$\bar{2}42$	+	−
57.7	in (+)	131	$\bar{4}20$	+	+
63.3	out (−)	$\bar{5}\bar{1}\bar{1}$	222	+	−
76.2	out (−)	$\bar{8}20$	$5\bar{1}\bar{1}$	+	−

among the symmetry-related reflections of the space group $C2/m$, i.e., $\Psi(hkl) = \Psi(\bar{h}\bar{k}\bar{l}) = \Psi(h\bar{k}l) = \Psi(\bar{h}k\bar{l})$, a diagram showing the linking between both the secondary and the coupling reflections involved can be constructed. This diagram is shown in Fig. 7.12. In Fig. 7.12a, eight non-equivalent equations of the type $S_1 \cdot S_2 \cdot S_3 = \pm 1$ can be derived which allow one to recognize the following phase relationships:

$$S(11\bar{3}) = S(511) = -S(531) = -S(13\bar{3}),$$

$$S(202) = -S(222) = -S(820) = -S(42\bar{4}), \qquad (7.23)$$

$$S(\bar{3}11) \cdot S(202) \cdot S(113) = +1.$$

The space group $C2/m$ requires two reflections, whose indices (hkl) belong to either the o o o, e e o or the o o e, e e o parity groups, to fix the origin of

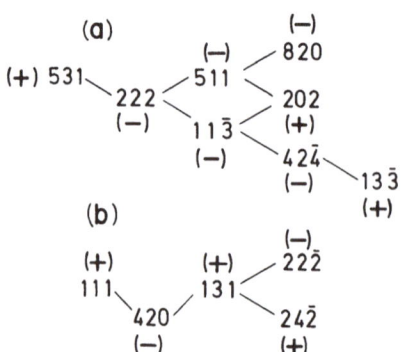

Fig. 7.12. Diagram for linked triplet phase relationships (after *Han* and *Chang* [7.43])

the unit cell. The notations e and o stand for even and odd values for h, k and l, respectively. One of the two required reflections can be found in the reflections above [for example, the reflection $(11\bar{3})$ belongs to the o o o parity group]. Then the relation set (7.23) leads to two solutions, namely, $\cdot S(202) = \pm 1$.

In Fig. 7.12b, the links between (111), (420), (131), $(22\bar{2})$ and $(24\bar{2})$ give the following phase relationships:

$$S(111) = S(1\,3\,1)\,,$$

$$S(420) = S(22\bar{2}) = -S(24\bar{2})\,, \tag{7.24}$$

$$S(\bar{3}11)\,S(1\,1\,1)\,S(420) = +1\,.$$

The combination of Figs. 7.12a and b leads to four solution sets (Table 7.4). Unfortunately, we cannot use any one of these reflections together with $(11\bar{3})$ for fixing the origin. We must, therefore, consider all four sets to find a reflection of the e e o parity group suitable for the determination of the origin.

If more individual phases are desired, more weak reflections need to be used as primary reflections to generate more three-beam diffraction line profiles. Fortunately, the direct methods available, for example, the MULTAN program [7.45], provide a means of developing more individual phases without carrying out further experiments. The four sets A, B, C and D of Table 7.4 serve as the starting phase set for the MULTAN program.

Since the MULTAN program selects only those reflections whose E values are greater than a given minimum E value, the reflections in the starting set that have a small value for E must be included manually. In addition, the ex-

Table 7.4. Possible starting phases for MULTAN

Index	Phases of the solution sets				E	Correct phase
	A	B	C	D		
$31\bar{1}$	−	−	+	+	0.13	−
111	+	−	+	−	0.90	+
$11\bar{3}$	−	−	−	−	2.39	−
$13\bar{3}$	+	+	+	+	0.95	+
131	+	−	+	−	1.25	+
202	+	+	−	−	1.63	+
222	−	−	+	+	0.93	−
$22\bar{2}$	−	+	−	+	2.24	−
$24\bar{2}$	+	−	+	−	1.29	+
$42\bar{4}$	−	−	+	+	2.75	−
420	−	+	−	+	1.55	−
511	−	−	−	−	2.45	−
531	+	+	+	+	1.13	+
820	−	−	+	+	2.31	−

Table 7.5. The parameters for MULTAN and the calculated results

	A	B	C	D	E	F
E_{min}	1.7	1.7	1.7	1.7	1.7	1.7
$E3_{min}$	2.7	2.7	2.7	2.7	2.7	2.5
No. of refl. used in MULTAN	225	225	225	225	205	296
No. of $\Sigma 2$ relations	3042	3042	3042	3042	2711	5836
No. of indeterminate phases	0	0	0	0	61	0
Percentage of correct phases	100%	60%	59%	50%	61%	75%
ABS FOM	1.1343	1.1332	0.9906	0.9918	0.8080	1.0206
RESID	15.60	15.69	25.79	25.79	22.15	28.00

perimentally determined phases of those reflections that are considered correct should be given high weighting factors, for example, 0.99, in the tangent refinements (7.3). As was discussed above, an additional reflection is necessary to fix the origin of the unit cell. The additional reflection in this case is $(410\bar{7})$ which is provided by the MULTAN program.

The parameters and results of the MULTAN calculation are given in Table 7.5. Using sets A, B, C and D of Table 7.4 as the starting sets, columns A, B, C and D of Table 7.5 show the corresponding minimum E values and $E3$ (the triple product of E for $\Sigma 2$-related reflections), the number of accepted reflections of the generated $\Sigma 2$ relations (7.21) and of the indeterminate phases, the percentage of correct phases, the absolute figure of merit (FOM), and the residual values (RESID). Columns E and F are the corresponding values determined from a normal MULTAN calculation without using known phases. The most appropriate set is therefore set A, which has the highest FOM and the lowest residual value. The structure determined using set A as the starting phases is shown in Fig. 7.13, projected on the ac plane.

The procedure described above provides a direct way of obtaining enough known phases to form a large starting set for MULTAN. The combination of

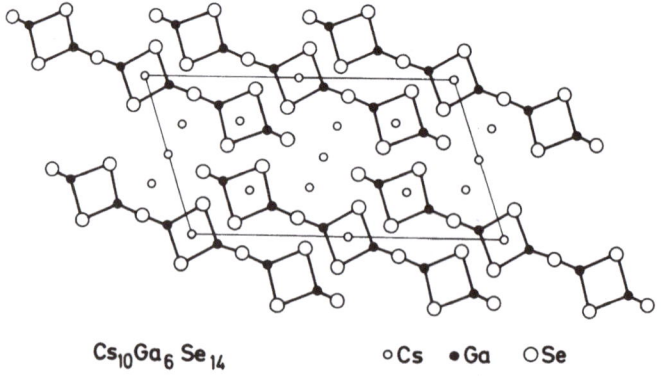

$Cs_{10}Ga_6Se_{14}$ ∘Cs   •Ga   ⊙Se

Fig. 7.13. Crystal structure of $Cs_{10}Ga_6Se_{14}$

the experimentally determined phases with the direct methods gives a new way of solving crystal structures.

7.1.5 Phase Determination for Non-Centrosymmetric Crystals

The dispersion surface at the exact three-beam point, as described by (6.3), depends on the term T_p defined in (7.10). Without considering the effect of anomalous scattering on the atomic scattering factors f of (4.22), the real and imaginary parts of the two terms in T_p satisfy

$$\operatorname{Re}\{F_{-G_1}F_{G_2}F_{G_1-G_2}\} = \operatorname{Re}\{F_{G_1}F_{-G_2}F_{G_2-G_1}\},$$
$$\operatorname{Im}\{F_{-G_1}F_{G_2}F_{G_1-G_2}\} = -\operatorname{Im}\{F_{G_1}F_{-G_2}F_{G_2-G_1}\}. \tag{7.25}$$

Therefore, the term T_p in (7.11) only depends on the sum of the real parts. This seems to imply that multiple diffraction provides no information about $\sin \Psi$. If the effect of anomalous scattering is introduced into the atomic scattering factors, the situation is changed. For simplicity, let us consider the case with two different atoms in a unit cell whose atomic scattering factors are f_1 and f_2. Both f_1 and f_2 can be written in the form

$$f = f^0 + f' + if'', \tag{7.26}$$

where the correction in f due to the anomalous scattering is considered. f' and f'' are the real and the imaginary parts of the correction, and f^0 is the uncorrected value. Supposing that the two atoms are located respectively at the origin of the unit cell and the position defined by the vector r_2 from the origin, the structure factor F_G can be written

$$F_G = (A_G + \Delta'_G) + i(B_G + \Delta''_G), \tag{7.27}$$

where

$$A_G = f_1^0 + f_2^0 \cos(2\pi g \cdot r_2), \quad \Delta'_G = -f_2'' \sin(2\pi g \cdot r_2),$$
$$B_G = f_1'' + f_2'' \cos(2\pi g \cdot r_2), \quad \Delta''_G = f_2^0 \sin(2\pi g \cdot r_2). \tag{7.28}$$

For a $-G$ reflection, the following relations hold:

$$A_{-G} = A_G, \quad B_{-G} = B_G, \quad \Delta'_{-G} = -\Delta'_G, \quad \Delta''_{-G} = -\Delta''_G. \tag{7.29}$$

After a few steps of manipulation, the real and imaginary parts of T_p for the three-beam, O, G and H, diffraction are obtained:

$$\operatorname{Re}\{T_p\} = 2[A_{G-H}(A_G A_H - B_G B_H - \Delta'_G \Delta'_H + \Delta''_G \Delta''_H) - B_{G-H}(A_G B_H$$
$$+ B_G A_H - \Delta'_G \Delta''_H - \Delta''_G \Delta'_H) + \Delta'_{G-H}(A_G \Delta'_H - \Delta'_G A_H - B_G \Delta''_H$$
$$+ \Delta''_G B_H) + \Delta''_{G-H}(\Delta''_G A_H - B_G \Delta'_H - A_G \Delta''_H + \Delta'_G B_H)], \tag{7.30}$$

$$Im\{T_p\} = 2[A_{G-H}(A_G B_H + B_G A_H - \Delta'_G \Delta''_H - \Delta''_G \Delta'_H) + B_{G-H}(A_G A_H$$
$$- B_G B_H - \Delta'_G \Delta'_H + \Delta''_G \Delta''_H) + \Delta'_{G-H}(A_G \Delta''_H - \Delta'_G B_H - \Delta''_G A_H$$
$$+ B_G \Delta'_H) + \Delta''_{G-H}(A_G \Delta'_H - \Delta'_G A_H - B_G \Delta''_H + \Delta''_G B_H)] \; .$$

Since the coordinates of the tie point on the dispersion surface depend on the real part of the eigenvalues and the linear absorption is determined by the imaginary part, the dispersion surface is affected mainly by the term $Re\{T_p\}$, which is related to $\cos\Psi$. This fact is shown in Fig. 7.14 for the artificial cases with Ψ equal to $11°$, $90°$, $-90°$ and $180°$. The case with $\Psi = 11°$ is the real case for the three-beam (000) (111) $(\bar{1}\bar{1}1)$ diffraction in GaAs with MoK_{α_1} radiation (in this case, the anomalous scattering effect was included in the atomic scattering factors). Figure 7.14 shows the intersection of the dispersion surface with the plane of incidence for the (111) reflection. The dispersion sheets A, B, C and D are the additional sheets, branch 3 of Fig. 7.1, for Ψ equal to $11°$, $90°$, $-90°$ and $180°$. It is clear that the dispersion sheet of branch 3 in these three-beam cases sweeps from a position close to curve A to one close to curve D as the phase Ψ varies from $0°$ to $180°$; it then returns to the position of curve A as Ψ increases from $180°$ to $360°$. The dispersion sheets corresponding to $\Psi = 90°$ and $-90°$ should coincide at the Lorentz

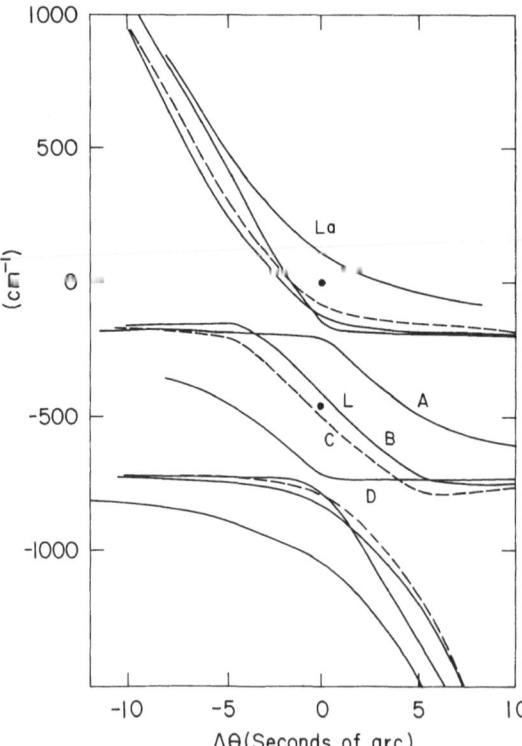

Fig. 7.14. Calculated dispersion sheets of the GaAs (000) (111) $(\bar{1}\bar{1}1)$ Borrmann diffraction (MoK_{α_1}) for various phase angles

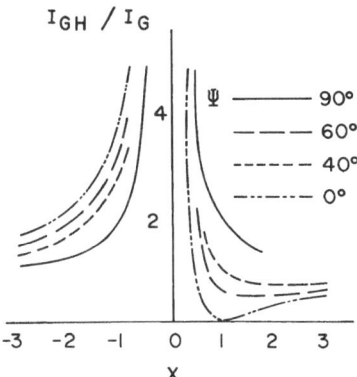

Fig. 7.15. Calculated integrated intensity profiles for various phase angles (after *Juretschke* [7.36])

point L if no absorption is involved. This fact was predicted by *Ewald* and *Heno* [7.1] [see also (4.77)]. In the present case, the small deviations of the two dispersion sheets from the L point, for Ψ equal to 90° and $-90°$, are mainly due to the imaginary part of T_p. These small deviations are, however, difficult to use in profile analysis for phase determination. The discussion between the case with $\Psi = 90°$ and that with $\Psi = -90°$, i.e., the determination of the enantiomorphy, can hardly be achieved from the analysis of the dispersion surface. Recent theoretical studies by *Juretschke* [7.36] show, according to (7.17), the phase dependence of the diffraction line profiles for non-centrosymmetric crystals (Fig. 7.15). Although the calculated line profiles in Fig. 7.15 indicate the difference between, say, $\Psi = 0°$ and $\Psi = 60°$, the distinction between $\Psi = 60°$ and $\Psi = -60°$ cannot be obtained from the line-profile analysis. In other words, the sign of sin Ψ is still indeterminate. Fortunately, Im $\{T_p\}$ is closely related to f''. The future trend to solving the phase problem for non-centrosymmetric crystals may lie in the analysis of either the linear absorption coefficients or the intensity profile, using anomalous scattering together with multiple diffraction.

7.2 Determination of Lattice Constants of Single Crystals

The possibility of using multiple diffraction for lattice constant determination was first recognized by *Kossel* [7.46] and *Renninger* [7.38]. Since there are two possible ways of generating multiple diffractions, i.e., the divergent-beam (Sect. 2.2.2) and the collimated-beam techniques (Sect. 2.2.1), we shall concentrate on these two methods in discussing the determination of lattice constants.

7.2.1 Divergent-Beam Photographic Methods

The Kossel conic (Fig. 2.10) of a reflection (hkl) obtained in a divergent-beam experiment can be described by the analytical expression

$$x'^2 + y'^2 = z'^2 \tan^2 \alpha \,, \tag{7.31}$$

where $\alpha = 90° - \theta_G$ is the half-apex angle of the Kossel cone. The angle θ_G is the Bragg angle of the reflection. The axis of the cone, i.e., the direction of the incident x-ray beam, is along the g direction, where g is the reciprocal lattice vector of the reflection. If g is not along the incident direction, a convenient new orthogonal coordinate system (x, y, z) with the z axis along the incident direction may be used to relate the conic to the incident beam. The relation between (x, y, z) and (x', y', z') is

$$x' = l_{11}x + l_{21}y + l_{31}z \,,$$
$$y' = l_{12}x + l_{22}y + l_{32}z \,, \tag{7.32}$$
$$z' = l_{13}x + l_{23}y + l_{33}z \,,$$

where l_{ij} is the cosine of the angle between the new i axis and the old j axis. By substituting (7.32) into (7.31), (7.32) can be written as

$$l_{13}x + l_{23}y + l_{33}z = \sqrt{\frac{x^2 + y^2 + z^2}{1 + \tan^2 \alpha}} \,. \tag{7.33}$$

If a film is placed normal to the z axis at a distance t, the conic on the film can be described by

$$L_1 x + L_2 y + L_3 t = \frac{\lambda}{2d} \sqrt{x^2 + y^2 + t^2} \,, \quad \text{with} \tag{7.34}$$

$$d = \frac{1}{g} \,. \tag{7.35}$$

The L's are the direction cosines of g with respect to the x, y and z axes such that

$$L_1^2 + L_2^2 + L_3^2 = 1 \,. \tag{7.36}$$

In reciprocal space, the Kossel plane (Fig. 2.12) of the (hkl) reflection takes a similar form:

$$L_1' x + L_2' y + L_3' z = \frac{1}{d} \,, \quad \text{where} \tag{7.37}$$

$$\sqrt{x^2 + y^2 + z^2} = \frac{2}{\lambda} \,. \tag{7.38}$$

$2/\lambda$ is the modulus of the double wavevector defined in (2.9). The L's are the direction cosines between g ($= ha^* + kb^* + lc^*$) and the reciprocal vectors a^*, b^* and c^*, i.e.,

$$L_1' = \frac{ha^*}{g}, \qquad L_2' = \frac{kb^*}{g}, \qquad L_3' = \frac{lc^*}{g}, \qquad (7.39)$$

where $a^* = 1/a$, $b^* = 1/b$, $c^* = 1/c$ and $g = 1/d$. The quantities a, b and c are lattice constants of the crystal. By substituting (7.39) into (7.37), (7.37) can be written

$$\frac{hx}{a} + \frac{ky}{b} + \frac{lz}{c} = \frac{1}{d^2} . \qquad (7.40)$$

For a non-coplanar coincidental diffraction (Sect. 2.1.2), $(g_i \times g_j) \cdot g_m \neq 0$, or

$$\begin{vmatrix} h_i & k_i & l_i \\ h_j & k_j & l_j \\ h_m & k_m & l_m \end{vmatrix} \neq 0 , \qquad (7.41)$$

and the linear equations (7.40) for the $(hkl)_j$ planes involved can be easily be solved. The relation between the lattice constants and the wavelength λ can be obtained. Thus, for a non-coplanar coincidental diffraction of a known wavelength λ_0, the lattice constants of the crystal can be determined. For a cubic crystal, a four-beam diffraction is sufficient to determine the lattice constants. However, three four-beam diffractions would be required for orthorhombic crystals.

Lonsdale [7.47] in 1947 used the four-beam (000) $(2\,2\,0)$ $(1\,3\,3)$ $(3\,1\,3)$ diffraction of CuK_{α_1} radiation to determine the lattice constant of diamond. The reflections $(2\,2\,0)$ $(1\,3\,3)$ $(3\,1\,3)$ provide, via (7.40), the following three relations:

$$2x + 2y = \frac{8}{a}, \qquad 3x + y + 3z = \frac{19}{a}, \qquad x + 3y + 3z = \frac{19}{a}, \qquad (7.42)$$

which lead to

$$x = \frac{2}{a}, \qquad y = \frac{2}{a}, \qquad z = \frac{11}{3a} . \qquad (7.43)$$

According to (7.38), we obtain

$$a = \frac{\sqrt{193}}{6} \lambda_0 . \qquad (7.44)$$

Hence, the lattice constant of diamond is determined by setting $\lambda_0 = 1.540562$ Å. Although this method furnishes an accuracy in a of

$\pm 0.5 \times 10^{-4}$ Å, the occurrence of this type of coincidental diffraction is fortuitous, which limits the applicability of this method.

Kossel photographs record the reflection conics for both characteristic and continuous radiation. The spatial distance in Kossel photographs represents the range of the wavelength of the radiation. Figure 7.16 shows the schematic representation of three sets of Kossel lines for K_{α_1} and K_{α_2} radiation. At the point O, the three reflection lines intersect one another for a wavelength λ_0. The image of these three lines may not be visible because λ_0 may not be a characteristic wavelength. By a linear interpolation using K_{α_1} and K_{α_2} reflection lines, λ_0 can be calculated. Employing the known λ_0 in the Lonsdale method, one can determine the lattice constants of the crystal [7.48, 49]. This method has been used to determine the lattice constant of a GaAs single crystal with an accuracy of one part in 10^5 [7.50, 51].

An alternative in determining lattice constants is to use two intersecting Kossel lines in a lens configuration (Fig. 7.17a) in which the center O of the Kossel pattern lies at the center of the lens [7.52]. The geometrical relation (Fig. 7.17b) between the source S', the center O and the length L of the lens gives

$$L = 2l \tan(\xi/2), \tag{7.45}$$

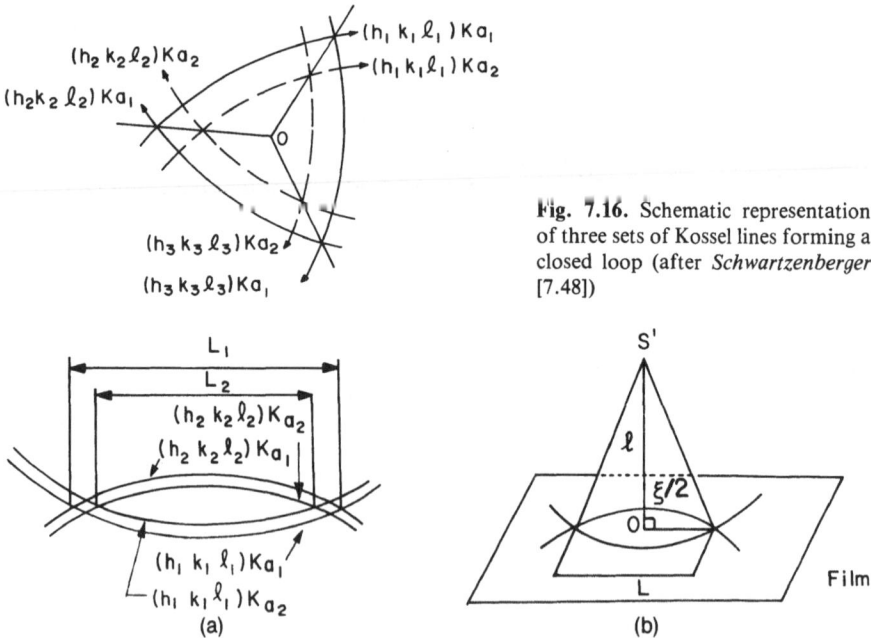

Fig. 7.16. Schematic representation of three sets of Kossel lines forming a closed loop (after *Schwartzenberger* [7.48])

Fig. 7.17. (a) Lens configuration of Kossel lines; (b) the geometric relation between source and center of the lens (after *Heise* [7.52])

where l is the source-film distance and ξ is the angle subtended by L at S'. Using two wavelengths, λ_1 and λ_2, we obtain

$$\frac{L(\lambda_1)}{L(\lambda_2)} = \frac{\tan(\xi_1/2)}{\tan(\xi_2/2)}. \tag{7.46}$$

The angles ξ can be calculated from the wavelengths of K_{α_1} and K_{α_2} and the indices (hkl) of these two intersecting lines. The drawback of this method is that the center of the lens must be very close to the center of the Kossel pattern. A spatial resolution of $\pm(0.2-1.0) \times 10^{-4}$ Å has been reported [7.52, 53]. Modification of this method can be achieved by using more Kossel lines for the lattice constant analysis [7.54].

7.2.2 Collimated-Beam Method

The azimuthal position ϕ of a multiple diffraction peak is, referring to (2.22, 25), a function of the wavelength of the radiation used and the crystal lattice constants. For a given known radiation, the lattice constant of the crystal can be determined from the peak positions of the multiple diffractions. Since the value of the angle ϕ depends on the choice of origin and reference vector (Sect. 2.3.1), the useful angles for the lattice constant determination are those independent of the choice of the origin. There exist two angles, $\phi_1 - \phi_2$ and β defined in (4.23), which satisfy this requirement. ϕ_1 and ϕ_2 are the azimuthal angles of two individual multiple diffractions. The difference between ϕ_1 and ϕ_2 obviously does not depend on the choice of the reference vector. If $\phi_2 = 0$ corresponds to the symmetry mirror of the diffraction pattern (see Fig. 2.6), ϕ_1 is also useful. The factor $\cos\beta$ in (4.25) is only a function of the wavelength, the lattice constants, and the Miller indices of the primary and secondary reflections.

The first lattice constant determination using the collimated-beam technique was carried out by *Renninger* [7.38] for a diamond crystal and CuK_{α_1} radiation. The azimuthal angles measured from the symmetry mirrors of the $(2\bar{2}2)$ multiple diffraction pattern for CuK_{α_1} were used. The lattice constant of diamond was determined to be 3.5668 Å with a precision of 10^{-4} Å. *Post* [7.55] and *Hom* et al. [7.56] repeated the same experiments with an improved experimental arrangement (Sect. 2.2.1). The angle β was determined experimentally from the difference in azimuth between the position IN and OUT of several multiple diffractions in diamond, silicon and germanium crystals. The primary reflection was (002) for all these crystals. CuK_{α_1} and CuK_{α_2} were employed for silicon and germanium, and for diamond, respectively.

Equation (2.25) has the following form for cubic crystals:

$$a = \frac{\lambda(p^2 - p_p g)}{2 p_n \cos \theta_G \cos \beta},$$ (7.47)

where the angle θ_G is the Bragg angle of the primary reflection and a is the lattice constant. Differentiating (7.47) leads to

$$\frac{\Delta a}{a} = (\tan \theta_G) \Delta \theta_G + (\tan \beta) \Delta \beta.$$ (7.48)

Since θ_G is the same for all multiple diffractions of a given primary reflection, the ratio $\Delta a/a$ is thus affected by the presence of the secondary reflection via the angle β. Multiple diffraction with a small angle β is therefore suitable for precise measurement of a. The consequence of having a small angle β, as discussed in Sect. 3.9, is the wider peak width of the diffraction.

Multiple diffractions with the secondary reflections $\{513\}$, $\{551\}$ and $\{117\}$, and $\{331\}$ for silicon, germanium, and diamond, respectively, were used in [7.55] for the measurements of the angle β. These angles are about $32.72°$, $11.33°$ for Si and Ge, and $15.78°$ (CuK_{α_1}) and $9.12°$ (CuK_{α_2}) for diamond. The corrections due to crystal absorption, beam divergence and the difference in refractive index between the crystal and air were considered in *Post*'s experiment [7.55]. The lattice constants of these three crystals determined in this way are listed in Table 7.6. Values obtained by other means are also given in this table. An error of about 10^{-5} Å was attained using this method.

The position of multiple diffraction in the divergent-beam and the collimated-beam techniques is influenced by the dynamical effect in multiple

Table 7.6. Lattice constants for silicon, germanium and diamond at 25 °C

Crystal	a [Å]	Reference
Silicon	5.430941	*Bond* [7.57]
	5.430938	*Deslates* and *Henins* [7.58]
	5.43094	*Segmuller* [7.59]
	5.430941	*Hom* et al. [7.55]
Germanium	5.6578	*Straumanis* and *Aka* [7.60]
	5.65775	*Smakula* and *Kalnajs* [7.61]
	5.65778	*Cooper* [7.62]
	5.65796	*Batchelder* and *Simmons* [7.63]
	5.657819	*Hom* et al. [7.55]
Diamond	3.56687	*Straumanis* and *Aka* [7.60]
	3.5668	*Renninger* [7.38]
	3.566986	*Hom* et al. [7.55]

diffraction, especially when a perfect crystal and strong reflections are involved. This can be seen from Fig. 6.30. The shift of the peak position for germanium and CuK_{α_1} is about 10^{-6} rad which is within the experimental error. For imperfect crystals, this dynamical effect need not be considered.

7.3 Determination of Lattice Mismatch in Thin Layered Materials

The advanced technology in crystal growth, such as chemical vapor deposition (CVD), liquid phase (LPE), vapor phase (VPE) and molecular-beam epitaxy (MBE) and metal organic chemical vapor deposition (MOCVD), provides new materials for semiconductors, light-emitting diodes, and for computer magnetic core memories. These new materials are sometimes affected by the presence of surface unevenness and lattice strains which occur during their preparation. One of the most common problems is the lattice mismatches in layered III – V, II – VI compounds and garnet materials.

Lattice mismatches result from either the difference in thermal expansion coefficients between the epitaxial layers and the substrates, or the compositional difference of the ternary and quaternary layers from the substrate material. The stresses due to lattice mismatches are harmful to the performance and lifetime of the electronic devices. Various x-ray methods, such as ABAC (automatic Bragg-angle control) [7.64] and double-crystal spectrometry [7.65], have been developed or adapted to characterize the lattice mismatches by detecting the shifts of layer diffraction peaks from substrate peaks.

Epitaxial layered materials usually possess a tetragonal unit cell, in comparison with the cubic crystals of the substrates, owing to small difference between the lattice mismatches parallel and perpendicular to the interfacial boundaries, $\Delta a_{\parallel}$ and $\Delta a_{\perp}$ [7.66 – 68]. $\Delta a_{\parallel}$ and $\Delta a_{\perp}$ are equal to $a_{\parallel} - a_s$ and $a_{\perp} - a_s$, respectively, where a_s is the undistorted lattice constant. An indirect observation of $\Delta a_{\parallel}$ and $\Delta a_{\perp}$ for the LPE $Ga_{1-x}Al_xAs$ layer on a GaAs substrate has been reported [7.68] using one symmetric and one inclined Bragg reflection. The parameter x is the aluminium concentration. Similar investigations have also been carried out by many others [7.69 – 71]. Because of the inclination of the asymmetric Bragg reflection with respect to the interfacial boundary, the information obtained about the lattice mismatches is limited and incomplete.

As mentioned at the beginning of this book, multiple diffraction provides three-dimensional information about the crystal lattice. In principle, multiple diffraction can be used to detect simultaneously the lattice mismatches, $\Delta a_{\parallel}$ and $\Delta a_{\perp}$, between the layered materials and the substrates. An attempt, utilizing a four-beam Borrmann diffraction, to investigate lattice mismatches was realized by *Chang* in 1978 [7.72] for a LPE GaAlAs/GaAs system. In 1980, *Chang* applied the divergent-beam technique (Sect. 2.2.2) to a six-beam Bragg diffraction for the LPE InGaAsP/InP system [7.72]. The lattice mismatches parallel and perpendicular to the interfacial boundary were simultaneously determined from the divergent-beam photographs. Independently, *Isherwood* et al. [7.73, 74] determined in the same year $\Delta a_{\parallel}$ and $\Delta a_{\perp}$ using two three-beam multiple diffractions, (000) (006) (511) and (000) (006) ($\bar{5}\bar{1}5$), for the GaAlAs/GaAs LPE system.

In *Chang*'s experiment, (001) InGaAsP/InP crystals were examined, using the multiple diffraction involving (000) (006) ($2\bar{2}4$) ($2\bar{2}2$) ($\bar{2}24$) ($\bar{2}22$) reflections. For a cubic ⟨001⟩-cut InP crystal, this six-beam multiple diffraction occurs when the crystal is first placed in position for the (006) reflection and is then rotated around the normal to the (006) planes to bring the direction ⟨1$\bar{1}$0⟩ to the plane of incidence of the (006) reflection. Figure 7.18 shows the geometry of this six-beam case in reciprocal space: Fig. 7.18a shows the plane of incidence for the (006) reflection and Fig. 7.18b is the top view of Fig. 7.18a. The side view of Fig. 7.18b is shown in Fig. 7.18c. θ_G and ϕ are the Bragg angle of the (006) reflection and the azimuthal angle of rotation, respectively, and β is the angle between the plane of incidence and the plane containing the six reciprocal lattice points. Because of the tetragonal unit cell for InGaAsP epitaxial materials, the corresponding six reciprocal lattice points of the epitaxial layer can no longer be brought simultaneously onto the surface of the Ewald sphere. Instead, only four points, either (000) (006) ($2\bar{2}4$) ($2\bar{2}2$) (set A) or (000) (006) ($\bar{2}24$) ($\bar{2}22$) (set B), can enter or leave the Ewald sphere together. The six-beam multiple diffraction for a cubic crystal is then decomposed into two four-beam cases for a tetragonal crystal. Since the recip-

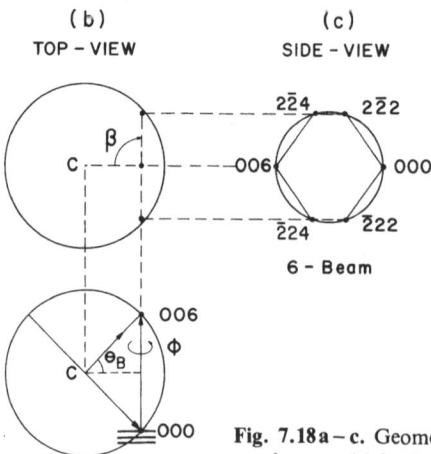

Fig. 7.18a–c. Geometry of the six-beam Bragg diffraction in reciprocal space: (a) in the plane of incidence, (b) top view of (a), (c) side view of (b) (after *Chang* [7.72])

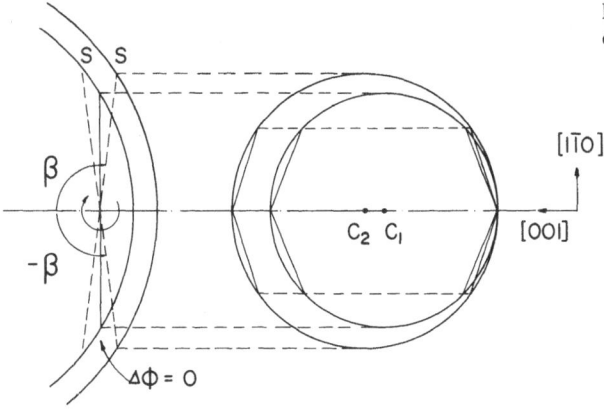

Fig. 7.19. Top and side views of Fig. 7.18a

rocal lattice points $(2\bar{2}4)$ and $(2\bar{2}2)$ lie on one side while $(\bar{2}24)$ and $(\bar{2}22)$ lie on the other side of the $\langle 006 \rangle$ rotation axis, the relative motion of the reciprocal lattice points of set A with respect to the Ewald sphere has an opposite sense to that of set B. During the rotation about the $\langle 006 \rangle$ axis, set A enters the Ewald sphere just after set B leaves, because of the relative motion. This situation corresponds to the position S for a negative angle β, shown in Fig. 7.19. The ordering of the sequence is reversed for a positive angle β, when set A leaves just after set B enters the Ewald sphere (Fig. 7.19).

According to (2.25), the angle β for a tetragonal system can be calculated as

$$\cos \beta = 2 \left(\frac{h^2 + k^2}{a_\parallel^2} + \frac{l^2 - lL}{a_\perp^2} \right) \Bigg/ \sqrt{\frac{h^2 + k^2}{a_\perp^2}} \sqrt{\frac{1}{\lambda^2} - \frac{L^2}{4a_\perp^2}} \,, \qquad (7.49)$$

if the rotation axis is $\langle 00L \rangle$ and the secondary reflection is (hkl). λ is the wavelength of the x-ray used. For InP and CuK_{α_1} (actually this multiple diffraction occurs for all wavelengths), β is $90°$, since $a_\parallel = a_\perp = a_s$. For the quaternary layer, the deviation $\Delta \beta$ from $90°$ can be determined to be

$$\Delta \beta = [(h^2 + k^2) \Delta a_\parallel + (l^2 - lL) \Delta a_\perp] a_s \sqrt{h^2 + k^2} \left(\frac{1}{\lambda^2} - \frac{L^2}{4a_s^2} \right), \qquad (7.50)$$

where a_s is 5.8696 Å, the lattice constant of InP. For these two particular four-beam cases,

$$\Delta \beta = 0.205 \, (\Delta a_\parallel - \Delta a_\perp), \qquad (7.51)$$

where $\Delta a_\perp$ can be determined from

$$\frac{\Delta a_\perp}{a_s} = -\cot \theta_G (\Delta \theta), \qquad (7.52)$$

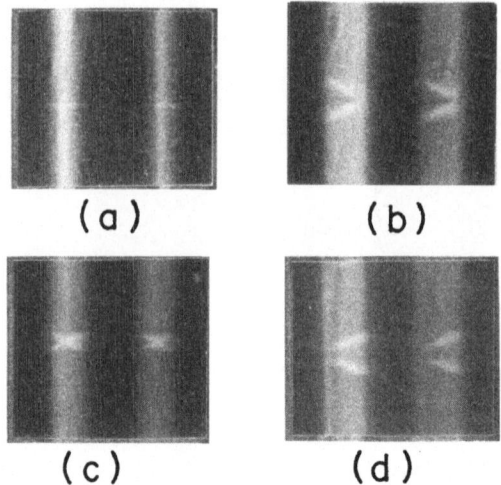

(a)   (b)

(c)   (d)

Fig. 7.20a–d. (006) reflection images in the vicinities of **(a)** the six-beam case for InP, and the four-beam cases for InGaAsP with X'_{As} equal to **(b)** 0.0105, **(c)** 0.0099, **(d)** 0.0091, where $X'_{Ga} = 0.0007$. CuK_{α_1} lines are to the left of the CuK_{α_2} lines (after *Chang* [7.72])

and the angular deviation $\Delta\theta$ from the Bragg angle θ_G of the (006) reflection. $\Delta\beta$ can be measured from the angular separation between the two four-beam reflection lines. $\Delta a_{\parallel}$ can therefore be determined.

Figure 7.20 shows the image of the six-beam case of the InP substrate and the images of the two four-beam cases of the quaternary InGaAsP within various arsenic concentrations, X'_{As}, in the liquid composition. The photographs were taken 150 cm from the crystal, with the section divergent-beam technique in a Bragg geometry (Sect. 2.2.2). The thicknesses of the InP substrate and the InGaAsP layer were about 530 and 5 μm.

Comparison of the observed images (Fig. 7.20) can be made with the calculated image (Fig. 7.21), with $\Delta a_{\parallel} = 0$. Since the horizontal and vertical axes represent respectively $\Delta(2\theta)$ and $\Delta(2\beta)$, the slopes of the four-beam reflection lines can be determined by

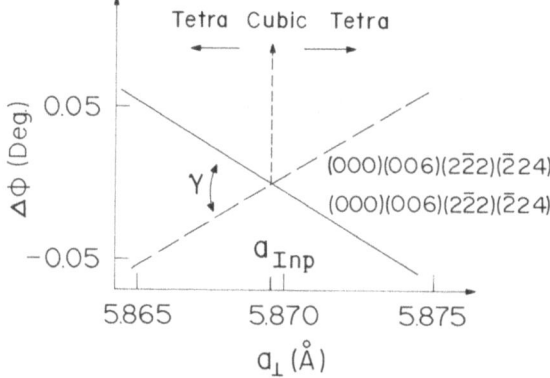

Fig. 7.21. Calculated azimuthal positions $\Delta\phi$ for two four-beam multiple diffractions against $a_{\perp}$ ($\Delta a_{\parallel} = 0$) (after *Chang* [7.72])

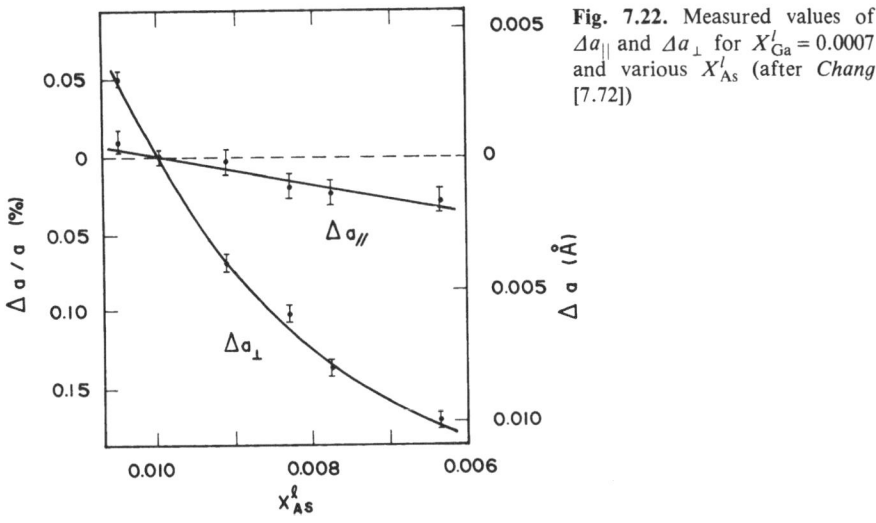

Fig. 7.22. Measured values of $\Delta a_{\parallel}$ and $\Delta a_{\perp}$ for $X_{Ga}^{l} = 0.0007$ and various X_{As}^{l} (after *Chang* [7.72])

$$\tan \frac{\gamma_B}{2} = \frac{\Delta \beta}{2(\Delta \theta)}, \qquad (7.53)$$

where γ_B is the angle between the two four-beam reflection lines. From (7.51 – 53), it is clear that the slope is a function of $\Delta a_{\parallel}$ and $\Delta a_{\perp}$ and is constant if $\Delta a_{\parallel}$ is linearly porportional to $\Delta a_{\perp}$.

A broad reflection band shown in Fig. 7.20 is common for the quaternary layer, since $a_{\perp}$ varies along the interface normal. $\Delta a_{\perp}$ and $\Delta a_{\parallel}$ were determined by measuring $\Delta \theta$ and $\Delta \beta$ at the middle of the reflection band. In Fig. 7.22, the measured values of $\Delta a_{\parallel}$ and $\Delta a_{\perp}$ are given as functions of X_{As}^{l}, with $X_{Ga}^{l} \simeq 0.0007$. They resemble the curves reported by *Oe* et al. [7.75] for X_{As}^{l} and X_{Ga}^{l} one order of magnitude higher. This method provides a means of determining simultaneously $\Delta a_{\parallel}$ and $\Delta a_{\perp}$ from a single divergent-beam photograph using multiple diffraction. It should be noted that it is very easy to align a (001)-cut crystal for the six-beam diffraction, because the two cleaved (110) and (1$\bar{1}$0) faces indicate clearly the orientation of the crystal. For (111)-oriented quaternary materials, other N-beam cases can be used for the same purpose.

The other similar method proposed by *Isherwood* et al. [7.73, 74] is actually an application to epitaxial materials of the experiment of *Spooner* and *Wilson* [7.51] discussed in Sect. 7.2.1. The most striking and interesting feature of *Isherwood*'s investigation is the reported images of hybrid multiple diffraction, i.e., combined diffraction from both substrate and epitaxial layer. Figure 7.23 is the diffraction image and Fig. 7.24 is the identification of the lines from the hybrid multiple diffraction. The subscripts x and o refer to the layer and the substrate respectively. The distances between the intersection points measure the angular separations.

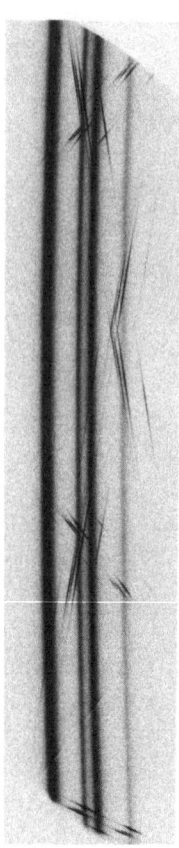

Fig. 7.23. (006) reflection image of $Ga_{0.34}Al_{0.66}As/GaAs$ for CuK_α (courtesy of *Isherwood* et al. [7.74])

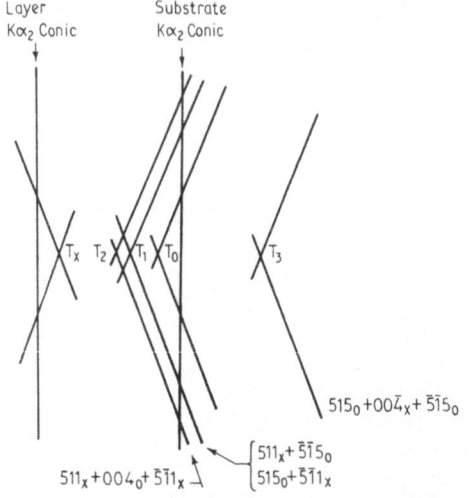

Fig. 7.24. Interpretation of the hybrid reflection lines of Fig. 7.23 (courtesy of *Isherwood* et al. [7.74])

7.4 Multi-Beam X-Ray Topography

X-ray topography is one of the most frequently used methods in the examination of crystal defects. There are many topographic techniques which have been developed for specific purposes [7.76]. In general, they can be classified into two types of topography, transmission and reflection. The former is usually employed for characterizing defects of a bulk crystal; the latter is very often adopted to reveal surface defects, such as hillocks, pyramids, terraces, and dislocations. In both cases, topographs taken from at least two non-coplanar reflections are necessary to determine the Burgers vector of a dislocation. This is, however, rather time consuming, because one has to align the crystal for at least two reflections. According to the nature of multiple diffraction of x-rays, simultaneous detection of diffraction images is feasible. In the following, the method proposed by *Chang* [7.77] is mentioned, to illustrate multi-beam x-ray topography.

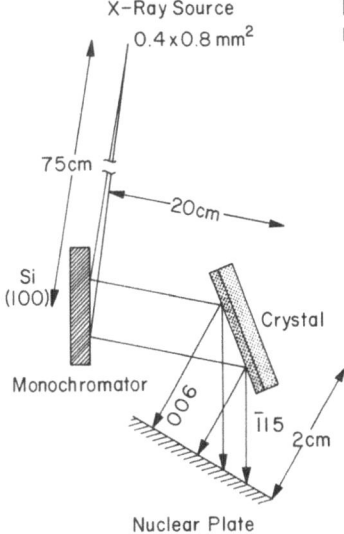

X-Ray Source
0.4 x 0.8 mm²

75cm

20cm

Si (100)

Crystal

Monochromator

006

Ī15   2cm

Nuclear Plate

A double-crystal arrangement, shown in Fig. 7.25, is employed. A fine focus copper tube and a (100) perfect silicon single crystal are used as x-ray source and monochromator, respectively. The focal size in the plane normal to Fig. 7.25 is 0.4×0.8 mm². The asymmetric (422) reflection of the silicon monochromator provides a broad, parallel and monochromatic CuK_{α_1} beam. The sample is a (001) LPE $In_{0.76}Ga_{0.24}As_{0.38}P_{0.62}/In_{0.97}Ga_{0.03}As_{0.08}P_{0.92}/InP$ double heterostructure. The three compositions represent the materials for the active layer, the buffer layer and the substrate, respectively. The thicknesses are 0.5, 2 and 300 μm for the active layer, the buffer layer and the substrate, respectively. The large (001) face has dimensions 7×8 mm². The buffer layer and the substrate are lattice matched. The lattice mismatches between the top active layer and the buffer layer are 0.0170 Å and 0.0018 Å less than the lattice constant of the substrate in directions normal and parallel to the interfacial boundary. These are determined by using the six-beam multiple diffraction method discussed in Sect. 7.3. The LPE layers cover only an area of 7×7 mm². The distances between the sample crystal, the monochromator, and a nuclear emulsion plate are indicated in the figure.

The sample is first aligned for the symmetric (006) Bragg reflection and then rotated about the (006) crystal surface normal to bring the ($\bar{1}11$), ($\bar{1}15$) and ($3\bar{3}3$) atomic planes into positions where they simultaneously satisfy Bragg's law. The geometry of this five-beam, (000) (006) ($\bar{1}11$) ($\bar{1}15$) ($3\bar{3}3$), multiple diffraction in reciprocal space for CuK_{α_1} is shown in Fig. 7.26. As can be seen from Fig. 7.25, the ($\bar{1}15$) reflection is an inclined Bragg reflection, which can be used to facilitate the crystal alignment for this five-beam diffraction. A scintillation counter is therefore placed at $\chi = 31.66°$ [in the plane of incidence of the (006) reflection] and $\gamma = 22.18°$ (in the plane normal to the

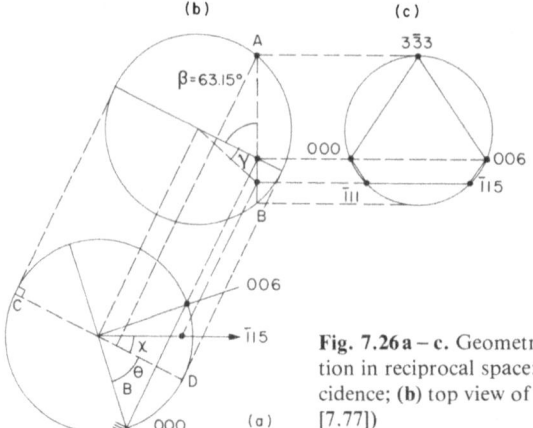

Fig. 7.26 a – c. Geometry of the five-beam multiple diffraction in reciprocal space: (a) a projection on the plane of incidence; (b) top view of (a); (c) side view of (b) (after *Chang* [7.77])

plane of incidence) to monitor the $(\bar{1}15)$ reflected beam. As discussed in Sect. 2.1, the equatorial plane, *CD* in Fig. 7.26, parallel to the (001) planes, represents the crystal surface. $(\bar{1}11)$ and $(3\bar{3}3)$ are referred to as transmitted and surface reflections, since their reciprocal lattice points are below the *CD* plane. These two diffracted beams cannot be detected from the same side of the (006) reflected beam. Only the $(\bar{1}15)$ and (006) reflected beams can be recorded simultaneously on a nuclear plate. This recording, shown in Fig. 7.27, is obtained with the nuclear plate placed perpendicular to the (006) reflected beam. The intensity variations of both the (006) and $(\bar{1}15)$ reflection images exhibit the effect of crystal bending and the interaction effect among the diffracted beams. The $(\bar{1}15)$ reflection is more sensitive to the crystal bending than (006) because the $(\bar{1}15)$ reflected beam is closer to the crystal surface. Although part of the $(\bar{1}15)$ image is missing due to these effects, the rest of the image reveals a one-to-one correspondence to the (006) image (Fig. 7.27). The enlarged ($\times 50$) (006) and $(\bar{1}15)$ images, shown in Fig. 7.28, give details about

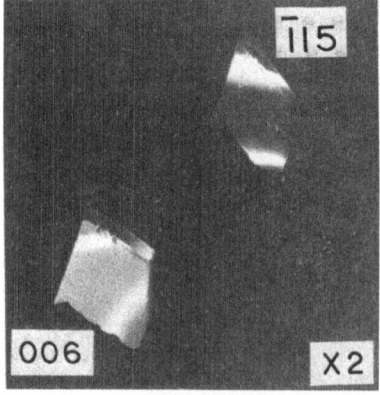

Fig. 7.27. Topographic images ($\times 2$) of the (006) (*lower left*) and $(\bar{1}15)$ (*upper right*) reflections (after *Chang* [7.77])

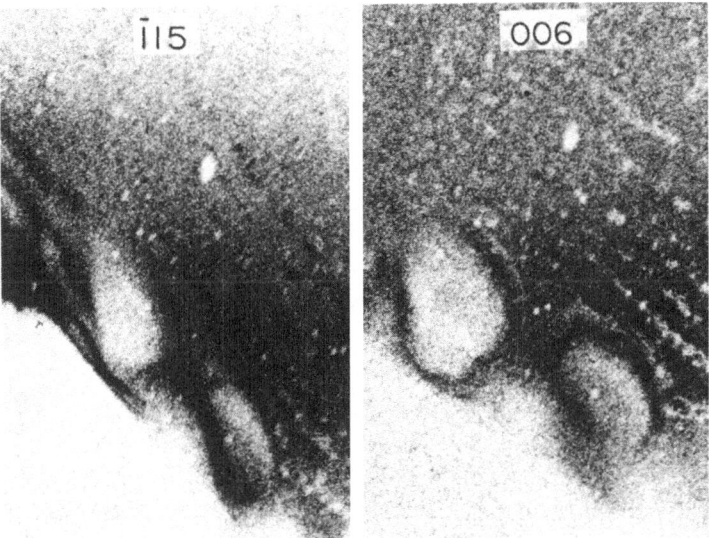

Fig. 7.28. The enlarged images (×50) of the upper portion of the crystal shown in Fig. 7.27 for the ($\bar{1}$15) and the (006) beams. Field width 0.15 mm (after *Chang* [7.77])

the border between the LPE region and the bare substrate portion. The relatively large round black contours and small white spots are due to the uneven distribution and the inhomogeneity of the InGaAsP melt, respectively.

This demonstration clearly shows that the use of multiple diffraction makes a simultaneous recording of more than one topograph of a crystal possible. This may facilitate the determination of Burgers vectors. For thinner crystals, depending on the radiation used, both transmitted and reflected beams can be utilized for this purpose. This method is equally applicable to transmission-type topography. It is also worth pointing out that to find many multiple diffractions according to Chap. 2 is not a problem. However, one should choose those cases where the diffracted beams are as close as possible to the crystal surface normal.

7.5 Multi-Beam X-Ray Interferometer

X-ray interferometers have, since their invention in 1965 by *Bonse* and *Hart*, been widely utilized for various applications, such as the precise determination of atomic scattering factors, lattice constants, x-ray wavelengths and the extended x-ray absorption fine-structure (EXAFS). The most commonly used interferometer is the triple Laue (LLL) type, in which two-beam transmission in three crystal leaves of the same crystal is employed as a means for beam

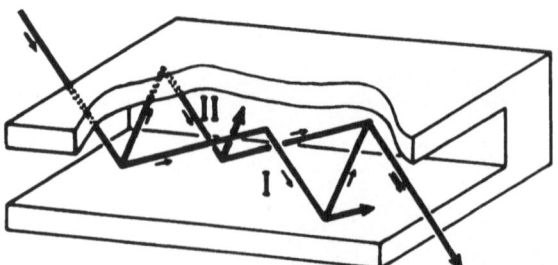

Fig. 7.29. Beam path of a three-beam x-ray interferometer (after *Graeff* and *Bonse* [7.79])

splitting, mirroring and recombination. However, this arrangement suffers crystal absorption and beam broadening due to the Borrmann-fan effect [7.78]. The alternative, triple-Bragg (BBB) geometry, avoids the absorption problem at the expense of having a rather ineffective beam-splitting ability, and beam broadening is still associated with the BBB interferometer.

A way of minimizing the absorption loss of the diffracted intensity and increasing the coherent splitting was recently proposed by *Graeff* and *Bonse* [7.79] utilizing simultaneous excitation of multiple diffraction. A three-beam x-ray interferometer was therefore designed for this purpose. Figure 7.29 shows schematically the beam splitting and recombination in a general three-beam wavevector-coplanar (Sect. 2.1.2) x-ray interferometer. Paths I and II are the two principal routes for the diffracted beams. It should be noted that each diffraction from a leaf of the crystal simultaneously generates three reflected beams, i.e., beams O, G and H. In Fig. 7.29, only one reflected beam is shown following each path. The reflections involved in path I are, in sequence, the G and $-G$ reflections and then the H and $-H$ reflections. In path II the reflected beams involved are, in sequence, the H and $-H$ and then the G and $-G$ reflections.

Referring to (4.214, 223), the wavefield E_M of an outgoing beam M is related to the wavefield E'_L of the incoming beam L by

$$E_M = \begin{pmatrix} E_{\sigma M} \\ E_{\pi M} \end{pmatrix} = \begin{pmatrix} E_{\sigma M}(\sigma) + E_{\sigma M}(\pi) \\ E_{\pi M}(\sigma) + E_{\pi M}(\pi) \end{pmatrix} = (M|L) \begin{pmatrix} E'_{\sigma L} \\ E'_{\pi L} \end{pmatrix} = (M|L)E'_L , \quad (7.54)$$

where the transition matrix has the form

$$(M|L) = \begin{pmatrix} c_{ML}^{\sigma\sigma} & c_{ML}^{\sigma\pi} \\ c_{ML}^{\pi\sigma} & c_{ML}^{\pi\pi} \end{pmatrix}, \quad \text{with} \quad (7.55)$$

$$c_{ML}^{\sigma\sigma} = \sum_{j=1}^{N_p} \frac{\mathrm{Cof}(\sigma M, \sigma L)_j}{\sum_{\sigma G} \mathrm{Cof}(\sigma G, \sigma G)_j}, \qquad c_{ML}^{\sigma\pi} = \sum_{j=1}^{N_p} \frac{\mathrm{Cof}(\sigma M, \sigma L)_j}{\sum_{\pi G} \mathrm{Cof}(\pi G, \pi G)_j},$$

$$\qquad\qquad\qquad\qquad\qquad\qquad\qquad\qquad\qquad\qquad\qquad (7.56)$$

$$c_{ML}^{\pi\sigma} = \sum_{j=1}^{N_p} \frac{\mathrm{Cof}(\pi M, \pi L)_j}{\sum_{\sigma G} \mathrm{Cof}(\sigma G, \sigma G)_j}, \qquad c_{ML}^{\pi\pi} = \sum_{j=1}^{N_p} \frac{\mathrm{Cof}(\pi M, \pi L)_j}{\sum_{\pi G} \mathrm{Cof}(\pi G, \pi G)_j}.$$

The interfering beams have therefore the following amplitudes: for the direct-diffracted beam O,

$$E_{O,\text{out}} = (O \mid G)(G \mid O)(O \mid H)(H \mid O)E_{O,\text{in}} \tag{7.57}$$

for path I, and

$$E_{O,\text{out}} = (O \mid H)(H \mid O)(O \mid G)(G \mid O)E_{O,\text{in}} \tag{7.58}$$

for path II.

Let us denote the matrix products of (7.57, 58) as the matrix A:

$$A_{\text{I,II}} = \begin{pmatrix} A^{\sigma\sigma} & \pm A^{\sigma\pi} \\ \pm A^{\pi\sigma} & A^{\pi\pi} \end{pmatrix} . \tag{7.59}$$

Because of the opposite signs in the off-diagonal elements in this matrix A_{I} and A_{II} represent elliptically polarized beams with opposite senses of rotation. If a phase shift Ψ is introduced in path II by a wedge-shaped material, the total amplitude of the interference beam is then

$$E_{O,\text{out}} = \begin{pmatrix} A^{\sigma\sigma}(1 + e^{i\phi}) & A^{\sigma\pi}(1 - e^{i\phi}) \\ A^{\pi\sigma}(1 - e^{i\phi}) & A^{\pi\pi}(1 + e^{i\phi}) \end{pmatrix} E_{O,\text{in}} . \tag{7.60}$$

If the incoming beam is σ polarized, the intensity of the outgoing beam is obtained as

$$I_{O,\text{out}} = [2 \mid A^{\sigma\sigma} \mid^2 (1 + \cos \Psi) + 2 \mid A^{\sigma\pi} \mid^2 (1 - \cos \Psi)] I_{O,\text{in}} . \tag{7.61}$$

The visibility Γ' of the interference fringes can be calculated:

$$\Gamma'_\sigma = \frac{I_{\max} - I_{\min}}{I_{\max} + I_{\min}} = \frac{\mid \mid A^{\sigma\sigma} \mid^2 - \mid A^{\sigma\pi} \mid^2 \mid}{\mid A^{\sigma\sigma} \mid^2 + \mid A^{\sigma\pi} \mid^2} . \tag{7.62}$$

Γ'_σ is maximal when $A^{\sigma\pi}$ vanishes. In an actual case, $A^{\sigma\pi}$ varies with the angle of incidence, and the visibility of the diffraction contrast is therefore poor. This indicates that the correlation between the σ- and π-polarized waves in three-beam non-coplanar diffraction destroys the image contrast. This type of multi-beam geometry is not well suited for x-ray interferometry.

The correlation between the σ- and π-polarized waves vanishes if the three-beam case is a wavevector-coplanar diffraction. As discussed in Chap. 2, this kind of three-beam diffraction occurs at a wavelength equal to the maximum wavelength λ_m $(= 1/r_0)$ for multiple diffraction, where r_0 is defined in (2.4). The expressions for the wavefield amplitudes and the diffracted intensities (7.60, 61) can be used for this case with the $\sigma\pi$ and $\pi\sigma$ terms equal to zero.

In the experiments of *Graeff* and *Bonse* [7.79], the reflections G, H and $G - H$ were chosen to be symmetry-related reflections such that $g_G = g_H =$

Table 7.7. Symmetry-related wavevector-coplanar N-beam diffractions

N	g_G	g_H	g_{G-H}	λ_m (Si) [Å]	λ_m (Ge) [Å]
3	220	202	02$\bar{2}$	3.32551	3.46450
12	422	24$\bar{2}$	2$\bar{2}$4	1.92000	2.00023
3	440	404	04$\bar{4}$	1.66276	1.73225
6	624	46$\bar{2}$	2$\bar{4}$6	1.25693	1.30946
6	660	606	06$\bar{6}$	1.10887	1.15483

g_{G-H}. This implies the existence of a three-fold symmetry linking the three reciprocal lattice points, O, G and H. It is therefore necessary to choose the reciprocal lattice points for G and H lying in the (1 1 1) planes. Table 7.7 gives a representative list of such combinations [7.79].

The wavevector-coplanar three-beam, (000) (440) (404), case in silicon for $\lambda_m = 1.66276$ Å was used in [7.79] to devise a three-beam interferometer. Figure 7.30 shows the beam path inside the interferometer. The crystal orientation of the interferometer is also indicated.

The continuous spectrum of a synchrotron provides ideal experimental conditions for realizing interference via three-beam wavevector-coplanar reflection, since the wavelength can be chosen exactly equal to λ_m. The experimental set-up for synchrotron radiation of *Graeff* and *Bonse* consisted of a grooved germanium crystal as the monochromator for selecting the appropriate wavelength, and the three-beam interferometer. Films were placed in position to record the resultant beam in the incident direction. Two detectors were used to detect the (440) and (404) reflected beams in the alignment of the interferometer. A lucite wedge, with a five degree vertex angle was put in one of the beam paths to introduce a phase lag. Figure 7.31 is the interference pattern of the lucite wedge obtained with a synchrotron source at $\lambda_m = 1.6628$ Å. The densitometric trace shows clearly the intensity maxima and minima. Because of the use of the totally reflected beams in the three-

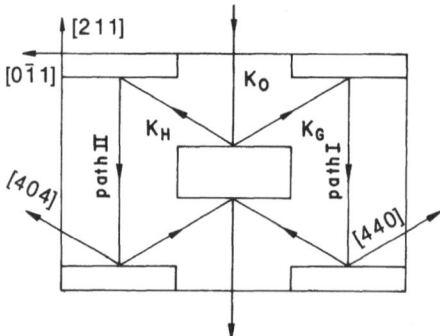

Fig. 7.30. Geometry of the wavevector-coplanar three-beam x-ray interferometer (after *Graeff* and *Bonse* [7.79])

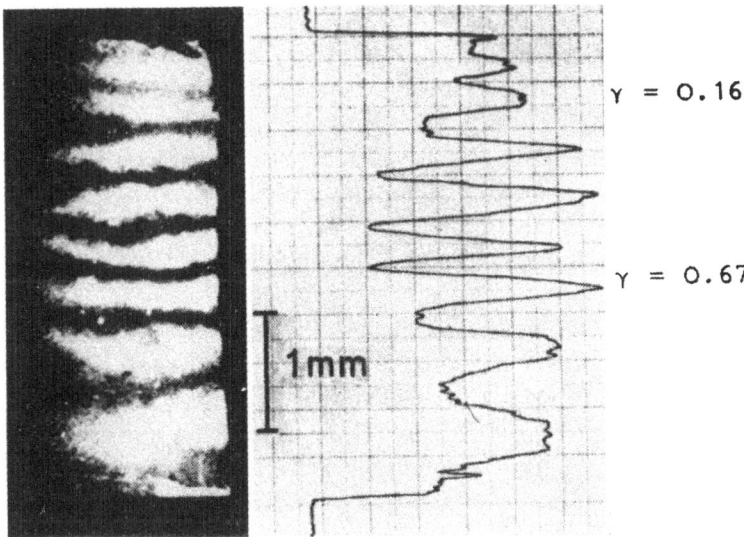

Fig. 7.31. Interference pattern of a lucite wedge and the corresponding densitometer trace (after *Graeff* and *Bonse* [7.79])

beam interferometer, the line broadening due to the Borrmann-fan effect in a transmission L L L interferometer is avoided. The spatial resolutions are 0.5 μm horizontal and 5 μm vertical, much higher than the 20 to 30 μm in L L L interferometers [7.80]. Furthermore, this three-beam interferometer can be adopted for neutron phase-contrast microscopy where, due to the lack of absorption of neutrons in crystals, the problem of complicated Pendellösung patterns in the Borrmann fan is overcome.

7.6 Monochromatization of X-Ray Beams

The interaction between the diffracted beams in multiple x-ray diffraction takes place, as discussed in Chap. 6, in a very small angular region, typically a few seconds to one or two minutes of arc. This characteristic of multiple diffraction immediately suggest the possibility of using multiple diffraction to monochromatize an incident x-ray beam. However, the considerations first made by *Kottwitz* [7.81] on using Umweg-type multiple reflection for high-resolution monochromatization originated from the use of high-flux thermal neutrons for diffraction studies. The energy of the neutrons ranges from 0.003 to 0.05 eV. This energy range corresponds to 1 − 5 Å for x-rays.

The degree of monochromaticity depends on the intrinsic diffraction peak width of a perfect crystal, the mosaic spread of the crystal and the beam divergence. The intrinsic width of a multiple diffraction peak, according to the discussion in Chap. 2, is proportional to the angular range in which strong diffracted beams are generated. This angular range usually corresponds to $\Delta k/k \simeq 10^{-5}$, i.e., a few seconds to one or two minutes of arc. This angle is independent of the beam divergence. For a mosaic crystal, the peak width is determined not only by the intrinsic peak width but also by the mosaic spread and the beam divergence. The width of mosaic spread is usually of the order of a few minutes to one degree, i.e., $\Delta \lambda/\lambda \simeq 10^{-4}$ to 10^{-2}. The beam divergence we are considering, in relation to the peak width of multiple diffraction (Sect. 3.9), is the vertical component.

Diffracted intensity is another important factor which needs to be considered in beam monochromatization. The diffracted intensity of a two-beam reflection is proportional to the square of the structure factor F for an imperfect crystal and to $|F|$ for a perfect crystal. It is similar in the case of multiple diffraction. The intensity of a multiple diffracted beam for an imperfect crystal can be, in some cases, as large as 10^3 times that for a perfect crystal. As usual, the intensity and the angular resolution requirements must be balanced against each other.

In addition to the intensity and resolution requirements, the overlapping of multiple diffraction peaks should also be considered, since this usually generates intensity doublets for the diffracted beam. This overlapping destroys the beam confinement. *Kottwitz* [7.81] calculated the peak positions for several well-resolved multiple diffractions for diamond-structure crystals, such as Ge, Si and diamond, and for hcp Be and Zn crystals. Experimental simulation of the (0003) and (0001) forbidden reflections in a beryllium mosaic crystal and of the (002) reflection in a germanium crystal were carried out by *Kottwitz* using neutrons.

The use of Umweg reflections from a bent perfect crystal for monochromatizing thermal neutrons was recently reported by *Mikula* et al. [7.82 – 85] It is known, from the dynamical theory of *Takagi* [7.86] for diffraction in an imperfect crystal, that the integrated intensity from a deformed crystal depends on the scalar product of the reciprocal lattice vector g of the reflection and the deformation vector u. In a three-beam reflection, for example, with the vectors g_1, g_2 and $g_1 - g_2$ as the reciprocal lattice vectors of the primary, the secondary and the coupling reflections, the product $g_1 \cdot u$ may be zero for the primary reflection and for its higher-order harmonics if the deformation vector is made to be perpendicular to g_1. Meanwhile, the products $g_2 \cdot u$ and $(g_1 - g_2) \cdot u$ then have non-zero values which may lead to a large increase in the Umweg diffraction intensity. In this manner, the diffracted intensity of the primary reflection and high-order harmonics is kept to a minimum, while the intensity of the Umweg is increased according to the degree of elastic deformation of the crystal.

According to *Vrána* et al. [7.85], the intensity of multiple diffraction in a elastically deformed crystal can be derived in the following way.

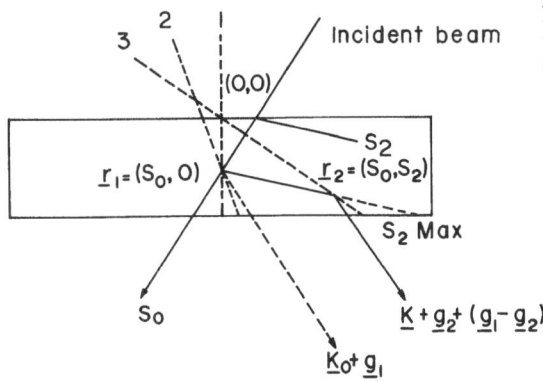

Fig. 7.32. Schematic diagram of multiple diffraction inside a crystal plate (after *Mikula* et al. [7.82])

Suppose that the incident neutron beam is successively and independently reflected by the secondary planes G_2 at the position r_1 and by the coupling planes $G_1 - G_2$ at r_2 (Fig. 7.32). The distance $|r_2 - r_1|$ is assumed to be larger than the extinction length l_x (Sect. 4.3.1). Let us define the incident wavevector k' so that it deviates from the exact three-beam vector k by Δk, i.e., $k' = k + \Delta k$. The reciprocal lattice vector g' of the slightly deformed reflection planes are described, according to *Takagi*'s theory of diffraction [7.86], by

$$g_j'(r) = g_j - \text{grad}\,[g_j \cdot u(r)] , \tag{7.63}$$

for the reflection j, where j can be G_1, G_2 and $G_1 - G_2$. For the incident neutron beam of wavevector k', which has successive reflections G_2 and $G_1 - G_2$, the following Bragg's laws must be fulfilled (Sect. 2.1.1):

$$k'(r_1) \cdot g_2'(r_1) = -\frac{|g_2'(r_1)|^2}{2} , \tag{7.64}$$

$$[k'(r_1) + g_2'(r_1)] \cdot [g_1'(r_1) - g_2'(r_2)] = -\frac{|g_1'(r_2) - g_2'(r_2)|^2}{2} , \tag{7.65}$$

where $r_2 = r_1 + S_2$ (Fig. 7.32). For a slight deformation, the terms $\text{grad}\,[g_2 \cdot u(r_1)]$ and $\text{grad}\,[(g_1 - g_2) \cdot u(r_2)]$ are very small in comparison with g_2 and $|g_1 - g_2|$. By substituting (7.63) into (7.65) and neglecting the second-order terms involving gradients, the following differential equations are obtained:

$$\Delta k \cdot g_2 = k \left. \frac{\partial(g_2 \cdot u)}{\partial S_2} \right|_{r_1} = A(S_0) ,$$

$$\Delta k \cdot g_1 = k \left[\left. \frac{\partial(g_1 \cdot u)}{\partial S_1} \right|_{r_1} + \left. \frac{\partial(g_2 \cdot u)}{\partial S_1} \right|_{r_1} - \left. \frac{\partial(g_2 \cdot u)}{\partial S_1} \right|_{r_2} \right] = B(S_0, S_2) , \tag{7.66}$$

where

$$\frac{\partial}{\partial S_1} = \frac{k_0 + g_1}{k_0} \cdot \mathrm{grad}, \qquad \frac{\partial}{\partial S_2} = \frac{k_0 + g_2}{k_0} \cdot \mathrm{grad}. \qquad (7.67)$$

The solution of (7.66) is

$$\Delta k(p, S_0, S_2) = \frac{g_1 \times g_2}{|g_1 \times g_2|} p + \frac{A g_2^2 - B(g_1 \cdot g_2)}{|g_1 \times g_2|^2} g_1 + \frac{B g_1^2 - A g_1 \cdot g_2}{|g_1 \times g_2|^2} g_2, \qquad (7.68)$$

where S_0 and S_2 are the path lengths of the beams in the directions k_0 and k_2. S_0 and S_2 are the distances from the entrance point on the crystal surface to r_1, and from the point at r_1 to the exit point on the lower crystal surface, respectively. S_2 is a function of S_0. The parameter p describes the collimation condition of the incident beam, which is independent of S_0 and S_2. This solution (7.68) is valid for cases in which g_1 is not parallel to g_2.

The integrated intensity of the diffracted beam in the three-beam case may be written in the form

$$I = \int P(\theta, \phi) d\theta d\phi = \int N(\Delta k) R(\Delta k) dV, \qquad (7.69)$$

where $N(\Delta k)$ is the density of the incident neutrons of the wavevector k' and $R(\Delta k)$ is the combined effective reflectivity of the three-beam diffraction. In the lamella model [7.87] $R(\Delta k)$ is unity for those Δk satisfying (7.68). Otherwise, $R(\Delta k)$ is zero. The term $d\theta d\phi$ is related to Δk, and P is the reflection power of the three-beam case, which is defined in (3.4). The volume element can be written in terms of the variables S_0, S_1 and p as:

$$dV = \left[\left(\frac{\partial \Delta k}{\partial S_0} \times \frac{\partial \Delta k}{\partial S_2} \right) \cdot \frac{\partial \Delta k}{\partial p} \right] dp \, dS_0 dS_2$$

$$= \frac{1}{|g_1 \times g_2|} \frac{\partial A(S_0)}{\partial S_0} \frac{\partial B(S_0, S_2)}{\partial S_2} dp \, dS_0 dS_2. \qquad (7.70)$$

This implies that simultaneous diffraction occurs in a deformed crystal if the conditions $\partial A / \partial S_0 \neq 0$ and $\partial B / \partial S_2 \neq 0$ are fulfilled. When the crystal is subject to vibrations, the displacement $u(r)$ along g_1 in the middle of the bent crystal may be written approximately as

$$u(r) = \hat{n} u_0 \cos\left(\frac{\pi g_1}{L|g_1|} \cdot r \right), \qquad \text{with} \qquad (7.71)$$

$$u_0 = \frac{L^2}{8R}, \qquad (7.72)$$

where $\hat{n}$ is the unit vector of the surface normal. L and R are the length of the crystal bar and the radius of curvature. By combining (7.64 – 70), the diffracted intensity has the simple form

$$I = \frac{G_0}{R^2} \int \frac{(g_1 \cdot k_2)^2}{|g_1 \times g_2|} \, dS_0 \, dS_2 \, , \tag{7.73}$$

where $k_2 = k_0 + g_2$ and G_0 contains all the constant factors. Since S_0 varies from zero to $(T/\cos \theta_{G_1})$ and S_2 depends on S_0, the integrated intensity is proportional to the square of the crystal thickness T. According to (7.72, 73), the intensity is also proportional to the square of the amplitude of $u(r)$. For a wavevector-coplanar three-beam case, the intensity is

$$I \simeq \frac{T^2}{R^2} \frac{1}{\cos \theta_{G_1} \sin (\alpha - \theta_{G_2})} \, , \tag{7.74}$$

where α is the angle between the entrance surface of the crystal and the atomic planes of the G_2 reflection, and θ_{G_1} and θ_{G_2} are the Bragg angles of the primary and the secondary reflections. The optimal condition for obtaining strong intensity is to have a small radius of curvature R, a thick crystal and a secondary reflection plane which has a small value for $\sin (\alpha - \theta_{G_2})$. However, this expression (7.74) is not valid for $\theta_{G_2} = \alpha$.

Figure 7.33 shows the (222) reflected intensity at the position of the four-beam case, (000) (222) (153) ($\bar{3}$11), in silicon for $\lambda = 1.54$ Å. (222) is the

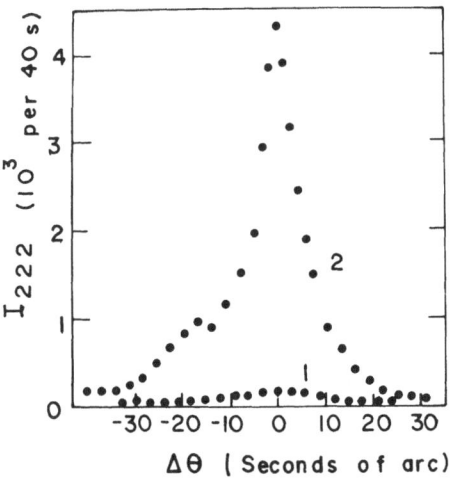

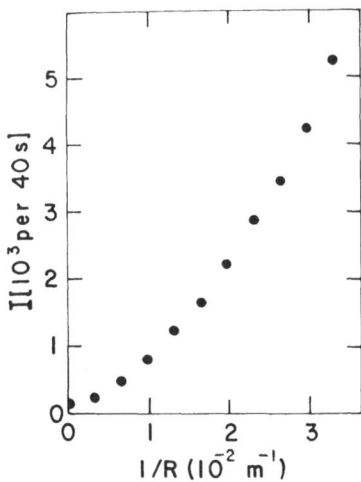

Fig. 7.33. The observed intensity distribution I_{222} of the simulated (222) reflection for (1) perfect nondeformed crystal and (2) bent crystal with $R = 33$ m (a Zn single crystal is used as the monochromator for neutrons) (courtesy of *Mikula* et al. [7.84])

Fig. 7.34. The dependence of the intensity I_{222} at $\Delta\theta = 0$ of Fig. 7.33 on $1/R$ (courtesy of *Mikula* et al. [7.84])

primary reflection. Curves 1 and 2 in this figure are the intensity distribution versus $\Delta\theta$, the angular deviation from the Bragg angle of the (222) reflection, for a perfect crystal and a deformed crystal with $R = 33$ m, respectively. The tremendous increase in the diffracted intensity is due to the fact that the angle $\alpha - \theta_{G_2}$ is about 3.25°. The variation of the intensity versus $1/R$ or the vibration amplitude is also shown (Fig. 7.34).

Transmission multiple diffraction is an alternative way of monochromatizing an incident beam in x-ray cases, since the anomalous transmission of x-rays at the exact N-beam point furnishes a highly collimated and monochromatic beam. As discussed in Chap. 6, the crystal absorption off the N-beam position limits the divergence of the transmitted beam through a crystal.

There is a special six-beam case, $(000)\ (0\bar{4}4)\ (2\bar{2}0)\ (\bar{2}02)\ (2\bar{4}2)\ (\bar{2}\bar{2}4)$, in which the six reciprocal lattice points form a regular hexagon (Fig. 7.35). According to *Joko* and *Fukuhara* [7.88], two of the twelve modes (2×6) of propagation have linear absorption coefficients μ almost equal to zero (4.164), i.e.,

$$\mu = 2\pi k (\chi^i_{000} - \chi^i_{\bar{2}02} - \chi^i_{\bar{2}\bar{2}4} + \chi^i_{0\bar{4}4}) \simeq 0 , \tag{7.75}$$

at the exact six-beam point. By using a rather thick perfect crystal, only these two modes can be effectively excited and the rest of the modes are suppressed by the crystal absorption. These two residual modes should, in principle, provide an intense, monochromatic and extremely parallel transmitted beam. Investigations on this effect have been carried by many researchers [7.89 – 92]. In the experiment of [7.92], intensity enhancement has been observed for germanium and CuK_{α_1}. However, the observed enhancement extended along the $\langle 0\bar{4}4\rangle$ direction, and the two strong reflections $(2\bar{2}0)$ and $(\bar{2}02)$ still showed two intense diffraction lines very close to the six-beam point. With the aid of a linearly polarized source such as synchrotron radiation, a sharp

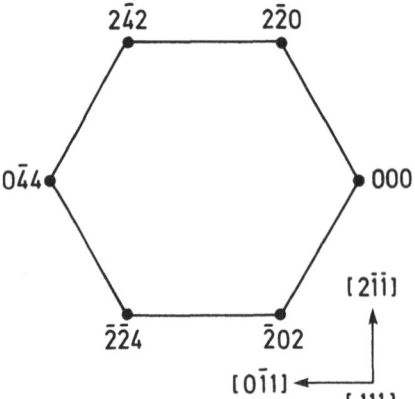

Fig. 7.35. Representation of the six-beam case in reciprocal space

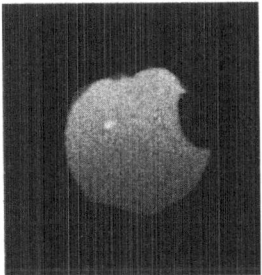

Fig. 7.36. Image of the direct (000) beam at the exact six-beam diffraction position for $\lambda = 1.54$ Å (after *Chang* [7.93])

Fig. 7.37. Calculated intensity distribution of Fig. 7.36 (after *Chang* [7.93])

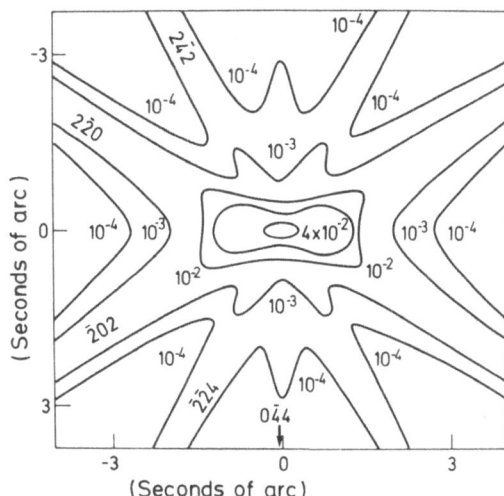

enhanced intensity spot with two seconds of arc beam divergence was recently observed on the direct (000) beam [7.93]. In this experiment, the synchrotron radiation from a storage ring was used as the source. A ⟨111⟩-cut single perfect crystal of silicon, 3 mm thick, was aligned for the wavelength 1.54 Å and six-beam diffraction. The image of the (000) transmitted beam is shown in Fig. 7.36. The shape of the direct image with a diameter corresponding to 20″ is due to the geometry of the beam collimator. The bright spot inside the direct beam is the exact six-beam diffraction image, which has an angular divergence of two seconds of arc. This is in agreement with the calculated intensity distribution given in Fig. 7.37.

The dynamical calculation, based on the procedure described in Sect. 5.2, gives the linear absorption coefficient μ_0, the excitation of mode and the accommodation δ for each mode at the exact six-beam diffraction position in Table 7.8. Only nine of the twelve modes are appreciably excited by the incident beam. This is because of the linear π polarization [along the (0$\bar{4}$4) direction] of the synchrotron radiation. For a thick crystal, only the first four modes contribute to the transmitted intensity. For a Si crystal of 3 mm thickness, only modes 2 and 4 are effective. The wavevector difference of the two modes is only 24.57 cm^{-1}.

Polarization of the incident beam plays, as discussed in Chap. 4, a very important role in the dynamical interaction of x-ray beams within a crystal, especially when a linearly polarized source like synchrotron radiation is used. Table 7.9 provides the relation between the incident polarization angle ϕ and the excitation of the four effective modes with $\mu < 1$ cm^{-1}. The angle ϕ is the angle between the π-polarization vector and the resultant electric field of the incident wave. The angle ϕ is zero for a π-polarized field and 90° for a σ-polarized field. If the incident wave is unpolarized, $\phi = 45°$. For the π polari-

Table 7.8. Linear absorption coefficient μ_0 along the direction of the incident beam, the percentage of mode excitation and the accommodation $k\delta$ for a π-polarized beam ($\phi = 0$)

Mode	$\mu_0\,[\text{cm}^{-1}]$	Excitation [%]	$k\delta\,[\text{cm}^{-1}]$
1	0.03	0.0	246.57
2	0.19	10.1	251.33
3	0.19	0.1	251.33
4	0.87	16.7	275.90
5	13.17	4.0	441.15
6	13.17	6.6	441.15
7	16.54	18.0	514.68
8	16.54	5.1	514.68
9	24.28	16.7	637.34
10	521.66	0.0	2083.26
11	543.48	6.1	2111.65
12	543.48	16.6	2111.65

Table 7.9. Relation between the incident polarization angle ϕ and the excitation of modes 1, 2, 3 and 4

$\phi\,[°]$	Excitation of modes [%]			
	1	2	3	4
0	0.0	10.1	0.1	16.7
10	0.5	9.8	0.8	16.2
20	1.9	9.0	2.8	14.7
30	4.2	7.7	5.8	12.5
40	6.9	6.0	9.5	9.8
45	8.3	5.2	11.5	8.3
50	9.8	4.3	13.4	6.9
60	12.5	2.7	17.1	4.2
70	14.7	1.4	20.2	1.9
80	16.2	0.6	22.1	0.5
90	16.7	0.3	22.8	0.0

zation ($\phi = 0°$), only modes 2 and 4 are strongly excited, while for the σ polarization modes 1 and 3 are effective. The transition from the excitations of modes 2 and 4 to that of modes 1 and 3 takes place when ϕ varies from 0° to 90°. For an unpolarized beam, these four modes are nearly equally excited. Since the four modes are related to the ($2\bar{2}0$) and ($\bar{2}02$) reflections, they are also effectively excited at positions slightly off the six-beam point along the ($2\bar{2}0$) and ($\bar{2}02$) lines. It is therefore very difficult to observe only a sharp tiny enhanced intensity at the exact six-beam point.

The polarization of the transmitted (000) beam depends, of course, on the polarization of the incident beam. This is of importance when the enhanced transmitted beam is used as a source for x-ray diffraction and imaging studies

Table 7.10. Relation between the incident polarization angle ϕ and the transmitted polarization angle β for different crystal thicknesses t

ϕ [°]	β [°]		
	$t = 0.1$ cm	$t = 0.2$ cm	$t = 0.3$ cm
0	-180.0	0.0	180.0
10	-138.3	5.8	158.4
20	-118.5	11.8	140.8
30	-108.9	18.3	127.7
40	-103.3	25.6	118.0
45	-101.2	29.7	114.0
50	-99.4	34.3	110.5
60	-96.5	44.7	104.4
70	-94.1	57.5	99.2
80	-92.0	72.9	94.5
90	-90.0	90.0	90.0

[7.94]. Suppose that the resultant transmitted wave of the (000) reflection has the form $|E|\exp(i\beta)$ where β is the polarization angle. Table 7.10 shows the relation between β and ϕ as the crystal thickness varies. For a linearly σ- or π-polarized incident wave, the (000) transmitted beam is also linearly σ or π polarized, respectively.

The observation of the intensity enhancement clearly demonstrates that the six-beam Borrmann diffraction provides a means of generating a monochromatic beam with an exceedingly low divergence. In addition, the single mode excitation in this case is a condition which may make it possible to lower the critical inversion density needed for net gain by stimulated emission, which is important for efforts towards realizing gamma-ray lasers [7.95].

7.7 Plasma Diagnosis

The position of a multiple diffraction peak in terms of $\Delta\theta$ and $\Delta\phi$, described in Chap. 2, is a function of λ/d, where d, the spacing between atomic planes of a given reflection, depends on the lattice constant. If the wavelength λ is known, the peak position provides information about the lattice constants. The determinations of lattice constants and 'mismatches' discussed in Sects. 7.2, 3 are examples. In turn, if d is known, the wavelength can be unambiguously determined, provided that the spectral resolution is good enough to resolve overlapped peaks. In other words, if a crystal with a known d is used as an analyzer, the spectrum of an x-ray source can be obtained. The idea of using multiple diffraction for spectroscopic study of x-ray sources has recently been adapted by *Fraenkel* to plasma diagnosis [7.96 – 98].

Hot plasma, for example, generated by a tokamak, usually emits non-uniform x-radiation. The determination of the spatial and temperature distributions of the highly ionized atoms inside a tokamak is always one of the main problems in plasma diagnosis. Since multiple diffraction furnishes high angular resolution (0.2 mrad) in the direction perpendicular to the plane of incidence, *Fraenkel* in 1980 suggested the use of this type of diffraction to image intense x-ray emitting sources, like hot plasma from a tokamak [7.97]. Very recently, he applied two Umweg reflections from germanium, (000) (200) (1$\bar{1}$1)/(11$\bar{1}$) and (000) (200) (111)/(1$\bar{1}$$\bar{1}$), to low-energy plasmas diagnosis. The indices after the slashes indicate the coupling reflections [7.96].

For illustration, the images of these two Umweg reflections from the plasma of sulphur produced with a 20 Joule Nd-laser pulse of three nanoseconds duration at $\lambda = 1.05$ μm is shown in Fig. 7.38. The horizontal direction indicates the wavelength variation. The vertical direction represents the azimuthal angle of rotation around the (200) plane normal. Two intense spots at the left of this figure are the two Umweg reflections at the wavelength 5.0385 Å of the helium-like resonance transition. Figure 7.39, the calculated image of Fig. 7.38, shows the relationship between the azimuthal angle and the wavelength for these two Umweg reflections. The intersection point of these two reflection lines corresponds to $\lambda = 5.06$ Å. Note that the calculated image resembles Fig. 7.21, except that the present case deals with the wavelength variation of the azimuthal angle of multiple diffraction, while the

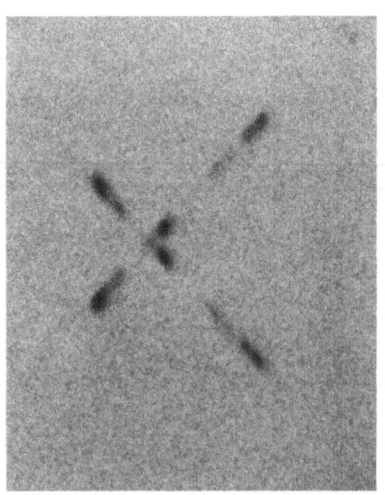

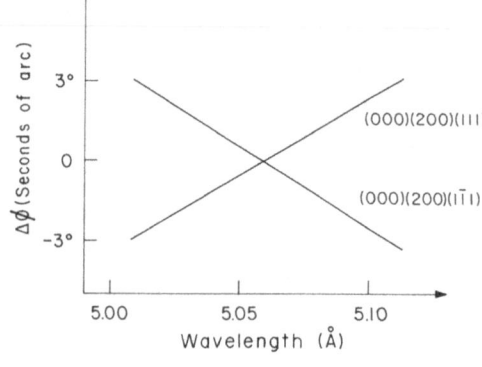

Fig. 7.38. Multiple diffraction images of the He-like and Li-like sulphur spectrum obtained from a laser-produced plasma. The vertical axis indicates the azimuthal angle and the horizontal axis is proportional to the wavelength (courtesy of *Fraenkel* [7.96])

Fig. 7.39. Calculated image of Fig. 7.38

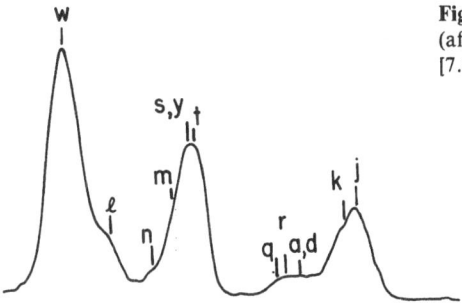

Fig. 7.40. Spectrum of highly ionized sulphur (after *Feldman* et al. [7.99]; courtesy of *Fraenkel* [7.96])

former deals with the lattice-parameter variation. The other spots in Fig. 7.38 correspond to the spectrum of sulphur described by *Feldman* et al. [7.99]. An intensity trace along one of the Umweg reflection lines is shown in Fig. 7.40. The peaks labeled w, y, j, k and l indicate the transitions at $\lambda = 5.0385$, 5.0862, 5.103, 5.100 and 5.049 Å, respectively. The fine structure of the spectrum also shows the splitting of the He-like resonance spots, i.e., peaks w and l. This could be attributed to the Doppler effect, resulting from the velocity difference between the two components w and l. Using the same method, the fine structures of He-like spectra of laser-produced titanium plasmas were also obtained [7.96].

Besides the common considerations on the film–source distance and on the use of high resolution films, the optimal experimental condition for unraveling the spectral information is the right choice of Umweg reflection of an appropriate crystal. From the calculated image shown in Fig. 7.39, it is clear that the optimal condition is to have an Umweg reflection such that the diffraction line is parallel or nearly parallel to the horizontal direction, i.e., $\partial(\Delta\phi)/\partial\lambda \simeq 0$. This consideration is similar to that for the precise determination of lattice constants and lattice mismatches discussed in Sects. 7.2, 3. The other alternative is to use energy-dispersive multiple diffraction [7.100] for plasma diagnosis.

7.8 Determination of Mosaic Spread of Crystals

The mosaic spread η of a crystal is a measure of the misorientation of the perfect crystal blocks within the crystal. For a perfect crystal, the mosaic spread is almost equal to zero. For a mosaic crystal, the mosaic spread can be as large as one degree. As discussed in Chap. 3, mosaic spread, assumed to have an isotropic Gaussian distribution, is usually introduced in the effective reflectivity of a given reflection defined in (3.15). Since the reflected intensity of a simple reflection G depends only on a single Gaussian distribution of the mosaic spread, it is straightforward to extract the mosaic spread of a crystal by measuring the full widths at half maximum, W_I and W_P, of the diffraction from the crystal studied and from a perfect crystal, respectively. This gives

$$\eta = \sqrt{W_I^2 - W_P^2}, \tag{7.76}$$

provided the beam divergence is much smaller than the mosaic spread.

An alternative way of determining the mosaic spread is to use multiple diffraction, which was first proposed by *Caticha-Ellis* [7.101]. According to (3.45), the reflectivity at the diffraction peak is equal to Q_{ij} with $\Delta\varepsilon = 0$. By substituting this reflectivity into (3.22, 26), the ratio between the diffracted power $P_1(0)$ of an N-beam case and the diffracted power $P_1'(0)$ of a two-beam case can be written as

$$R_\eta = \frac{P_1(0)}{P_1'(0)}, \tag{7.77}$$

which is a function of η. Although an analytical expression for η in terms of R_η is very complicated, numerical calculation facilitates the extraction of the information about η from the comparison of the measured R_η with the calculated one. Figure 7.41 depicts η determined for a germanium crystal as a function of the direction cosines γ of the secondary reflections in the cases with (222) and (111) as the primary reflections for MoK_{α_1} radiation [7.101]. The negative and positive γ's indicate that the secondary reflections involved are Bragg and Laue reflections, respectively. Since the germanium crystal used had considerable surface damages due to saw cut, the mosaic spread has its highest value at the superficial layer and decreases as the crystal depth increases. This fact is clearly demonstrated in the figure.

The mosaic spread has its maximum value when γ approaches zero, namely, when the diffracted beam of the secondary reflection is parallel to the crystal surface. When γ increases, the direction of the secondary beam becomes closer to the crystal normal. This signifies that the x-ray beam penetrates more into the crystal. In other words, the deeper layers are exposed by the secondary beam. The corresponding η values are therefore decreased. The

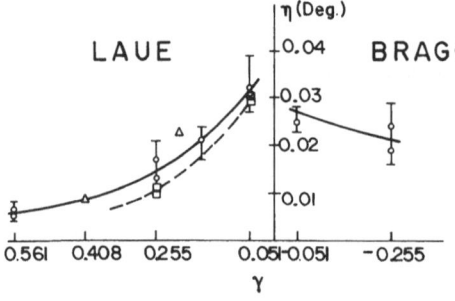

Fig. 7.41. Measured mosaic spread η versus the direction cosine γ (after *Caticha-Ellis* [7.101])

curve for Bragg-Bragg diffractions (Sect. 2.1) varies less drastically than that for Bragg-Laue diffractions, because the Bragg-reflected secondary beams penetrate the crystal less than the Laue-diffracted beams. As the direction cosine is close to -1, the mosaic spread approaches an asymptotic value which is almost independent of the secondary reflections involved in the multiple diffraction. This asymptotic value is approximately equal to the vertical beam divergence due to the fact that the width of multiple diffraction depends mainly on the vertical divergence δ_v, instead of the mosaic spread, for $\eta > \delta$ (Sect. 3.9). This conclusion has also been pointed out by *Prager* [7.102].

In conclusion, this method determines the variation of mosaic spread along the crystal surface normal; this quantity is not obtainable using a simple two-beam reflection. Furthermore, once the crystal is aligned for multiple diffraction, all the diffraction peaks and dips provide values for η at the corresponding azimuthal angles. In the same way, the anisotropic distribution of mosaic spread can, in principle, be determined.

The use of three-beam multiple diffraction for the determination of mosaic spread has been also applied to aluminium single crystals using neutron diffraction [7.103]. This method, however, cannot be used in the case when the primary reflection is a forbidden reflection, because (7.77) has a singularity for $P_1'(0) = 0$.

7.9 Multi-Beam X-Ray Standing-Wave Excited Fluorescence Technique for Surface Studies – A Proposed Method

An x-ray standing wave is the result of the dynamical interaction of the wavefields of the incident and the diffracted waves within a perfect crystal lattice. The excitation of the atoms in the crystal by this standing wave causes the atoms to emit radiation, the x-ray fluorescence. As discussed in Sects. 4.9 and 7.1, the intensity of the fluorescence is directly related to the photoelectric absorption of the atoms, the positions of the atoms and the reflection phases. Since the absorption is a function of the angles of incidence, the fluorescence intensity is also dependent on this angle.

The first measurement of the intensity of the fluorescence of a two-beam diffraction from a perfect crystal was carried out by *Batterman* [7.104] in 1962. Since then, this technique has been continuously developed and now the positions of the impurity atoms at the surface of semiconductor single crystals can be detected [7.105 – 110]. There are two possible ways for the crystal to accommodate the impurity atoms: interstitial and substitutional. If the impurity atoms are located inbetween the host lattice sites, they are interstitial. If the impurity atoms replace the host atoms, the impurity atoms are substitutional. If the x-ray employed has an energy sufficient to excite the im-

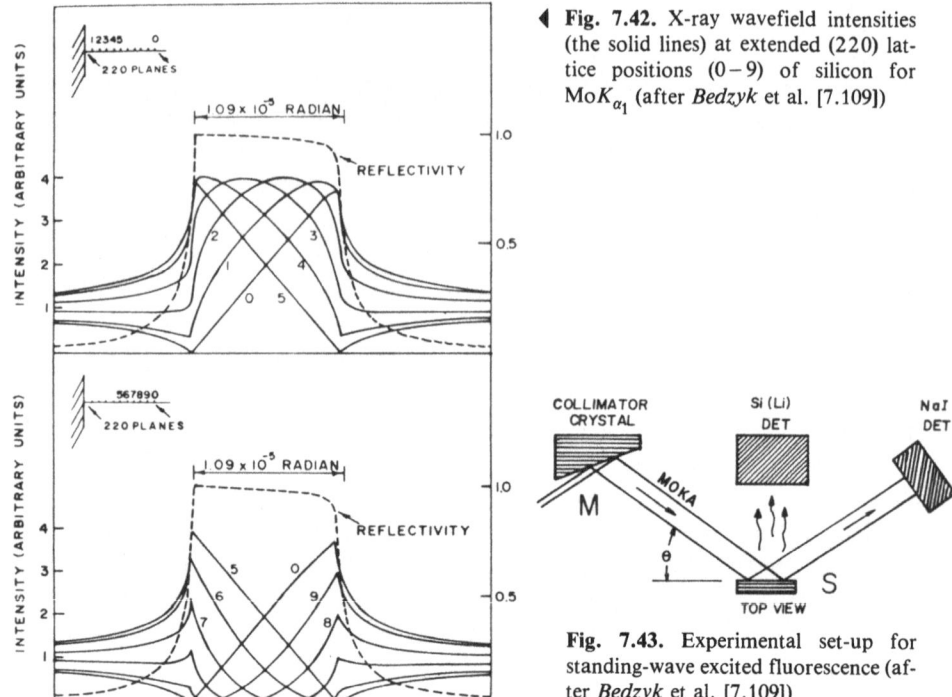

◀ **Fig. 7.42.** X-ray wavefield intensities (the solid lines) at extended (220) lattice positions (0 – 9) of silicon for MoK_{α_1} (after *Bedzyk* et al. [7.109])

Fig. 7.43. Experimental set-up for standing-wave excited fluorescence (after *Bedzyk* et al. [7.109])

purity atoms, the intensity distribution of the fluorescence from the impurity depends on the relative positions of the impurity atoms in the host crystal lattice. This intensity distribution is proportional to the wavefield intensity I_F, defined in (4.246), of the two-beam diffraction used. Figure 7.42 shows the wavefield intensities at various positions between the (220) reflection planes of a silicon crystal for MoK_{α_1} radiation. The thickness of the impurity layer is less than the extinction length so that distortion of the fluorescence curves (4.246) due to the absorption can be avoided. By comparing the observed fluorescence intensity distribution with the calculated curves (Fig. 7.42), the relative position of the impurity atom between the lattice planes of the host crystal can be determined.

Figure 7.43 is a schematic representation of a typical experimental set-up. An asymmetrically cut crystal M serves as the monochromator, and the crystal S is a silicon sample. A scintillation counter and a lithium-drifted silicon solid-state detector are used to monitor the Bragg-reflected beam and the fluorescence, respectively. Since a two-beam reflection can only provide information about the relative positions of the impurities with respect to one set of atomic planes, two two-beam Bragg reflections should be used to determine the two-dimensional distribution of the impurity atoms at the crystal surface. Recently, *Golovchenko* et al. [7.110] used (111) and (220) reflections of silicon to determine the positions A and B of the bromine atom relative to the (111) and (220) rows of Si atoms (Fig. 7.44). Figure 7.45 shows the experimentally obtained bromine fluorescence and reflected intensities of the (111) reflection. Similar results were also obtained for the (220) reflection. The relative positions A and B were determined as 2.56 ± 0.03 Å and 1.75 ± 0.02 Å.

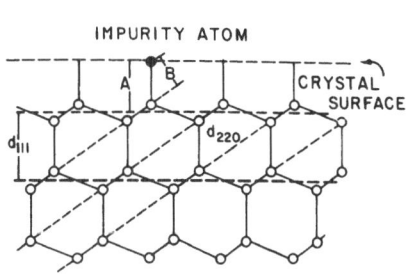

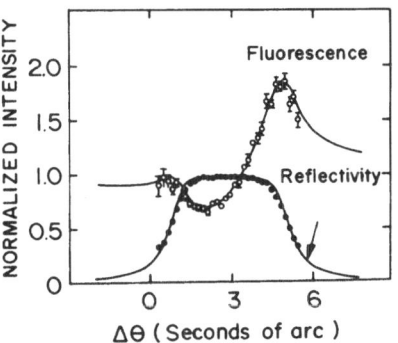

Fig. 7.44. Projection of a (111)-cut silicon crystal on the ($1\bar{1}0$) plane (after *Golovchenko* et al. [7.110])

Fig. 7.45. Bromine fluorescence and reflectivity $|E_{111}/E_{000}|^2$ versus $\Delta\theta$ for (111) Bragg reflection on a silicon (111) surface (after *Golovchenko* et al. [7.110])

In principle, this information can also be obtained using the three-beam, (000) (111) (220), reflection with (111) as the symmetric Bragg reflection and (220) as the inclined Bragg reflection. The following procedure is proposed. The crystal is first set for the (111) reflection and is then rotated around the reciprocal lattice vector of the (111) reflection to bring the (220) reflection also in position to diffract the incident beam. The variation of the fluorescence intensity versus the azimuthal rotation angle $\Delta\phi$ for a given $\Delta\theta$ is a function of the parameters A and B. For illustration, Fig. 7.46 shows the calculated fluorescence intensities versus $\Delta\phi$ at $\Delta\theta = 4.5''$ for $A = 2.56$ Å and $B = 1.920, 1.750, 1.344, 0.576, 0.192$ Å. The angle $\Delta\theta$ is the angular deviation from the exact Bragg angle of the (111) reflection. The curves labeled 1 to 5 correspond, respectively, to the various positions of B mentioned above. By comparing Fig. 7.46 with experimental results, the parameters A and B can be determined. Furthermore, the effective layer which contributes to the fluorescence intensity is not as thin as a monoatomic layer. Since the extinction length for this particular case of diffraction is of the order of a few microns, the effective layer still contains a few hundred atomic planes. The three-dimensional information about the positions of the impurity atoms within these few hundred atomic planes is needed. The multi-beam standing-wave analysis just mentioned may serve this purpose. However, in this case the phase invariants involved in multi-beam diffractions should be taken into account in analyzing the fluorescence intensities.

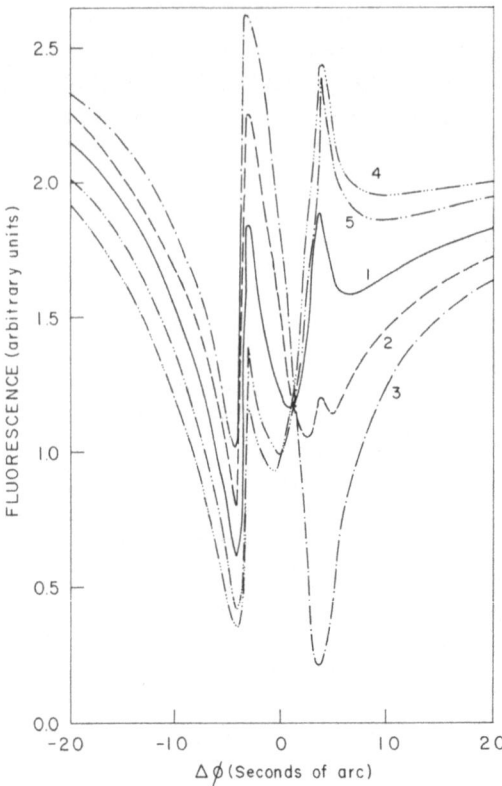

Fig. 7.46. Calculated wavefield intensities at various lattice positions between (111) planes for the three-beam, (000) (111) (220), diffraction in silicon for MoK_{α_1}

7.10 Possible Future Trend of Development

Possible future development of the multiple diffraction techniques can be foreseen along the discussion of their applications given in the previous sections of this chapter. The development naturally includes continuous studies in attacking the unsolved problems associated with the applications mentioned and the creation of new ideas for using multiple diffraction in materials characterization. In the following, the possible future lines of development are outlined briefly.

a) Crystal-structure determination. Continuous research on the application of multiple diffraction to the solution of the x-ray phase problem for non-centro-symmetric crystals will be the first step in the future development of the technique for determining crystal structures. The possibility of generalizing this technique and solving the structures of large molecules is the final goal. To pursue this goal, the use of radiation sources with continuous spectra, like synchrotron radiation, is indispensable.

b) Characterization of semiconductor materials. The structural characterization of semiconductor materials needs various techniques which complement each other. X-ray diffraction methods serve as appropriate techniques for studying defects and dislocations at the crystal surface and in the crystal bulk. Since perfect semiconductor crystals are increasingly involved nowadays in electronic technology, the methods for determining lattice constants and lattice mismatches should take into account the dynamical effects of diffraction. The development of multi-beam topography to the stage where it provides a sensitive and easy way to determine the Burgers vectors of dislocations is needed. The use of a real-time x-ray topographic arrangement [7.111, 112] may be helpful.

Determination of the concentration and the positions of impurity atoms in crystals may be achieved from the analysis of the reflected intensities and the peak positions of multiple diffractions. However, the perfection of the crystal surface should be checked before the analysis of the reflected intensities. Moreover, determination of the distribution of the covalent electron density is another subject for which multiple diffraction techniques may be used. In this case, multiple diffraction with a forbidden reflection as the primary reflection should be employed.

c) Applications in x-ray optics. Components such as monochromators, beam splitters, interferometers and lenses are often necessary for studies related to x-ray optics. The coherent interaction in N-beam diffraction provides an additional means of devising these components. The anomalous transmission in N-beam Borrmann diffraction gives an extremely parallel x-ray source for possible use in x-ray lithography and x-ray laser development.

d) Surface studies. In addition to the standing-wave technique for solving the surface registration problem [7.110], Bragg surface multiple diffraction may be utilized to determine surface structures by detecting the surface diffracted beams. This may be complementary to the method using two-beam x-ray glancing reflection for the study of surface reconstruction [7.113].

References

Chapter 1

1.1 E. Wagner: Phys. Z. **21**, 94 (1923)
1.2 O. Berg: Wiss. Veröff. (Siemen's conference) **5**, 89 (1926)
1.3 G. Mayer: Z. Kristallogr. **66**, 585 (1928)
1.4 M. Renninger: Z. Kristallogr. **106**, 141 (1937); Naturwissenschaften **25**, 43 (1937)
1.5 A. Pabst: Am. Mineral. **24**, 566 (1939)
1.6 C. J. Davisson, F. E. Haworth: Phys. Rev. **66**, 351 (1944)
1.7 W. N. Lipscomb: Acta Crystallogr. **2**, 193 (1949)
1.8 G. Borrmann, W. Hartwig: Z. Kristallogr. **121**, 401 (1965)
1.9 G. Borrmann: Phys. Z. **42**, 157 (1941); Z. Phys. **127**, 297 (1950)
1.10 H. Bethe: Ann. Phys. (Leipzig) **87**, 55 (1928)
1.11 H. Raether: Z. Phys. **78**, 527 (1932)
1.12 J. A. Darbyshire, E. R. Cooper: Proc. Roy. Soc. **A152**, 104 (1935)
1.13 S. Miyake, K. Kambe: Acta Crystallogr. **7**, 216 (1954)
1.14 K. Kambe, S. Miyake: Acta Crystallogr. **7**, 218 (1954)
1.15 K. Kambe: J. Phys. Soc. Jpn. **12**. 13, 25 (1957)
1.16 J. C. Slater: Rev. Mod. Phys. **30**, 197 (1958)
1.17 R. W. James: Solid State Phys. **15**, 53 (1962)
1.18 P. P. Ewald: J. Phys. Soc. Jpn. Suppl. B-II **17**, 48 (1962); Rev. Mod. Phys. **37**, 48 (1965)
1.19 B. W. Battermann, H. Cole: Rev. Mod. Phys. **36**, 681 (1964)
1.20 G. Borrmann: Z. Kristallogr. **120**, 143 (1964)
1.21 A. Authier: In *Advances in Structure Research by Diffraction Methods*, ed. by. R. Brill, R. Mason (Pergamon, Oxford 1970)
1.22 N. Kato: In *X-Ray Diffraction*, ed. by L. V. Azaroff (McGraw-Hill, New York 1974)
1.23 W. H. Zachariasen: *Theory of X-Ray Diffraction in Crystals* (Wiley, New York 1945)
1.24 R. W. James: *The Optical Principle of the Diffraction of X-Rays* (Bell, London 1950)
1.25 M. von Laue: *Röntgenstrahl-Interferenzen* (Akademische, Frankfurt 1960)
1.26 L. H. Schwartz, J. B. Cohen: *Diffraction from Materials* (Academic, New York 1977)
1.27 P. G. Pinsker: *Dynamical Scattering of X-Rays in Crystals*, Springer Ser. Solid-State Sci., Vol. 3 (Springer, Berlin, Heidelberg, New York 1977)
1.28 M. von Laue, W. Friedrich, P. Knipping: Ann. Phys. (Leipzig) **41**, 971 (1913)
1.29 C. G. Darwin: Philos. Mag. **27**, 315, 675 (1914)
1.30 P. P. Ewald: Ann. Phys. (Leipzig) **49**, 1, 117 (1916); Ann. Phys. (Leipzig) **54**, 519 (1917); Z. Kristallogr. **97**, 1 (1937)
1.31 M. von Laue: Ergebn. Exakten Naturwiss. **10**, 133 (1931)
1.32 N. Kato, R. A. Lang: Acta Crystallogr. **12**, 787 (1959)
1.33 N. Kato: Z. Naturforsch. **15a**, 369 (1960); Acta Crystallogr. **14**, 526, 627 (1961); J. Appl. Phys. **39**, 2225, 2231 (1968)
1.34 P. Penning, D. Polder: Acta Crystallogr. **17**, 950 (1964)
1.35 S. Takagi: Acta Crystallogr. **15**, 1311 (1962); J. Phys. Soc. Jpn. **27**, 1239 (1969)
1.36 D. Taupin: Bull. Soc. Fr. Minéral. Cristallogr. **87**, 469 (1964)
1.37 N. Kato: Acta Crystallogr. **16**, 276, 282 (1963); J. Phys. Soc. Jpn. **18**, 1785 (1963); J. Phys. Soc. Jpn. **19**, 67, 971 (1964)
1.38 K. Kambe: Z. Naturforsch. **20a**, 770 (1965)
1.39 M. Wilkins: Phys. Status Solidi **13**, 529 (1966)
1.40 F. Balibar, A. Authier: Phys. Status Solidi **21**, 413 (1967)

1.41 F. N. Chukhovskii, A. A. Shtolberg: Phys. Status Solidi **41**, 815 (1970)
1.42 A. M. Afanas'ev, V. G. Kohn: Acta Crystallogr. **A27**, 421 (1971)
1.43 A. Howie, M. J. Whelan: Proc. Roy. Soc. **A263**, 217 (1961)
1.44 U. Bonse: Z. Phys. **177**, 385 (1964)
1.45 N. Kato: Acta Crystallogr. **A32**, 453, 458 (1976); **A35**, 9 (1979); **A36**, 171 (1980)
1.46 N. Kato: Acta Crystallogr. **A36**, 763, 770 (1980)
1.47 G. Molière: Ann. Phys. (Leipzig) **35**, 272, 297 (1939); **36**, 265 (1939)
1.48 H. Wagenfeld: Phys. Rev. **144**, 216 (1966)
1.49 Y. H. Ohtsuki: J. Phys. Soc. Jpn. **19**, 2285 (1964); **20**, 374 (1965)
1.50 M. Ashkin, M. Kuriyama: J. Phys. Soc. Jpn. **21**, 1549 (1966)
1.51 Y. H. Ohtsuki, S. Yanagawa: J. Phys. Soc. Jpn. **21**, 326, 502 (1966)
1.52 M. Kuriyama: J. Phys. Soc. Jpn. **23**, 1369 (1967); Phys. Status Solidi **24**, 743 (1967);
      J. Phys. Soc. Jpn. **25**, 846 (1968); Acta Crystallogr. **A25**, 682 (1969); **A26**, 56 (1970); **A27**,
      273 (1971); **A28**, 588 (1972)
1.53 J. P. Hannon, G. T. Trammell: Phys. Rev. **169**, 315 (1968); Phys. Rev. **186**, 306 (1969)
1.54 A. M. Afanas'ev, Y. Kagan: Acta Crystallogr. **A24**, 163 (1968)
1.55 P. H. Dederichs: Phys. Kondens. Mater. **5**, 347 (1966); Phys. Status Solidi **23**, 377 (1967)
1.56 H. Sano, K. Ohtaka, Y. H. Ohtsuki: J. Phys. Soc. Jpn. **27**, 1254 (1969)
1.57 P. P. Ewald: Z. Kristallogr. **A97**, 1 (1937)
1.58 E. Lamla: Ann. Phys. (Leipzig) **36**, 194 (1939)
1.59 N. Kato: Acta Crystallogr. **11**, 885 (1958)
1.60 E. J. Saccocio, A. Zajac: Acta Crystallogr. **18**, 478 (1965); Phys. Rev. **A139**, 255 (1965)
1.61 G. Hildebrandt: Phys. Status Solidi **24**, 245 (1967); Acta Crystallogr. **A25**, 209 (1969);
      Phys. Status Solidi **(a)15**, K83 (1973)
1.62 T. Joko, A. Fukuhara: J. Phys. Soc. Jpn. **22**, 597 (1967)
1.63 P. P. Ewald, Y. Heno: Acta Crystallogr. **A24**, 5 (1968)
1.64 Y. Heno, P. P. Ewald: Acta Crystallogr. **A24**, 16 (1968)
1.65 A. M. Afanas'ev, V. G. Kohn: Acta Crystallogr. **A32**, 2, 308 (1976)
1.66 A. Dalisa, A. Zajac, C. H. Ng: Phys. Rev. **168**, 859 (1968)
1.67 W. Uebach, G. Hildebrandt: Z. Kristallogr. **129**, 1 (1969)
1.68 P. Penning, D. Polder: Philips Res. Rep. **23**, 1 (1968)
1.69 P. Penning: Advan. X-Ray Anal. **10**, 67 (1967); Philips Res. Rep. **23**, 12 (1968)
1.70 Yu S. Terminasov, L. V. Tuzov: Usp. Fiz. Nauk. **83**, 223 (1964) [English transl.: Sov. Phys.-
      Usp. **7**, 734 (1964)]
1.71 S. Balter, R. Feldman, B. Post: Phys. Rev. Lett. **27**, 307 (1971)
1.72 R. Feldman, B. Post: Phys. Status Solidi **(a)12**, 273 (1972)
1.73 T. C. Huang, B. Post: Acta Crystallogr. **A29**, 35 (1973)
1.74 T. C. Huang, M. H. Tillinger, B. Post: Z. Naturforsch. **28a**, 600 (1973)
1.75 M. Umeno: Phys. Status Solidi **(a)2**, K203 (1970)
1.76 M. Umeno, G. Hildebrandt: Phys. Status Solidi **(a)31**, 583 (1975)
1.77 M. Umeno: Phys. Status Solidi **(a)37**, 561 (1976); **(a)38**, 701 (1976)
1.78 T. I. Borodina, V. I. Iveronova, A. A. Katznelson, T. K. Runova: Sov. Kristallogr. **20**, 490
      (1975) [English transl.: Sov. Phys.-Crystallogr. **20**, 300 (1975)]
1.79 S. A. Kshevetsky, I. P. Mikhailyuk: Sov. Kristallogr. **21**, 381 (1976) [English transl.: Sov.
      Phys.-Crystallogr. **21**, 209 (1976)]
1.80 A. M. Afanas'ev, V. G. Kohn: Phys. Status Solidi **(a)28**, 61 (1975); Acta Crystallogr. **A33**,
      178 (1977)
1.81 V. D. Kos'mik, S. A. Kshevetsky, M. L. Kshevetskaya, I. P. Mikhailyuk, M. V.
      Ostapovich: Sov. Kristallogr. **21**, 899 (1976) [English transl.: Sov. Phys.-Crystallogr. **21**,
      514 (1976)]
1.82 B. Post, S. L. Chang, T. C. Huang: Acta Crystallogr. **A33**, 90 (1977)
1.83 S. L. Chang: Z. Naturforsch. **37a**, 501 (1982)
1.84 B. Post, P. W. Wang, T. Hom: Z. Naturforsch. **37a**, 528 (1982)
1.85 O. Pacherová, R. Bubáková: Z. Naturforsch. **37a**, 617 (1982)
1.86 R. M. Moon, C. G. Shull: Acta Crystallogr. **17**, 805 (1964)
1.87 W. H. Zachariasen: Acta Crystallogr. **18**, 705 (1965)
1.88 W. H. Zachariasen: Acta Crystallogr. **16**, 1139 (1965)

1.89 S. Caticha-Ellis: Acta Crystallogr. **A25**, 666 (1969)
1.90 S. L. Chang: Acta Crystallogr. **A38**, 41 (1982)
1.91 R. Colella: Acta Crystallogr. **A30**, 413 (1974)
1.92 V. G. Kohn: Phys. Status Solidi (a)**54**, 375 (1979)
1.93 S. L. Chang: Acta Crystallogr. **A35**, 543 (1979); **A38**, 516 (1982)
1.94 K. Lonsdale: Philos. Trans. **A240**, 219 (1947)
1.95 B. J. Isherwood, C. A. Wallace: Nature **212**, 173 (1966)
1.96 F. J. Spooner, C. G. Wilson: J. Appl. Crystallogr. **6**, 132 (1973)
1.97 B. Post: J. Appl. Crystallogr. **8**, 452 (1975)
1.98 T. Hom, W. Kiszenik, B. Post: J. Appl. Crystallogr. **8**, 457 (1975)
1.99 B. R. Brown, M. A. G. Halliwell, B. J. Isherwood: J. Microscopy **118**, 375 (1980)
1.100 B. J. Isherwood, B. R. Brown, M. A. G. Halliwell: J. Cryst. Growth **54**, 449 (1981)
1.101 S. L. Chang: Appl. Phys. Lett. **34**, 239 (1979); **37**, 819 (1980); Acta Crystallogr. **A37**, 876 (1981)
1.102 M. Hart, A. R. Lang: Phys. Rev. Lett. **7**, 120 (1961)
1.103 B. Post: Phys. Rev. Lett. **39**, 760 (1977); Acta Crystallogr. **A35**, 17 (1979)
1.104 L. D. Chapman, D. R. Yoder, R. Colella: Phys. Rev. Lett. **46**, 1578 (1981)
1.105 S. L. Chang: Appl. Phys. **A26**, 221 (1981); Phys. Rev. Lett. **48**, 163 (1982)
1.106 H. J. Juretschke: Phys. Rev. Lett. **48**, 1487 (1982); Phys. Lett. **92A**, 183 (1982)
1.107 M. F. C. Ladd, R. A. Palmer (eds.): *Theory and Practice of Direct Methods in Crystallography* (Plenum, New York 1980)
1.108 S. L. Chang, F. S. Han: Acta Crystallogr. **A38**, 414 (1982)
1.109 F. S. Han, S. L. Chang: Acta Crystallogr. **A39**, 98 (1983)
1.110 S. L. Chang: Phys. Bl. **39**, 46 (1983)
1.111 P. Mikula, J. Kulda, M. Vrána, R. T. Michalec, J. Vávra: Nucl. Instrum. Meth. **197**, 563 (1982)
1.112 S. L. Chang: Appl. Phys. Lett. **40(9)**, 793 (1982)
1.113 W. Graeff, U. Bonse: Z. Phys. **B27**, 19 (1977)
1.114 S. L. Chang: J. Appl. Phys. **53(4)**, 2988 (1982)
1.115 B. W. Battermann: Appl. Phys. Lett. **1**, 68 (1962); Phys. Rev. **133**, A759 (1964)
1.116 J. A. Golovchenko, J. R. Patel, D. R. Kaplan, P. L. Cowan, M. J. Bedzyk: Phys. Rev. Lett. **49**, 560 (1982)
1.117 B. S. Fraenkel: Appl. Phys. Lett. **41(3)**, 234 (1982)
1.118 W. Kossel: Nachr. Akad. Wiss. Göttingen, Math. Phys. **1**, 229 (1935)
1.119 W. Kossel: Ann. Phys. (Leipzig) **25**, 512 (1935)
1.120 W. Kossel: Ann. Phys. (Leipzig) **26**, 533 (1936)
1.121 W. Kossel: Ergebn. Exakten Naturwiss. **16**, 295 (1937)
1.122 W. Kossel, H. Voges: Ann. Phys. (Leipzig) **23**, 677 (1935)
1.123 R. Kohler, W. Mohling, H. Peibst: Phys. Status Solidi **41**, 75 (1970); **61(b)**, 173 (1974)
1.124 S. Kishino, K. Kohra: Jpn. J. Appl. Phys. **10**, 551 (1971)
1.125 T. Bedynska: Phys. Status Solidi (a)**19**, 365 (1973)
1.126 S. Kishino: J. Phys. Soc. Jpn. **31**, 1168 (1971)
1.127 S. Kishino, A. Noda, K. Kohra: J. Phys. Soc. Jpn. **33**, 158 (1972)
1.128 T. Bedynska: Phys. Status Solidi (a)**25**, 405 (1974)
1.129 D. A. O'Connor: Proc. Phys. Soc. **91**, 917 (1967)
1.130 H. W. Schürmann: Z. Phys. **194**, 425 (1966)
1.131 K. Haruta: J. Appl. Phys. **38**, 3312 (1967)
1.132 M. Kuriyama, T. Miyakawa: J. Appl. Phys. **40**, 1967 (1969)
1.133 A. R. Hutson, J. H. McFee, D. L. White: Phys. Rev. Lett. **7**, 235 (1961)
1.134 D. L. White: J. Appl. Phys. **33**, 2547 (1962)
1.135 D. G. Carlson, A. Segmüller, E. Mosekilde, H. Cole, J. A. Armstrong: Appl. Phys. Lett. **18**, 330 (1971)
1.136 D. G. Carlson, A. Segmüller: Phys. Rev. Lett. **28**, 175 (1972)
1.137 Ishibashi, M. Kitamura, A. Odajima: Phys. Lett. **44A**, 371 (1973)
1.138 S. D. LeRoux, R. Colella, R. Bray: Phys. Rev. Lett. **35**, 230 (1975); **37**, 1056 (1976)
1.139 G. Hildebrandt: J. Phys. **E15**, 1140 (1982)
1.140 H. Fricke, V. Gerold: J. Appl. Phys. **30**, 661 (1959)

1.141 K. Thomas, A. Franks: J. Appl. Phys. **30**, 649 (1959)
1.142 V. I. Betekhtin, A. I. Slutsker: Sov. Phys. – Solid State **4**, 94 (1962)
1.143 U. Bonse, M. Hart: Appl. Phys. Lett. **7**, 238 (1965)

Chapter 2

2.1   G. Borrmann, W. Hartwig: Z. Kristallogr. **121**, 401 (1965)
2.2   K. Lonsdale: Philos. Trans. **A240**, 219 (1947)
2.3   R. D. Burbank: Acta Crystallogr. **19**, 957 (1965)
2.4   P. R. Prager: Acta Crystallogr. **A27**, 563 (1971)
2.5   N. F. M. Henry, K. Lonsdale (eds.): *International Tables for X-Ray Crystallography*, Vol. 1 (Kynoch, Birmingham 1968)
2.6   M. Renninger: Z. Krystallogr. **106**, 141 (1937); Naturwissenschaften **25**, 43 (1937)
2.7   R. S. Williamson, I. Fankuchen: Rev. Sci. Instrum. **30**, 908 (1959)
2.8   H. Cole, F. W. Chambers, H. M. Dunn: Acta Crystallogr. **15**, 138 (1962)
2.9   For example, E. W. Nuffield: *X-Ray Diffraction Methods* (Wiley, New York 1966)
2.10   W. H. Zachariasen: Acta Crystallogr. **17**, 805 (1964)
2.11a H. L. Yakel, I. Fankuchen: Acta Crystallogr. **15**, 1188 (1962)
2.11b H. L. Yakel, I. Fankuchen, J. W. Jeffery, A. Whitaker: Acta Crystallogr. **19**, 963 (1965)
2.12   A. Santoro, M. Zocchi: Acta Crystallogr. **21**, 293 (1966)
2.13   E. Rutherford, E. N. da C. Andrade: Philos. Mag. **28**, 263 (1914)
2.14   W. H. Bragg: Nature **43**, 31 (1914)
2.15   H. Seemann: Ann. Phys. (Leipzig) **51**, 391 (1916)
2.16   H. Seemann: Ann. Phys. (Leipzig) **53**, 461 (1917)
2.17   H. Seemann: Phys. Z. **20**, 169 (1919)
2.18   W. Gerlach: Phys. Z. **22**, 557 (1921)
2.19   W. Gerlach: Verh. Phys. Med. Ges. Würzburg **56**, 55 (1921)
2.20   H. Seemann: Ann. Phys. (Leipzig) **6**, 793 (1930); **7**, 633 (1930)
2.21   H. Seemann, O. Kantorowicz: Naturwissenschaften **18**, 526 (1930)
2.22   B. Hess: Z. Kristallogr. **97**, 197 (1937)
2.23   B. Hess: Z. Kristallogr. **104**, 294 (1942)
2.24   W. Linnik: Nature **124**, 946 (1929)
2.25   W. Linnik: Z. Phys. **61**, 220 (1930)
2.26   S. Kikuchi: Proc. Imp. Acad. Jpn. **4**, 354 (1928)
2.27   W. Kossel: Nachr. Akad. Wiss. Göttingen, Math. Phys. **1**, 229 (1935)
2.28   W. Kossel: Ann. Phys. (Leipzig) **25**, 512 (1936)
2.29   W. Kossel: Ann. Phys. (Leipzig) **26**, 533 (1936)
2.30   W. Kossel: Ergebn. Exakten Naturwiss. **16**, 295 (1937)
2.31   W. Kossel, H. Voges: Ann. Phys. (Leipzig) **23**, 677 (1935)
2.32   H. Voges: Ann. Phys. (Leipzig) **27**, 702 (1936)
2.33   G. Borrmann: Naturwissenschaften **23**, 591 (1935)
2.34   G. Borrmann: Ann. Phys. (Leipzig) **27**, 669 (1936)
2.35   R. Castaing, A. Guinier: C. R. Acad. Sci. Paris **232**, 1948 (1951)
2.36   J. L. McCall, K. L. Strabel, J. S. Duerr: J. Appl. Phys. **38**, 2695 (1967)
2.37   B. A. Newman, S. Weissmann: J. Appl. Crystallogr. **1**, 139 (1968)
2.38   T. Imura, S. Weismann, J. J. Slade Jr.: Acta Crystallogr. **15**, 786 (1962)
2.39   H. Yakowitz: J. Appl. Phys. **37**, 4455 (1966)
2.40   B. J. Isherwood, C. A. Wallace: Acta Crystallogr. **A27**, 119 (1971)
2.41   T. Ellis, L. G. Naani, A. Shrier, S. Weissmann, G. E. Padawer, N. Hosokawa: J. Appl. Phys. **35**, 3364 (1964)
2.42   T. Fujiwara: J. Sci. Hiroshima Univ. **A7**, 179 (1937)
2.43   T. Fujiwara: J. Sci. Hiroshima Univ. **A9**, 233 (1939)
2.44   D. Onoyama: J. Sci. Hiroshima Univ. **A9**, 125 (1939)
2.45   T. C. Huang, B. Post: Acta Crystallogr. **A29**, 35 (1973)
2.46   F. J. Spooner, C. G. Wilson: J. Appl. Crystallogr. **6**, 132 (1973)
2.47   S. L. Chang: Appl. Phys. Lett. **34**, 239 (1979); **37**, 819 (1980); Acta Crystallogr. **A37**, 876 (1981)

2.48 S. A. Kshevetsky, I. P. Mikaulyuk: Sov. Kristallogr. **21**, 381 (1976) [English transl.: Sov. Phys.-Crystallogr. **21**, 209 (1976)]
2.49 S. L. Chang, S. Caticha-Ellis: Acta Crystallogr. **A34**, 825 (1978)
2.50 F. S. Han, S. L. Chang: J. Appl. Crystallogr. **15**, 570 (1982)
2.51 W. C. Hamilton (ed.): *International Tables for X-Ray Crystallography*, Vol. IV (Kynoch, Birmingham 1974)
2.52 W. R. Busing, H. A. Levy: Acta Crystallogr. **22**, 457 (1967)
2.53 P. de Meester: Acta Crystallogr. **A36**, 734 (1980)
2.54 R. E. Hanneman, R. E. Ogilvie, A. Modrezjewski: J. Appl. Phys. **33**, 1429 (1962)
2.55 R. Tixier, C. Waché: J. Appl. Crystallogr. **3**, 466 (1968)
2.56 J. Frazer, G. Arrhenius: In *Proceedings of the IVth Congress on X-Ray Optics and Micro-analysis* (Hermann, Paris 1966) p. 516
2.57 W. G. Morris: J. Appl. Phys. **39**, 1813 (1968)
2.58 E. T. Peters, R. E. Ogilvie: Trans AIME **233**, 89 (1965)
2.59 K. J. H. Mackay: In *Proceedings of the IVth Congress on X-Ray and Microanalysis* (Hermann, Paris 1966) p. 544
2.60 P. L. Ryder, H. Halbig, W. Pitsch: In *Proceedings of the VIIth Congress on X-Ray Optics and Microanalysis*, ed. by G. Moellenstedt, K. H. Gaukle (Springer, Berlin, Heidelberg, New York 1969) p. 388
2.61 B. K. Vainshtein: *Modern Crystallography I*, Springer Ser. Solid-State Sci., Vol. 15 (Springer, Berlin, Heidelberg, New York 1981)

Chapter 3

3.1 M. von Laue: Ann. Phys. (Leipzig) **41**, 971 (1913)
3.2 C. G. Darwin: Philos. Mag. **27**, 315, 675 (1914)
3.3 W. H. Zachariasen: *Theory of X-Ray Diffraction in Crystals* (Wiley, New York 1945)
3.4 R. M. Moon, C. G. Shull: Acta Crystallogr. **17**, 805 (1964)
3.5 W. H. Zachariasen: Acta Crystallogr. **18**, 705 (1965)
3.6 S. Caticha-Ellis: Acta Crystallogr. **A25**, 666 (1969)
3.7 D. Unangst, W. Melle: Acta Crystallogr. **A31**, 234 (1975)
3.8 C. B. R. Parente, S. Caticha-Ellis: Jpn. J. Appl. Phys. **13**, 1501 (1974)
3.9 S. L. Chang: Acta Crystallogr. **A37**, 876 (1981)
3.10 S. L. Chang: Acta Crystallogr. **A38**, 41 (1982)
3.11 S. Caticha-Ellis: Jpn. J. Appl. Phys. **14**, 603 (1975)
3.12 R. Colella, A. Merlini: Phys. Status Solidi **18**, 157 (1966)

Chapter 4

4.1 C. G. Darwin: Philos. Mag. **27**, 315, 675 (1914)
4.2 P. P. Ewald: Ann. Phys. (Leipzig) **49**, 1, 117 (1916); **54**, 519 (1917); Z. Kristallogr. **97**, 1 (1937)
4.3 M. von Laue: Ergebn. Exakten Naturwiss. **10**, 133 (1931)
4.4 H. Wagenfeld: Acta Crystallogr. **A24**, 170 (1968)
4.5 S. Miyake, Y. H. Ohtsuki: Acta Crystallogr. **A30**, 103 (1974)
4.6 Z. G. Pinsker: *Dynamical Scattering of X-Rays in Crystals*, Springer Ser. Solid-State Sci., Vol. 3 (Springer, Berlin, Heidelberg, New York 1977)
4.7 T. Joko, A. Fukuhara: J. Phys. Soc. Jpn. **22**, 597 (1967)
4.8 K. Kambe: J. Phys. Soc. Jpn. **12**, 13, 25 (1957)
4.9 P. Penning: Advan. X-Ray Anal. **10**, 67 (1967); Philips Res. Rep. **23**, 12 (1968)
4.10 P. P. Ewald: Z. Kristallogr. **A97**, 1 (1937)
4.11 E. Lamla: Ann. Phys. (Leipzig) **36**, 194 (1939)
4.12 N. Kato: J. Phys. Soc. Jpn. **7**, 397 (1952)
4.13 M. Born: *Optik*, 2nd ed. (Springer, Berlin, Heidelberg, New York 1965)
4.14 E. J. Saccocio, A. Zajac: Acta Crystallogr. **18**, 478 (1965); Phys. Rev. **A139**, 255 (1965)
4.15 A. M. Afanas'ev, V. G. Kohn: Phys. Status Solidi **(a)28**, 61 (1975); Acta Crystallogr. **A33**, 178 (1977)
4.16 M. Kohler: Ann. Phys. (Leipzig) **18**, 265 (1933)

4.17 A. Authier: J. Phys. Radium **23**, 961 (1962)
4.18 N. Kato, T. Katagawa, T. Saka: Kristallofrafiya **16**, 1110 (1971)
4.19 S. L. Chang: Acta Crystallogr. **A35**, 543 (1979); **A38**, 516 (1982)
4.20 S. Kishino, K. Kohra: Jpn. J. Appl. Phys. **10**, 551 (1971)
4.21 T. Bedynska: Phys. Status Solidi **(a)19**, 365 (1973)
4.22 Y. H. Ohtsuki: J. Phys. Soc. Jpn. **19**, 2285 (1964); **20**, 374 (1965)
4.23 B. Okkerse: Philips Res. Rep. **17**, 464 (1962)
4.24 R. W. James: *The Optical Principle of the Diffraction of X-Rays* (Bell, London 1950)
4.25 H. Wagner: Z. Phys. **146**, 127 (1956)
4.26 R. W. James: Solid State Phys. **15**, 53 (1962)
4.27 N. Kato: Z. Naturforsch. **15a**, 369 (1960); Acta Crystallogr. **14**, 526, 627 (1961); J. Appl. Phys. **39**, 2225, 2231 (1968)
4.28 N. Kato, R. A. Lang: Acta Crystallogr. **12**, 787 (1959)
4.29 H. Jeffreys, B. Jeffreys: *Methods of Mathematical Physics* (Cambridge University Press, Cambridge 1956) p. 503

Chapter 5

5.1   W. H. Zachariasen: *Theory of X-Ray Diffraction in Crystals* (Wiley, New York 1945)
5.2   Z. G. Pinsker: *Dynamical Scattering of X-Rays in Crystals*, Springer Ser. Solid-State Sci., Vol. 3 (Springer, Berlin, Heidelberg, New York 1977)
5.3   I. Waller: Ann. Phys. **79**, 261 (1929)
5.4   C. G. Darwin: Philos. Mag. **43**, 800 (1922)
5.5   F. Miller: Phys. Rev. **47**, 209 (1935)
5.6   H. J. Juretschke: Phys. Rev. Lett. **48**, 1487 (1982); Phys. Lett. **92A**, 183 (1982)
5.7   S. L. Chang: Appl. Phys. **A26**, 221 (1981); Phys. Rev. Lett. **48**, 163 (1982)
5.8   G. Molière: Ann. Phys. (Leipzig) **35**, 272, 297 (1939); **36**, 265 (1939)
5.9   H. Wagenfeld: Phys. Rev. **144**, 216 (1966)
5.10 Y. H. Ohtsuki: J. Phys. Soc. Jpn. **19**, 2285 (1964); **20**, 374 (1965)
5.11 M. Ashkin, M. Kuriyama: J. Phys. Soc. Jpn. **21**, 1549 (1966)
5.12 Y. H. Ohtsuki, S. Yanagawa: J. Phys. Soc. Jpn. **21**, 326, 502 (1966)
5.13 M. Kuriyama: J. Phys. Soc. Jpn. **23**, 1369 (1967); Phys. Status Solidi **24**, 743 (1967); J. Phys. Soc. Jpn. **25**, 846 (1968); Acta Crystallogr. **A25**, 682 (1969); **A26**, 56 (1970); **A27**, 273 (1971); **A28**, 588 (1972)
5.14 J. P. Hannon, G. T. Trammell: Phys. Rev. **169**, 315 (1968); **186**, 306 (1969)
5.15 A. M. Afanas'ev, Y. Kagan: Acta Crystallogr. **A24**, 163 (1968)
5.16 P. H. Dederichs: Phys. Kondens. Mater. **5**, 347 (1966); Phys. Status Solidi **23**, 377 (1967)
5.17 H. Sano, K. Kthaka, Y. H. Ohtsuki: J. Phys. Soc. Jpn. **27**, 1254 (1969)
5.18 H. Cole, N. R. Stemple: J. Appl. Phys. **33**, 2227 (1962)
5.19 P. P. Ewald, Y. Heno: Acta Crystallogr. **A24**, 5 (1968)
5.20 T. Bedynska: Phys. Status Solidi **(a)19**, 365 (1973)
5.21 R. S. Martin, J. H. Wilkinson: Numer. Math. **12**, 349 (1968)
5.22 F. R. Gantmacher: *The Theory of Matrices* (Chelsea, New York 1960) p. 23
5.23 M. Kohler: Berliner Ber. Phys. Math. Kl. XIX, S1 (1935)
5.24 H. Yoshioka: J. Phys. Soc. Jpn. **12**, 618 (1957)
5.25 L. I. Schiff: *Quantum Mechanics* (McGraw-Hill, New York 1968)
5.26 G. Hildebrandt, J. D. Stephenson, H. Wagenfeld: Z. Naturforsch. **28a**, 588 (1973)
5.27 A. M. Afanas'ev, V. G. Kohn: Phys. Status Solidi **(a)28**, 61 (1975); Acta Crystallogr. **A33**, 178 (1977)
5.28 S. Kishino, K. Kohra: Jpn. J. Appl. Phys. **10**, 551 (1971)
5.29 S. Kishino: J. Phys. Soc. Jpn. **31**, 1168 (1971)
5.30 S. Kishino, A. Noda, K. Kohra: J. Phys. Soc. Jpn. **33**, 158 (1972)
5.31 T. Bedynska: Phys. Status Solidi **(a)25**, 405 (1974)
5.32 J. C. Slater: Rev. Mod. Phys. **30**, 197 (1958)
5.33 J. Laval: Rev. Mod. Phys. **30**, 222 (1958)
5.34 D. A. O'Connor: Proc. Phys. Soc. **91**, 917 (1967)
5.35 G. Borrmann: Phys. Z. **42**, 157 (1941); Z. Phys. **127**, 297 (1950)

5.36 H. W. Schürmann: Z. Phys. **194**, 425 (1966)
5.37 K. Haruta: J. Appl. Phys. **38**, 3312 (1967)
5.38 M. Kuriyama, T. Miyakawa: J. Appl. Phys. **40**, 1697 (1969)
5.39 A. R. Hutson, J. H. McFee, D. L. White: Phys. Rev. Lett. **7**, 237 (1961)
5.40 D. L. White: J. Appl. Phys. **33**, 2547 (1962)
5.41 R. Kohler, W. Mohling, H. Peibst: Phys. Status Solidi **41**, 75 (1970); **61(b)**, 173 (1974)
5.42 D. G. Carlson, A. Segmüller, E. Mosekilde, H. Cole, J. A. Armstrong: Appl. Phys. Lett. **18**, 330 (1971)
5.43 D. G. Carlson, A. Segmüller: Phys. Rev. Lett. **28**, 175 (1972)
5.44 O. Ishibashi, M. Kitamura, A. Odajima: Phys. Lett. **44A**, 371 (1973)
5.45 S. D. LeRoux, R. Colella, R. Bray: Phys. Rev. Lett. **35**, 230 (1975); **37**, 1056 (1976)

Chapter 6

6.1 S. L. Chang, B. Post: Acta Crystallogr. **A31**, 832 (1975)
6.2 N. F. M. Henry, K. Lonsdale (eds.): *International Tables for X-Ray Crystallography*, Vol. 1 (Kynoch, Birmingham 1968)
6.3 P. B. Hirsch, G. N. Ramachandran: Acta Crystallogr. **3**, 187 (1950)
6.4 S. L. Chang: Acta Crystallogr. **A38**, 41 (1982)
6.5 G. Borrmann, W. Hartwig: Z. Kristallogr. **121**, 401 (1965)
6.6 G. Hildebrandt: Phys. Status Solidi **24**, 245 (1967); Acta Crystallogr. **A25**, 209 (1969); Phys. Status Solidi **(a)15**, K83 (1973)
6.7 T. Joko, A. Fukuhara: J. Phys. Soc. Jpn. **22**, 597 (1967)
6.8 P. P. Ewald, Y. Heno: Acta Crystallogr. **A24**, 5 (1968)
6.9 Y. Heno, P. P. Ewald: Acta Crystallogr. **A24**, 16 (1968)
6.10 W. Uebach, G. Hildebrandt: Z. Kristallogr. **129**, 1 (1969)
6.11 P. Penning: Advan. X-Ray Anal. **10**, 67 (1967); Philips Res. Rep. **23**, 12 (1968)
6.12 S. Balter, R. Feldman, B. Post: Phys. Rev. Lett. **27**, 307 (1971)
6.13 R. Feldman, B. Post: Phys. Status Solidi **(a)12**, 273 (1972)
6.14 B. Post, P. W. Wang, T. Hom: Z. Naturforsch. **37a**, 528 (1982)
6.15 G. Hildebrandt: Phys. Status Solidi **15**, K131 (1966)
6.16 W. Uebach: Z. Naturforsch. **28a**, 1214 (1973)
6.17 M. Umeno, G. Hildebrandt: Phys. Status Solidi **(a)31**, 583 (1975)
6.18 P. P. Ewald: Ann. Phys. (Leipzig) **49**, 1, 117 (1916); **54**, 519 (1917); Z. Kristallogr. **97**, 1 (1937)
6.19 B. Post: Phys. Rev. Lett. **39**, 760 (1977); Acta Crystallogr. **A35**, 17 (1979)
6.20 A. A. Katsnelson, V. I. Kissin, N. A. Polyakova: Sov. Crystallogr. **14**, 965 (1969)
6.21 A. A. Katsnelson, T. I. Borodina, V. I. Kissin: Phys. Status Solidi **(a)3**, 105 (1970)
6.22 A. A. Katsnelson, V. I. Iveronova, T. I. Borodina, I. G. Sapkova: Phys. Status Solidi **(a)11**, 39 (1972)
6.23 A. A. Katsnelson, V. I. Iveronova, T. I. Borodina, T. K. Runova: Phys. Status Solidi **(a)28**, 365 (1975)
6.24 B. Post, S. L. Chang, T. C. Huang: Acta Crystallogr. **A33**, 90 (1977)
6.25 T. Tuomi, K. Naukarinen: Phys. Rev. **B24**, 6125 (1981)
6.26 S. L. Chang: Appl. Phys. Lett. **40(9)**, 793 (1982)
6.27 R. Colella: Acta Crystallogr. **A30**, 413 (1974)
6.28 V. G. Kohn: Phys. Status Solidi **(a)54**, 375 (1979)
6.29 S. L. Chang: Acta Crystallogr. **A35**, 543 (1979); **A38**, 516 (1982)
6.30 W. Graeff, U. Bonse: Z. Phys. **B27**, 19 (1977)
6.31 S. L. Chang: Acta Crystallogr. **A34**, 238 (1978)
6.32 S. L. Chang: Phys. Status Solidi **(a)47**, 717 (1978)
6.33 N. Kato: In *X-Ray Diffraction*, ed. by. L. V. Azaroff (McGraw-Hill, New York 1974)
6.34 C. Kittel: *Introduction to Solid State Physics* (Wiley, New York 1971) p. 697
6.35 M. Born, E. Wolf: *Principles of Optics* (Pergamon, Oxford 1970) p 381

Chapter 7
7.1 P. P. Ewald, Y. Heno: Acta Crystallogr. **A24**, 5 (1968)
7.2 M. F. C. Ladd, R. A. Palmer (eds.): *Theory and Practice of Direct Methods in Crystallography* (Plenum, New York 1980)
7.3 D. Harker, J. S. Kasper: Acta Crystallogr. **3**, 374 (1948)
7.4 D. Sayre: Acta Crystallogr. **5**, 60 (1952)
7.5 H. Hauptmann, J. Karle: *Solution of the Phase Problem I The Centrosymmetric Crystal*, American Crystallographic Association Monograph No. 3 (Polycrystal Service, Pittsburgh, 1953)
7.6 M. M. Woolfson: *Direct Methods in Crystallography* (Oxford University Press, Oxford 1961)
7.7 J. Karle, I. L. Karle: Acta Crystallogr. **8**, 1 (1966)
7.8 B. K. Vainshtein, V. M. Fridkin, V. L. Indenbom: *Modern Crystallography*, Vol. II, Springer Ser. Solid-State Sci. Vol. 21 (Springer, Berlin, Heidelberg, New York 1982)
7.9 W. Cochran, M. M. Woolfson: Acta Crystallogr. **8**, 1 (1955)
7.10 W. Cochran: Acta Crystallogr. **8**, 473 (1955)
7.11 G. Germain, P. Ain, M. M. Woolfson: Acta Crystallogr. **A27**, 368, 1040 (1971)
7.12 W. Cochran, A. S. Douglas: Proc. Roy. Soc. London **A227**, 486 (1957)
7.13 A. L. Patterson: Z. Kristallogr. **A90**, 517 (1935)
7.14 M. G. Rossmann, D. M. Blow: Acta Crystallogr. **15**, 24 (1962); **16**, 39 (1963); **17**, 1474 (1964)
7.15 M. G. Rossmann, D. C. Hodgkin: *The Molecular Replacement Method*, ed. by. M. G. Rossmann (Gordon and Breach, New York 1972)
7.16 M. F. Perutz: Proc. Roy. Soc. London **A225**, 264 (1954)
7.17 S. Ramaseshan: *Advanced Methods of Crystallography*, ed. by G. N. Ramachandran (Academic, New York 1964)
7.18 Y. Okaya, R. Pepinsky: *Computing Methods and the Phase Problem in X-Ray Crystal Analysis*, ed. by R. Pepinsky, J. M. Robertson, J. C. Speakman (Pergamon, Oxford 1961) p. 273
7.19 W. N. Lipscomb: Acta Crystallogr. **2**, 193 (1949)
7.20 R. S. Williamson: Ph. D. Thesis, Polytechnic Institute of Brooklyn (1957) (unpublished)
7.21 S. Miyake, K. Kambe: Acta Crystallogr. **7**, 216 (1954)
7.22 K. Kambe, S. Miyake: Acta Crystallogr. **7**, 218 (1954)
7.23 K. Kambe: J. Phys. Soc. Jpn. **12**, 13, 25 (1957)
7.24 H. Bethe: Ann. Phys. (Leipzig) **87**, 55 (1928)
7.25 M. Hart, A. R. Lang: Phys. Rev. Lett. **7**, 120 (1961)
7.26 N. Kato, R. A. Lang: Acta Crystallogr. **12**, 787 (1959)
7.27 B. Post: Phys. Rev. Lett. **39**, 760 (1977); Acta Crystallogr. **A35**, 17 (1979)
7.28 H. Jagodzinski: Acta Crystallogr. **A36**, 104 (1980)
7.29 L. D. Chapman, D. R. Yoder, R. Colella: Phys. Rev. Lett. **46**, 1578 (1981)
7.30 S. L. Chang: Appl. Phys. **A26**, 221 (1981); Phys. Rev. Lett. **48**, 163 (1982)
7.31 B. Post: Acta Crystallogr. **A39**, 711 (1983)
7.32 P. P. Gong, B. Post: Acta Crystallogr. **A39**, 719 (1983)
7.33 R. Hoier, K. Marthinsen: Acta Crystallogr. **A39**, 854 (1983)
7.34 S. L. Chang: Acta Crystallogr. **A38**, 41 (1982)
7.35 R. Hoier, A. Aanestad: Acta Crystallogr. **A37**, 787 (1981)
7.36 H. J. Juretschke: Phys. Rev. Lett. **48**, 1487 (1982); Phys. Lett. **92A**, 183 (1982)
7.37 K. Hümmer, H. W. Billy: Acta Crystallogr. **A38**, 841 (1982)
7.38 M. Renninger: Z. Kristallogr. **106**, 141 (1937); Naturwissenschaften **25**, 43 (1937)
7.39 E. Wagner: Phys. Z. **21**, 94 (1923)
7.40 S. L. Chang: Phys. Status Solidi **(a)65**, 553 (1981)
7.41 H. Cole, F. W. Chambers, H. M. Dunn: Acta Crystallogr. **15**, 138 (1962)
7.42 S. L. Chang, F. S. Han: Acta Crystallogr. **A38**, 414 (1982)
7.43 F. S. Han, S. L. Chang: Acta Crystallogr. **A39**, 98 (1983)
7.44 H. J. Deiseroth, F. S. Han: Angew. Chem. **93**, 1011 (1981)
7.45 P. Main, M. M. Woolfson, L. Lessinger, S. Germain, J. P. Declerq: *MULTAN, A System of Computer Programs for the Automatic Solution of Crystal Structures* (University of York 1974, 1978, 1980)

7.46 W. Kossel: Ann. Phys. (Leipzig) **25**, 512 (1936)
7.47 K. Lonsdale: Philos. Trans. **A240**, 219 (1947)
7.48 D. R. Schwartzenberger: Philos. Mag. **47**, 1242 (1959)
7.49 J. P. Hannon, G. T. Trammell: Phys. Rev. **169**, 315 (1968); Phys. Rev. **186**, 306 (1969)
7.50 B. J. Isherwood, C. A. Wallace: Nature **212**, 173 (1966)
7.51 F. J. Spooner, C. G. Wilson: J. Appl. Crystallogr. **6**, 132 (1973)
7.52 B. H. Heise: J. Appl. Phys. **33**, 938 (1962)
7.53 E. T. Peters, R. E. Ogilvie: Trans. AIME **233**, 89 (1965)
7.54 W. G. Morris: J. Appl. Phys. **39**, 1813 (1968)
7.55 B. Post: J. Appl. Crystallogr. **8**, 452 (1975)
7.56 T. Hom, W. Kiszenik, B. Post: J. Appl. Crystallogr. **8**, 457 (1975)
7.57 W. L. Bond: Acta Crystallogr. **13**, 814 (1960)
7.58 R. Deslattes, A. Henins: Phys. Rev. Lett. **31**, 972 (1973)
7.59 A. Segmüller: Adv. X-Ray Anal. **13**, 155 (1970)
7.60 M. E. Straumanis, E. Z. Aka: J. Am. Chem. Soc. **73**, 5643 (1951)
7.61 A. Smakula, J. Kalnajs: Phys. Rev. **99**, 1937 (1955)
7.62 A. Cooper: Acta Crystallogr. **15**, 578 (1962)
7.63 D. N. Batchelder, R. O. Simmons: J. Appl. Phys. **36**, 2864 (1965)
7.64 G. A. Rozgonyi, T. J. Ciesielka: Rev. Sci. Instrum. **44**, 1053 (1973)
7.65 E. Estop, A. Izrael, M. Sauvage: Acta Crystallogr. **A32**, 627 (1976)
7.66 T. Hattanda, A. Takeda: Jpn. J. Appl. Phys. **12**, 1104 (1973)
7.67 H. Nagai: J. Appl. Phys. **45**, 3789 (1974)
7.68 K. Ishida, J. Matsui, T. Kamejima, I. Sakuma: Phys. Status Solidi (a)**31**, 255 (1975)
7.69 G. H. Olsen, R. T. Smith: Phys. Status Solidi (a)**31**, 739 (1975)
7.70 M. C. Rowland, D. A. Smith: J. Cryst. Growth **38**, 143 (1977)
7.71 W. Hagen: J. Cryst. Growth **43**, 739 (1978)
7.72 S. L. Chang: Appl. Phys. Lett. **34**, 239 (1979); **37**, 819 (1980); Acta Crystallogr. **A37**, 876 (1981)
7.73 B. R. Brown, M. A. G. Halliwell, B. J. Isherwood: J. Microscopy **118**, 375 (1980)
7.74 B. J. Isherwood, B. R. Brown, M. A. G. Halliwell: J. Cryst. Growth **54**, 449 (1981)
7.75 K. Oe, Y. Shinoda, K. Sugiyama: Appl. Phys. Lett. **33**, 962 (1978)
7.76 B. K. Tanner: *X-Ray Diffraction Topography* (Pergamon, Oxford 1977)
7.77 S. L. Chang: J. Appl. Phys. **53**(4), 2988 (1982)
7.78 B. W. Battermann, H. Cole: Rev. Mod. Phys. **36**, 681 (1964)
7.79 W. Graeff, U. Bonse: Z. Phys. **B27**, 19 (1977)
7.80 M. Ando, S. Hosoya: In *Proceedings of the 6th International Conference on X-Ray Optics and Microanalysis* (University of Tokyo Press, Tokyo 1972) p. 63
7.81 D. A. Kottwitz: Acta Crystallogr. **A24**, 117 (1968); Phys. Rev. **175**, 1056 (1968); Acta Crystallogr. **A25**, 459 (1969)
7.82 P. Mikula, R. T. Michalec, M. Vrána, J. Vávra: Acta Crystallogr. **A35**, 962 (1979)
7.83 P. Mikula, M. Vrána, R. T. Michalec, J. Kulda, J. Vávra: Phys. Status Solidi (a)**60**, 549 (1980)
7.84 P. Mikula, J. Kulda, M. Vrána, R. T. Michalec, J. Vávra: Nucl. Instrum. Meth. **197**, 563 (1982)
7.85 M. Vrána, P. Mikula, R. T. Michalec, J. Kulda, J. Vávra: Acta Crystallogr. **A37**, 459 (1981)
7.86 S. Takagi: Acta Crystallogr. **15**, 1311 (1962); J. Phys. Soc. Jpn. **27**, 1239 (1969)
7.87 R. Michalec, B. Chalupa, L. Sedláková, P. Mikula, V. Petrzilka, J. Zelenka: J. Appl. Crystallogr. **7**, 588 (1974)
7.88 T. Joko, A. Fukuhura: J. Phys. Soc. Jpn. **22**, 597 (1967)
7.89 T. C. Huang, M. H. Tillinger, B. Post: Z. Naturforsch. **28a**, 600 (1973)
7.90 M. Umeno, G. Hildebrandt: Phys. Status Solidi (a)**31**, 583 (1975)
7.91 M. Umeno: Phys. Status Solidi (a)**37**, 561 (1976); (a)**38**, 701 (1976)
7.92 S. A. Kshevetsky, I. P. Mikhailyuk: Sov. Kristallogr. **21**, 381 (1976) [English transl.: Sov. Phys.-Crystallogr. **21**, 209 (1976)]
7.93 S. L. Chang: Appl. Phys. Lett. **40**(9), 793 (1982)
7.94 M. Umeno, G. Hildebrandt: Acta Crystallogr. **A31**, S3, S253 (1975)

7.95   G. C. Baldwin, J. C. Solem, V. I. Gol'danskii: Rev. Mod. Phys. **53**, 687 (1981)
7.96   B. S. Fraenkel: Appl. Phys. Lett. **41(3)**, 234 (1982)
7.97   B. S. Fraenkel: Appl. Phys. Lett. **36**, 341 (1980)
7.98   B. S. Fraenkel: X-Ray Spectrom. **9**, 189 (1980)
7.99   U. Feldman, G. A. Doschek, D. J. Nagel, R. D. Cowan, R. R. Whitlock: Ap. J. **192**, 213 (1974)
7.100 C. S. G. Cousins, L. Gerward, J. Staun Olsen: Phys. Status Solidi **(a)48**, 113 (1978)
7.101 S. Caticha-Ellis: Acta Crystallogr. **A25**, 666 (1969)
7.102 P. R. Prager: Acta Crystallogr. **A27**, 563 (1971)
7.103 C. B. R. Parente, S. Caticha-Ellis: Jpn. Appl. Phys. **13**, 1506 (1974)
7.104 B. W. Batterman: Appl. Phys. Lett. **1**, 68 (1962); Phys. Rev. **133**, A759 (1964)
7.105 B. W. Batterman: Phys. Rev. Lett. **22**, 703 (1969)
7.106 J. A. Golovchenko, B. W. Batterman, W. L. Brown: Phys. Rev. **B10**, 4239 (1974)
7.107 S. K. Anderson, J. A. Golovchenko, G. Mair: Phys. Rev. Lett. **37**, 1141 (1976)
7.108 P. L. Cowan, J. A. Golovchenko, M. F. Robbins: Phys. Rev. Lett. **44**, 1680 (1980)
7.109 M. J. Bedzyk, W. M. Gibson, J. A. Golovchenko: J. Vac. Sci. Technol. **20**, 634 (1982)
7.110 J. A. Golovchenko, J. R. Patel, D. R. Kaplan, P. L Cowan, M. J. Bedzyk: Phys. Rev. Lett. **49**, 560 (1982)
7.111 J. Chikawa, I. Fujimoto: Appl. Phys. Lett. **13**, 387 (1968)
7.112 W. Hartmann: In *X-Ray Optics*, ed. by H. J. Queisser, Topics Appl. Phys., Vol. 22 (Springer, Berlin, Heidelberg, New York 1977) p. 191
7.113 P. Eisenberger, W. C. Marra: Phys. Rev. Lett. **46**, 1081 (1981)

Subject Index

B.K. Vainshtein
Modern Crystallography I
Symmetry of Crystals. Methods of Structural Crystallography
1981. 272 figures, some in color.
XVII, 399 pages. (Springer Series in Solid-State Sciences, Volume 15)
ISBN 3-540-10052-0

Contents: Crystalline State. – Fundamentals of the Theory of Symmetry. – Geometry of the Crystalline Polyhedron and Lattice. – Structure Analysis of Crystals. – Bibliography. – References. – Subject Index.

"This book is noteworthy for two reasons. Each subject is discussed in depth and with elegance. However, the discussion for each subject is extended into more areas than is generally found in crystallography texts. This will make it essential for everyone's bookself."
American Crystallographic Association

B.K. Vainshtein, V.M. Fridkin, V.L. Indenbom
Modern Crystallography II
Structure of Crystals
1982. 345 figures. XVII, 433 pages (Springer Series in Solid State Sciences, Volume 21)
ISBN 3-540-10517-4

This second volume of **Modern Crystallography** describes the ideal atomic structure of crystals, the real structure of crystals with its various disturbances, and the electron structure and lattice dynamics. The fundamentals of the theory of chemical bonding between atoms are given, and the geometric representations in the theory of crystalline structure and crystal chemistry, as well as the lattice energy, are considered. The important classes of crystalline structures in inorganic and organic compounds and also the structure of polymers, liquid crystals, biological crystals, and macromolecules are described.
The elements of the electron theory of crystal lattices, which help to classify crystals by their energy spectrum, are presented. Lattice dynamics and phase transitions are discussed. Concepts of the real structure of crystals with its various thermodynamic equilibrium and nonequilibrium disturbances are described. All types of defects are analyzed and described mathematically, with special emphasis on dislocation theory.

A.A. Chernov
Modern Crystallography III
Crystal Growth
With contributions by E.I. Givargizov, K.S. Bagdasarov, V.A. Kuznetsov, L.N. Demianets, A.N. Lobachev
1984. 244 figures. XX, 517 pages (Springer Series in Solid State Sciences, Volume 36)
ISBN 3-540-11516-1

The third in a four-part treatment of modern crystallography, this volume concentrates on the phenomena and techniques of crystal growth. The basic concepts are systematically presented and related to all facets of crystal growth. Recent results are reviewed and discussed by a team of experts. The first one-volume treatment of the main growth techniques used in industry today, this book is written on a level that affords the reader a basic understanding of the physical, chemical and technological principles involved.

Z.G. Pinsker
Dynamical Scattering of X-Rays in Crystals
1978. 124 figures, 12 tables. XII, 511 pages (Springer Series in Solid State Sciences, Volume 3)
ISBN 3-540-08564-5

"....Pinsker's text, ... is a masterful treatment. Not only are the basics presented in a thorough and comprehensible fashion but also most significant modern developments are revealed..." *American Scientist*

Springer-Verlag
Berlin
Heidelberg
New York
Tokyo

X-Ray Microscopy

Proceedings of the International Symposium, Göttingen, Federal Republic of Germany, September 14–16, 1983

Editors: **G.Schmahl, D.Rudolph**

1984. 262 figures. IX, 345 pages
(Springer Series in Optical Sciences,
Volume 43)
ISBN 3-540-13271-6

Contents: Introduction. – X-Ray Sources. – X-Ray Optics. – X-Ray Detectors. – X-Ray Microscopes. – Applications of X-Ray Microscopy. – X-Ray Holography. – List of Contributors.

B. K. Agarwal

X-Ray Spectroscopy

An Introduction

1979. 188 figures, 31 tables. XIII, 418 pages
(Springer Series in Optical Sciences,
Volume 15)
ISBN 3-540-09268-4

"... Even with its high density of information, I found the book easily readable. It begins with a classical description of X-ray production and then introduces only simple quantum mechanical treatments as they are necessary (although references direct one to more detailed treatments). This general procedure is used throughout, providing a rather physical interpretation to the more complicated issues. Much of the jargon is presented in a clear, concise manner so that the book also serves as a fine reference text. This book would serve well as a graduate-level textbook." *J. Am. Chem. Soc.*

Optical Data Processing

Applications

Editor: **D.Casasent**

1978. 170 figures, 2 tables. XIII, 286 pages
(Topics in Applied Physics, Volume 23)
ISBN 3-540-08453-3

Contents: *D. Casasent, H. J. Caulfield:* Basic Concepts. – *B. J. Thompson:* Optical Transforms and Coherent Processing Systems – With Insights From Cristallography. – *P. S. Considine, R. A. Gonsalves:* Optical Image Enhancement and Image Restoration. – *E. N. Leith:* Synthetic Aperture Radar. – *N. Balasubramanian:* Optical Processing in Photogrammetry. – *N. Abramson:* Nondestructive Testing and Metrology. – *H. J. Caulfield:* Biomedical Applications of Coherent Optics. – *D. Casasent:* Optical Signal Processing.

Synchrotron Radiation

Techniques and Applications

Editor: **C.Kunz**

1979. 162 figures, 28 tables. XVI, 442 pages
(Topics in Current Physics, Volume 10)
ISBN 3-540-09149-1

Contents: *C. Kunz:* Introduction – Properties of Synchrotron Radiation. – *E. M. Rowe:* The Synchrotron Radiation Source. *W. Gudat, C. Kunz:* Instrumentation for Spectroscopy and other Applications. – *A. Kotani, Y. Toyozawa:* Theoretical Aspects of Inner-Level Spectroscopy. – *K. Codling:* Atomic Spectroscopy. – *E. E. Koch, B. F. Sonntag:* Molecular Spectroscopy. – *D. W. Lynch:* Solid-State Spectroscopy.

Springer-Verlag Berlin Heidelberg New York Tokyo